Sustainable Development Goals Series

The **Sustainable Development Goals Series** is Springer Nature's inaugural cross-imprint book series that addresses and supports the United Nations' seventeen Sustainable Development Goals. The series fosters comprehensive research focused on these global targets and endeavours to address some of society's greatest grand challenges. The SDGs are inherently multidisciplinary, and they bring people working across different fields together and working towards a common goal. In this spirit, the Sustainable Development Goals series is the first at Springer Nature to publish books under both the Springer and Palgrave Macmillan imprints, bringing the strengths of our imprints together.

The Sustainable Development Goals Series is organized into eighteen subseries: one subseries based around each of the seventeen respective Sustainable Development Goals, and an eighteenth subseries, "Connecting the Goals", which serves as a home for volumes addressing multiple goals or studying the SDGs as a whole. Each subseries is guided by an expert Subseries Advisor with years or decades of experience studying and addressing core components of their respective Goal.

The SDG Series has a remit as broad as the SDGs themselves, and contributions are welcome from scientists, academics, policymakers, and researchers working in fields related to any of the seventeen goals. If you are interested in contributing a monograph or curated volume to the series, please contact the Publishers: Zachary Romano [Springer; zachary.romano@springer.com] and Rachael Ballard [Palgrave Macmillan; rachael.ballard@palgrave.com].

Abhinav Alakshendra · Amrita Datta ·
Bhim Reddy
Editors

Development, Inclusion and Sustainability

Issues and Perspectives

Springer

Editors
Abhinav Alakshendra
Department of Urban and Regional Planning
University of Florida
Gainesville, FL, USA

Amrita Datta
Department of Liberal Arts
Indian Institute of Technology
Hyderabad, Telangana, India

Bhim Reddy
Institute for Human Development
New Delhi, India

ISSN 2523-3084 ISSN 2523-3092 (electronic)
Sustainable Development Goals Series
ISBN 978-981-97-6862-2 ISBN 978-981-97-6863-9 (eBook)
https://doi.org/10.1007/978-981-97-6863-9

This Springer imprint is published by the registered company Springer Nature Singapore Pte Ltd.
The registered company address is: 152 Beach Road, #21-01/04 Gateway East, Singapore 189721, Singapore

Contents

Editors and Contributors

About the Editors

Abhinav Alakshendra is Associate Professor in the Department of Urban and Regional Planning at the University of Florida, Gainesville, USA. He is also the Director, Center for International Design and Planning at the University. He holds a Ph.D. in Economics from Kansas State University, Manhattan, USA. His research interests are mainly focused on the areas of urbanization, international development, and applied microeconomics in Southeast Asia. He is a recipient of many prestigious research grants including UKAID, Land and Housing Institute, Korea, and Florida Department of Transportation, and has many research publications to his credit published in international journals.

Amrita Datta is Assistant Professor of Development Studies at the Department of Liberal Arts, Indian Institute of Technology (IIT) Hyderabad and a Visiting Fellow at the Institute for Human Development (IHD), New Delhi. Amrita's research interests are in the areas of migration and mobilities, agrarian change and rural development, women's work, and gender and development. Based on empirical research, Amrita's book, *Migration and Development in India: The Bihar Experience* (2022), studies out-migration from rural Bihar in the context of neoliberal economic development in India. Amrita has published in several edited volumes and journals such as the *Journal of Development Studies*, *Asian Journal of Women's Studies*, *Children's Geographies*, *Indian Journal of Labour Economics*, and the *Economic and Political Weekly*. Amrita has a Ph.D. in Development Studies from the International Institute of Social Studies, Erasmus University Rotterdam, and an M.Phil. in the same subject from the University of Cambridge.

Bhim Reddy is Senior Fellow at the Institute for Human Development, New Delhi. He is a social anthropologist who is engaged in mixed-methods development research. His doctoral research was focussed on rural-urban labour migration, agrarian political economy and social change. His current interests are in the areas of rural transformation, farm technologies and work, urbanisation and socio-spatial inequalities; youth, migration and employment; gender inequality, intersections and violence; drinking cultures and alcohol policy. Dr. Reddy has been on the editorial board of the *Indian Journal of Human Development* as its Associate Editor since 2015.

Contributors

Sarthi Acharya Institute for Human Development, New Delhi, India

Shiva Raj Adhikari Central Department of Economics, Tribhuvan University, Kirtipur, Nepal

Abhinav Alakshendra Department of Urban and Regional Planning, University of Florida, Gainesville, USA

Jesse Anderson Senior Planning Project Manager, Orlando, USA

Manoj Bandan Balsamanta Institute for Human Development, New Delhi, India

Amrita Datta Department of Liberal Arts, Indian Institute of Technology Hyderabad, Kandi, India

Tanuka Endow Institute for Human Development, New Delhi, India

Ajit K. Ghose Institute for Human Development, New Delhi, India

Palashpriya Halder Department of Economics, Vijaygarh Jyotish Ray College, Kolkata, India

S. R. Hashim Government of India, Union Public Service Commission, New Delhi, India

Abhishek Kumar Independent Researcher, New Delhi, India

Ziming Li Department of Public Finance and Taxation, School of Economic and Management, Wuhan University, Wuhan, China

Shengfeng Lu Department of Public Finance and Taxation, School of Economic and Management, Wuhan University, Wuhan, China

Shipra Maitra Institute for Human Development, New Delhi, India

Rajarshi Majumder Department of Economics, University of Burdwan, Burdwan, West Bengal, India

Sunil Kumar Mishra Institute for Human Development, New Delhi, India

Ishita Mukhopadhyay Department of Economics, University of Calcutta, Kolkata, India

Jhilam Ray Department of Economics, Jadavpur University, Kolkata, West Bengal, India

Bhim Reddy Institute for Human Development, New Delhi, India

Christopher Silver University of Florida, Gainesville, USA

Introduction

1

Abhinav Alakshendra, Amrita Datta and Bhim Reddy

1.1 Development, Inclusion and Sustainability: The Interconnections

The growing social, economic, and environmental challenges of our times have posed some pressing concerns in the development path of the world. In the post-pandemic world—marked by polycrisis—progress on a majority of Sustainable Development Goals, in particular, in the areas of poverty, hunger and climate, has been slow, or has stalled or reversed (UNDESA 2023). It is widely established in the mainstream literature that while economic growth is necessary, it is not sufficient for development. Human development is beyond consumption, income, and wealth; it is about enlarging people's choices and the expansion of human capabilities, critically in the dimensions of health and education (UNDP 1990; Sen 2003).

A. Alakshendra (✉)
Department of Urban and Regional Planning, University of Florida, Gainesville, USA
e-mail: alakshendra@ufl.edu

A. Datta
Department of Liberal Arts, Indian Institute of Technology Hyderabad, Kandi, India
e-mail: amritadatta@la.iith.ac.in

B. Reddy
Institute for Human Development, New Delhi, India
e-mail: bhim@ihdindia.org

Inclusion and sustainable development have gained wide traction in this context. The policy literature, for long, has emphasised, that development *must be* made sustainable without compromising the ability of future generations to meet their own needs (Brundtland Report, 1987). However, *business as usual* has led to unprecedented climate change and global warming that threaten the planet. The current climate emergency is a 'direct consequence of our carbon-heavy land-use and agriculture, transport, buildings and industrial processes and our polluting energy sources' (UNEP 2019). Now that the urban population in the world has become more than half of the total population, a particular focus needs to be given to urban development in the Global South, where the rate of urban growth is expected to accelerate in the next two decades or so. The challenges of inclusion and sustainability will be complex and multifaceted in developing countries as they urbanise without clear-cut strategies. Sustainable development is not merely about upholding and preserving natural resources but ensuring that the communities thrive economically and socially. To achieve a truly sustainable and equitable future, linking sustainability with inclusivity and human development is imperative.

Economic sustainability emphasizes fostering economic development while minimizing negative environmental and societal impacts. It involves creating and nurturing a resilient and

A. Alakshendra et al. (eds.), *Development, Inclusion and Sustainability*,
Sustainable Development Goals Series, https://doi.org/10.1007/978-981-97-6863-9_1

just economy that is a must to reduce economic disparities. It has been established that a strong economy is essential for improving living standards by employment generation and providing decent work. The decent work framework evolved at the ILO to encompass 'productive work in which rights are protected, which generates an adequate income, with adequate social protection' and is embedded in SDG 8 that promotes inclusive and sustainable economic growth, employment and decent work for all (UNDESA 2023; ILO 1999).

The underlying goals can only be achieved by creating an economy which creates decent work and livelihoods, and promotes equity. One of the essential components of sustainable development is social inclusion, as it ensures that the gains from development and economic growth are shared equitably among the stakeholders. Excessive inequality adversely impacts growth (Stiglitz 2016), while employment generation and good jobs can address inequality (Nayyar 2014). The key indicators are having equal access to economic opportunity, equality, empowerment of all groups, and easy access to critical services. Social inclusion is based on human rights principles, and society thrives when inequalities are reduced, participation is enhanced, and the leading causes of exclusions are tackled. The social pillar of sustainability addresses all individuals' well-being and quality of life, ensuring that social justice, equity, and inclusivity are prioritized. It recognizes that sustainable development must consider the needs and aspirations of present and future generations. Sustainable development is deeply interconnected with inclusion, and achieving a sustainable future requires balancing their interactions.

1.1.1 This Book

The book addresses the overarching theme of promoting inclusive and sustainable development. Divided into 12 chapters, it contains selected and revised papers drawn from two international conferences—'Growth, Equality and Sustainability: Perspectives and Policies for a Better World' and 'Sustainable and Inclusive Urban Development in India: Learning from International Experiences', organised by the Institute for Human Development (IHD) in collaboration with the World Bank and the Department of Urban and Regional Planning, University of Florida, respectively. The 12 chapters are divided into four sections.

In the first section, the two chapters deal with employment and labor markets. The first chapter by Ajit Ghose and Abhishek Kumar uses nationally representative Indian data for a macro-analysis of employment trends, while the second chapter by Tanuka Endow and Sunil Kumar Mishra, 2024 is based on a micro-study that examines the potential of the manufacturing sector to absorb rural labor in an empirical context of deagrarianisation. Ghose and Kumar, based on rigorous data analysis, find that employment conditions worsened quite substantially during 2011-2017 in India, which was, in fact, a period of high growth. They argue that growth has been exclusionary and, hence, not accompanied by development. They explain that the most important reason behind such a development is that growth has been services-led on the one hand and growth of capital-and-skill-intensive manufacturing and services on the other, which is not suitable for a labor-surplus developing country like India. The authors conclude that such a growth process—where the benefits of growth disproportionately accrue to the rich—cannot be sustainable. The second chapter shifts the gaze to micro-level processes of industrialisation and structural change in the context of subaltern urbanisation in India. Based on an enterprise survey, with a focus on labor in two small and medium towns of Bihar, a backward and ruralized state, Endow and Mishra find that despite major policy initiatives to push industrialization, manufacturing units were in decline, while trading enterprises—the services sector—was a source of urban growth. The authors argue that the slow transfer of labor from agriculture to industry is one of the main reasons for the backwardness of the state. Further, poor governance and insufficient infrastructure have also contributed to the low productivity of existing manufacturing units.

The second section, comprising four chapters, deals with capabilities (education and health) and their roles in intergenerational mobility, contributing to poverty reduction and inclusive development. In the third chapter, Shiva Raj Adhikari analyses the dynamics of impoverishment impact of health care in the case of Nepal. In the literature on poverty, there is an overemphasis on studies related to its measurement, particularly in the context of developing countries. Poverty dynamics—how individuals and households transition in and out of poverty have received less attention. There is now an increasing acknowledgement in the academic and policy literature that poverty research needs to move beyond static analysis and focus on poverty dynamics (Addison et al. 2000). Adhikari's chapter goes beyond conventional cross-section approaches and constructs a synthetic panel dataset to assess poverty dynamics. The authors find that chronic poverty exists in all regions, and is higher among marginalized ethnic and Dalit (occupational caste) groups. The author argues that different anti-poverty policies are required to address chronic or transitory poverty. While instrumental policies such as credit, access to services, social safety net programmes, or even remittances can stabilize short-term income fluctuations may be more appropriate to address transitory poverty, structural or long term interventions such as the development of basic infrastructure, increasing social and political inclusion, redistribution of assets, increasing rates of capital accumulation among others are required to address chronic poverty.

The two papers on India examine intergenerational transfer of education among various social groups and find that upward mobility in education is relatively higher compared to occupational mobility. Chapter four by Rajarshi Majumder and Jhilam Ray also looks at the possible linkages of stickiness/mobility with several micro and macro-economic indicators to comprehend how stickiness acts as a barrier to development or how it can scale down and, in this context, strongly underline the need for inclusionary policies. In chapter five, Palashpriya Halder and Ishita Mukhopadhyay, in their chapter, examine whether inequality in human capital in today's generations reflects very unequal opportunities that individuals inherit from their parents due to their belonging to a particular social group. They argue that historical discrimination and social exclusion have a long-run effect. They find that mother's education, even in lower classes, plays a very important role in their child's capability enhancement. Their chapter posts an overall positive picture, and finds a high level of intergenerational mobility in education among most social groups. This is not surprising and may be explained by the high enrolment in recent years, and the prevalence of first-generation learners in India. However, the positive results on intergenerational mobility in education are not matched in other social and economic aspects. The analysis in the previous chapter by Majumder and Ray suggests that occupational and wealth rigidities have increased in the same time period. Overall, the results of both chapters resonate with other studies on social mobility in India (Krishna 2019; Iversen et al. 2017).

Based on the China Health and Nutrition Survey, in chapter six, Shengfeng Lu, Ziming Li and Abhinav Alakshendra investigate the impact of public goods provision on social mobility, particularly on upward intergenerational occupational mobility in China. The authors find that provisioning of public goods at the *local level* matters; it is associated with higher intergenerational mobility, particularly in the more deprived rural areas. In turn, the authors highlight the significance of equitable distribution mechanisms of public goods and their role in achieving a more inclusive society in developing countries. They also argue that the local governments should focus more on enlarging the coverage of public goods beneficiaries at the micro level than on smoothing the regional gap or rural-urban gap in terms of the total amount of public goods provision at the macro level.

The fourth section, containing four chapters, deals with various aspects of urbanization, which have increasingly become very crucial for sustainable and inclusive development. In chapter seven, Christopher Silver draws on a historical framework to focus on the urban problem of 'sinking cities'. Using a case from Indonesia, Silver traces the evolution of urban flooding in Jakarta, and unpacks the complexities of mitigation. He argues that while extreme cases of flooding have led to increased policy attention—at the local, national and global levels—paradoxically, this has also been accompanied by policy paralysis that contributes to inaction. This is further amplified by the conflicting positions of multiple stakeholders – experts, politicians, community advocates and development interests. Silver concludes that a more sustainable and equitable urban development may be possible by undertaking multiple mitigation strategies that address the fundamental problem of water pollution, and, at the same time, reduce overreliance of subsurface water.

Using a large household survey in two cities in India (Delhi and Patna), chapter eight, by Bhim Reddy and Manoj Bandan Balsamanta deals with the structure of urban inequality and infrastructural violence. The authors find that spatial and social locations largely overlap and intersect in cities, which, in turn, manifest in multiple deprivations and produce horizontal inequalities. According to the authors, the contemporary nature of urbanization appears to be more exclusive, with limited scope for residential spaces for the poor. In chapter nine, Jesse Anderson and Abhinav Alakshendra map the level of urban services and amenities—such as sanitation, drinking water, health services and housing quality—to the people of a city in eastern India (Patna). The authors report significant spatial inequalities in the access and availability of services in peripheral areas of the city compared to central areas. In particular, urban slums in the peripheral regions are worse off in terms of availability and access of public services in the urban areas. In chapter ten, S. R. Hashim raises some of the important issues that pose a challenge to sustainable urbanization in India, which is likely to increase in future. Among many issues of sustainability, this chapter particularly focuses on migration, poverty and urban crimes. The author argues for planning with a vision, provision of good infrastructure and amenities and facilitating interaction between various satellite units of the agglomeration which can go a long way in making these agglomerations work as efficient urban units. Of course, other complementary policies for bettering the conditions of the poor are needed.

The last section contains two diverse policy-focused chapters. In chapter eleven, Shipra Matira gives an overview of the smart cities policies in India and argues for making urban planning and policies inclusive and equitable. The author argues for coordinated development incentives with strong citizen and beneficiary involvement for building competitive yet inclusive cities. The last chapter by Sarthi Acharya, illustrates how the Sustainable Development Goals have been integrated within the national planning process of several countries. Based on an overview of the experiences of different countries, Acharya argues that in addition to institutional support, monitoring and evaluation, addressing local implementation challenges, having different stakeholders on board is critical to achieve results. Thus, while the goals and processes are not always the same for all countries, a particular country can certainly draw important lessons from the experiences of other countries, which can be very helpful in formulating policies for inclusive development.

References

Addison T, Hulme D, Kanbur R (2000) Poverty Dynamics. Oxford University Press, Oxford

UNDP (1990) Human Development Report 1990: Concept and Measurement of Human Development. United Nations Development Programme, New York

ILO (1999) Report of the Director-General: Decent Work. International Labour Office, Geneva

Iversen V, Krishna A, Sen K (2017) Rags to riches? Intergenerational occupational mobility in India. Econ Polit Wkly 52(44):107–114

Krishna A (2019) Obstacles to social mobility in India—and the way forward. Curr Hist 118(807):123–129

Nayyar D (2014) Why employment matters: Reviving growth and reducing inequality. Int Labour Rev 153(3):351–364

Sen A (2003) Development as capability expansion. In: Fukuda-Parr S, et al Readings in Human Development. Oxford University Press, New Delhi and New York

Stiglitz JE (2016) Inequality and Economic Growth. In: Jacobs M, Mazzucato M (eds) Rethinking capitalism. Wiley-Blackwell, New Jersey, pp 134–155

UNDESA (2023) The Sustainable Development Goals Report: A Reader. United Nations Department of Economic and Social Affairs, New York

UNEP – United Nations Environment Programme (2019). The UN Environment Programme and the climate emergency. Accessed at: https://www.unep.org/unga/our-position/unep-and-climate-emergency#:~:text=UNEP%20calls%20for%20greater%20ambition,adaptation%20and%20mitigation%20techniques

2 India's Deepening Employment Crisis in the Time of Rapid Economic Growth

Ajit K. Ghose and Abhishek Kumar

2.1 Introduction

The observed pattern of change in employment conditions in India in recent periods has justifiably aroused widespread concern. Employment conditions, it turns out, actually worsened quite substantially during 2011–2017, a period of high economic growth.[1] Apparently, growth has been exclusionary and hence unaccompanied by development.

In this paper, we take a close look into the evolution of employment conditions during a slightly longer period: 2011–2018.[2] We also do a review of employment trends over a much longer period (1999–2018) so as to place the developments during 2011–2018 in a context of longer-term developments. Furthermore, we examine the links between the observed employment trends and the pattern of growth of India's economy. We have three distinct but related objectives: one, to make an assessment of the extent of deterioration in employment conditions during 2011–2018; two, to determine if the worsening of employment conditions during 2011–2018 represents a wholly new development or a continuation (and perhaps intensification) of longer-term trends; and, three, to develop an understanding of why the employment conditions have been worsening in a period of high economic growth.

Our empirical assessment of the employment trends is based on a database built from the unit-level information available from four surveys conducted by the National Sample Survey Organisation (NSSO): the 55th, 61st and 68th Rounds (conducted respectively during 1999–2000, 2004–05 and 2011–2012) of The Employment and Unemployment Survey (EUS) and the recent Periodic Labour Force Survey (PLFS) conducted during 2018–2019.[3] Since

This chapter draws from two publications: Ghose, A. K., & Kumar, A. (2021). India's deepening employment crisis in the time of rapid economic growth. *The Indian Journal of Labour Economics*, *64*(2), 247–279 and a working paper published by the Institute for Human Development, with the same title. Permission has been obtained from the Institute for Human Development, New Delhi and the Indian Journal of Labour Economics to republish this article here.

A. K. Ghose
Institute for Human Development, New Delhi, India

A. Kumar (✉)
Independent Researcher, New Delhi, India
e-mail: mseabhishek@gmail.com

[1] See Kannan and Raveendran (2019) and Mehrotra and Parida (2019).

[2] Data from a more recent survey have now become available.

[3] PLFS was first conducted during 2017–18 and then again during 2018–19. We have examined the data from both surveys but have used only the data from the 2018–19 survey in our analysis in this paper. The data from the 2017–18 survey, however, are presented in the Appendix 2 (Tables). And a brief analysis of the developments between the two years is presented in Appendix 1.

A. Alakshendra et al. (eds.), *Development, Inclusion and Sustainability*, Sustainable Development Goals Series, https://doi.org/10.1007/978-981-97-6863-9_2

these surveys generally underestimate population, we derive key ratios and proportions (such as labour force participation rate, employment rate, and so on) from the surveys and use these together with estimates of population based on data available from the Population Censuses to derive the absolute numbers relating to labour force and employment.

As the NSSO surveys cover July-to-June years (e.g., July 2011–June 2012) and use a reference period of the preceding year, we take the survey data (i.e., the ratios and proportions that we shall be using) to refer to 1 July of the first year (e.g., data from the 2011–2012 survey to refer to 1 July 2011).[4] So, we use census-based estimates of adult (aged 15 or more years) populations (of rural male, rural female, urban male and urban female) on 1 July of the years 1999, 2004, 2011 and 2018 together with the ratios and proportions derived from the surveys to estimate the absolute numbers.[5]

We should mention a few other points about the data used in the paper. The first is that we use estimates of employment according to: Usual Principal Status (UPS), Usual Subsidiary Status (USS) and Usual Principal and Subsidiary Status (UPSS). Persons are employed according to UPS if they have been engaged in economically gainful activities for the major part (180+days) of the reference year. Persons are employed according to USS if they have been engaged in economically gainful activities for a minor part (30–180 days) of the reference year. Persons are employed according to UPSS if they are employed either according to UPS or according to USS, i.e., if they have been engaged in economically gainful activities for at least 30 days in the reference year. Persons engaging in economically gainful activities for less than 30 days in the reference year are not counted as employed; they could be either unemployed or out of the labour force.

We shall refer to the first two categories of employment respectively as *full-time* employment and *part-time* employment. Admittedly, these terms represent approximations rather than accurate descriptions (we only know that persons employed according to UPS have worked for more than 180 days in a year, for example), but their use helps avoid repeated use of UPS and USS as prefixes. Employment according to UPSS, which is our primary focus in this paper, will be referred to simply as employment. As for unemployment, we shall consider only that according to UPSS: persons in unemployment are those who did not work even for 30 days in the reference year but were looking or available for employment for at least 30 days. We can, therefore, simply use the terms unemployment and unemployed without adding any prefixes. The justification for using only this definition of unemployment is the following. Some of the persons who are unemployed according to UPS—persons who did not work for the major part of the reference year but were looking or available for employment for the major part—usually are found to be employed according to USS. Clearly, these persons would prefer *full-time* employment and can thus be considered as underemployed. But we cannot consider them as unemployed unless we choose to exclude *part-time* employment altogether from our analysis, which we do not do.

A second point we need to mention is that we have slightly modified the data available from the 61st Round of the EUS as we have very good reasons to believe that it erroneously overestimated a particular kind of employment—unpaid family work in rural areas.[6] Both

[4]The NSSO Surveys actually use several reference periods (a year, a week and each day of a week) thereby generating four different definitions, and hence four different estimates, of employment and unemployment. For a detailed discussion, see Ghose (2016), Box 1.1 (p. 4). In this paper, however, we shall be using only the data collected by using the reference period of one year.

[5]The issues involved in the choice of a reference date for the survey data are discussed by Nath and Basole (2020), who argue that this should be 1 June of the first year.

[6]See Ghose (2013) and Kapsos et al (2014) for discussions of these reasons.

the unmodified and modified estimates for 2004 are given in Appendix Table 2.27, the note to which explains the nature and method of the modification.

Finally, the analysis of the pace and pattern of economic growth relies on estimates of output (value added at constant prices) at national and sector levels available from the National Accounts Statistics produced by the National Statistical Office. These data refer to financial years (April-March) and we can reasonably consider them to hold for the mid-point of each of the financial years (e.g., consider the data for 2011–2012 to be actually for 1 September 2011). This means that there is a slight anomaly between the data on employment (which refer to 1 July) and the data on output (which refer to 1 September), but we do not believe this introduces serious distortions. However, there is another problem that might be more serious: there is a problem of comparability of the national accounts data for the period prior to 2011 with that for the period after 2011. For, a new series using a new base year—2011–2012—and also an altered methodology of estimation was launched in 2014. Detailed data, with 2011–2012 as the base year, are yet to be available for the entire period prior to 2012. Under the circumstances, we can only use the old series (with 2004–2005 as the base year) for the period 1999–2011 and the new series for the period 2011–2018.

The paper is structured as follows. In the section that follows, we seek to empirically establish the core employment trends and some remarkable associated developments. As our objective is to analyse rather than to merely describe, our discussion of the trends is selective rather than comprehensive.[7] In the next section, we examine the relationship between the core employment trends and output growth at the level of the aggregate economy as also in the broad sectors in an effort to understand why high economic growth was associated with worsening employment conditions. The final section states the main conclusions and offers some reflections on the growth process in India's economy, which produced the employment outcomes that we observe.

2.2 Evolution of Employment Conditions, 1999–2018

2.2.1 The Main Trends

There is little room for doubting that employment conditions in India worsened substantially between 2011 and 2018 (see Tables 2.1 and 2.2).[8] The average annual growth of *full-time* employment was a miserable 0.7%. And *part-time* employment recorded a dramatic decline (by 10% per annum). Thus, total employment showed near-zero growth. At the same time, the adult non-student population—the pool of potential workers—was growing at 1.8% per annum. So, the employment rate (defined with reference to non-student population) declined sharply, from 61.6% in 2011 to 54.5% in 2018. And the open unemployment rate recorded a sharp increase from just 2.2% in 2011 to 5.7% in 2018.

This substantial worsening of employment conditions during 2011–2018, however, was to a large extent the end-result of the trends that had set in right from the beginning of the millennium. The growth of *full-time* employment was decelerating throughout the period 1999–2018; it declined from 1.9% during 1999–2004 to 1.1% during 2004–2011 and then to 0.7% during 2011–2018. The growth of *part-time* employment was also decelerating throughout the period 1999–2018 though the deceleration

[7] Thus, we do not consider employment trends separately in organised and unorganised sectors, in rural and urban areas, of males and females, and of persons belonging to different caste/religious groups.

[8] Employment conditions showed a little deeper worsening during 2011–2017 than during 2011–2018. This is explained by the fact that the conditions improved a little between 2017 and 2018. See Appendix 1 for a brief discussion of the trends between 2017 and 2018.

Table 2.1 Employment: the basic trends-1

	Number of persons (millions)				Growth rate (per cent p.a.)		
	1999	2004	2011	2018	1999–04	2004–11	2011–18
Full-time employment	355.1	389.9	421.4	441.7	1.9	1.1	0.7
Part-time employment	28.4	33.4	35.1	16.3	3.4	0.7	−10.4
Employment	383.4	423.3	456.5	458.2	2.0	1.1	0.1
Non-student population	588.2	656.4	741.2	841.2	2.2	1.7	1.8
Population	639.1	717.5	842.8	968.0	2.3	2.3	2.0

Note The figures relate to persons aged 15 years or more
Source Authors' estimates based on data in Appendix Table 2.23

Table 2.2 Employment: the basic trends-2

	1999	2004	2011	2018
Employment rate (%)	65.2	64.5	61.6	54.5
Unemployment rate (%)	2.3	2.4	2.2	5.7

Note The employment rates are defined with reference to non-student population
Source Authors' estimates based on data in Appendix Table 2.23

between 2011 and 2018 was shockingly large. Till 2011, the growth of total employment was decelerating just as rapidly as the growth of *full-time* employment (it declined from 2% during 1999–2004 to 1.1% during 2004–2011) but then, after 2011, decelerated more sharply because of the drastic decline in *part-time* employment. Thus, the employment rate declined gradually from 65.2% in 1999 to 61.6% in 2011 and then sharply to 54.5% in 2018.

It is clear that decelerating employment growth was the basic trend for the entire period 1999–2017 and that the employment conditions were in fact deteriorating throughout this period (as evidenced by the declining employment rate). But the period 2011–2018 did witness a sharpening of the decelerating trend and, consequently, a deeper worsening of the employment conditions.[9] This period also witnessed a new development—a sharp rise in open unemployment. These are the trends and developments that we need to explore more fully and we do this below.

2.2.2 The Decelerating Employment Growth

Hidden behind the decelerating growth of employment was a growing education bias in job creation throughout the period 1999–2018 (Table 2.3). During 1999–2011, employment of persons with no education declined in absolute terms and the decline was faster during 2004–2011 than during 1999–2004. The growth of employment of persons with up-to-primary level education also decelerated very sharply: from a high 3.7% during 1999–2004 to an insignificant 0.5% during 2004–2011. In the more recent period (2011–2018), employment declined quite rapidly not only for persons with no education but also for persons with up-to-primary level education. Thus, progressive exclusion of the less educated from employment was an important part of the story of decelerating employment growth. The other part was decelerating growth of employment of persons with above-primary level education throughout 1999–2018.

A stark view of the nature and consequences of the education bias in job creation emerges if we divide the employed into just two groups: the less educated—those with either no schooling or up-to-primary level education—and the educated—those with above-primary level

[9] It is possible, indeed likely, that the shock delivered to the economy by the sudden demonetisation of 2016 had contributed to the worsening of the employment conditions during 2011–2017 though it is hard to establish this in empirical terms. However, see Lahiri (2020) for a discussion of the shock and some evidence of its adverse effect.

Table 2.3 The education bias: employment growth by education category

	Not literate	Up to primary	Up to secondary	Higher secondary and above
Total employment				
1999–2004	−1.2	3.7	3.5	6.9
2004–2011	−2.2	0.5	3.2	5.9
2011–2018	−2.8	−2.6	2.3	3.4

Source Authors' estimates based on data in Appendix Table 2.23

Table 2.4 The education bias and employment growth

	Number of persons (in millions)				Change in number (in millions)		
	1999	2004	2011	2018	1999–2004	2004–2011	2011–2018
Employment of							
Less educated	256.1	264.0	244.4	202.1	7.9	−19.6	−42.3
Educated	127.4	159.3	212.1	256.1	31.9	52.9	44.0

Note Less educated—persons with 0–5 years of education; educated—persons with more than 5 years of education
Source Authors' estimates based on data in Appendix Table 2.23

education.[10] We can then see very clearly that, throughout the period since 1999, a process of shrinking employment opportunities for the less educated has existed alongside a process of expanding employment opportunities for the educated (Table 2.4). Moreover, while the pace of decline of employment of the less educated has been accelerating over time, the pace of expansion of employment of the educated has been decelerating.

The decelerating growth of employment in the economy has been the combined outcome of these two parallel processes. During 1999–2004, employment increased for both the less educated and the educated, but it increased by just 8 million for the less educated and by 32 million for the educated; overall, employment increased by 40 million (i.e., by 8 million per year on average). During 2004–2011, employment of the less educated declined by 20 million while that of the educated increased by 53 million; overall, employment increased by 33 million (i.e., by less than 5 million per year on average). During 2011–2018, employment of the less educated declined by 42 million while that of the educated increased by 44 million so that, overall, employment increased by just two million (or by less than 0.3 million per year on average).

What is remarkable in all this is not the education-biased employment growth per se; we expect the share of the educated in employed population to be rising since the share of the educated in non-student population has also been rising. The remarkable facts are: (i) that, for the less educated, the decline in the number in employment was much faster than the decline in non-student population, and (ii) that, for the educated, the increase in the number in employment was significantly slower than the increase in non-student population. Both facts get reflected in declining employment rates for

[10] The division, of course, is unavoidably arbitrary. We could define "the educated" as those with secondary-and-above level of education or as those with higher-secondary-and above level of education. We have checked that use of these alternative definitions would not substantially alter the arguments and conclusions. In Indian conditions, however, we judge it appropriate to regard persons with middle-level education as educated. Detailed information is available in the Appendix Tables.

Table 2.5 Level of education and employment rate (percentages)

	1999	2004	2011	2018
Not literate	62.2	59.7	55.2	45.3
Up to primary	67.5	67.2	64.2	57.8
Up to secondary	67.3	67.4	64.8	58.8
Higher secondary and above	69.2	69.6	65.4	58.0

Note Employment rates are defined with respect to non-student populations
Source Authors' estimates based on data in Appendix Table 2.23

Table 2.6 Unemployment

	1999	2004	2011	2018
All unemployed (number in million)	9.2	10.4	10.3	27.9
Less educated	1.5	1.9	1.6	3.2
Educated	7.7	8.5	8.7	24.7
Unemployment rate (%)	2.3	2.4	2.2	5.7
Less educated	0.6	0.7	0.7	1.6
Educated	5.6	5.0	3.9	8.8

Source Authors' estimates based on data in Appendix Table 2.23

all education categories (Table 2.5).[11] And the second fact explains the rise in the unemployment rate, as we shall see below.

2.2.3 The Rise in Open Unemployment

Why did the rate of open unemployment increase so sharply during 2011–2018 while it had remained stable at a low level over a long period up to 2011 (Table 2.6)? In 2018, just as in all the preceding years, open unemployment was confined very largely to the educated. Persons with above-primary education accounted for 89% of all unemployed in 2018. The corresponding figures were 84% in 1999, 82% in 2004 and 85% in 2011.[12] The increase in open unemployment between 2011 and 2018, therefore, could only have resulted from a sharp increase in unemployment of the educated. That this was the case is quite clear from Fig. 2.1. In all the years, the rate of unemployment is observed to be rising with rising level of education. But in 2018, there was a spike in unemployment of persons with above-middle-level education.

To see the trends in starker terms, we can consider, once again, just the two categories—the less educated and the educated. The unemployment rate for the less educated was 0.7% in 2011 and 1.6% in 2018 while, for the educated, it was 3.9% in 2011 and 8.8% in 2018 (Table 2.6). Between 1999 and 2011, it is worth noting, the unemployment rate for the less educated had remained virtually constant; it was 0.6% in 1999, 0.7% in 2004 and 0.7% in 2011. On the other hand, the unemployment rate for the educated had shown a declining trend during this period; it was 5.6 in 1999, 5.0 in 2004 and 3.9% in 2011. The overall unemployment rate had remained stable at just over 2% throughout 1999–2011.

It is a remarkable fact that, for the less educated, the number in unemployment had shown only a small increase (of less than 2 million) between 2011 and 2018 even though a large number of them (42 million) had actually suffered loss of employment. It suggests that less educated persons, when confronted with loss of employment, went out of the labour force rather than join the ranks of the unemployed. Thus, for the less educated, labour force growth simply adjusted to employment growth so that

[11] Noticeably, the employment rate dispersion across education levels was rising. For example, the gap between the "not literate" and the "higher secondary and above" increased from 7 percentage points in 1999 to 13 percentage points in 2018.

[12] The unemployed have always been not just "educated" but "young and educated". The unemployment rate has always been high for the "educated persons aged between 15 and 29 years" but insignificant for the "educated persons aged 30 or more years". See Ghose (2016), Box 2.1 (p. 27). Much of the unemployment, therefore, is "educated youth" unemployment. This should not come as a surprise. We should expect the fresh entrants into the labour force to always face much higher unemployment than those who have been in the labour force for some time.

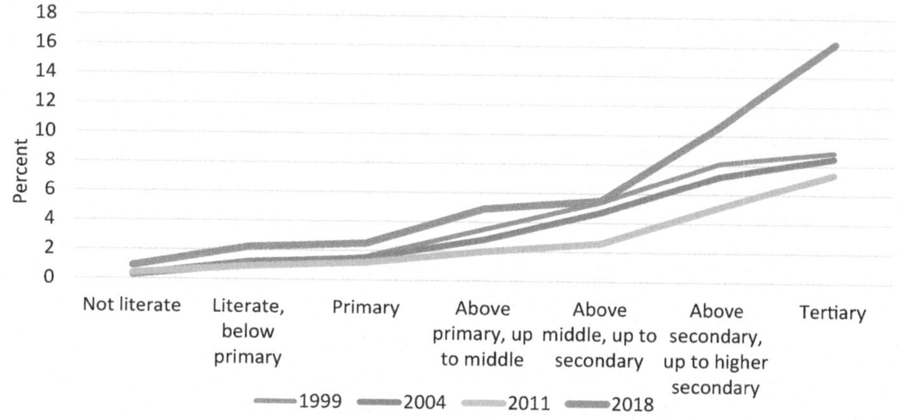

Fig. 2.1 Unemployment rate by education category. *Source* (data) Authors' estimates based on data in Appendix Table 2.23

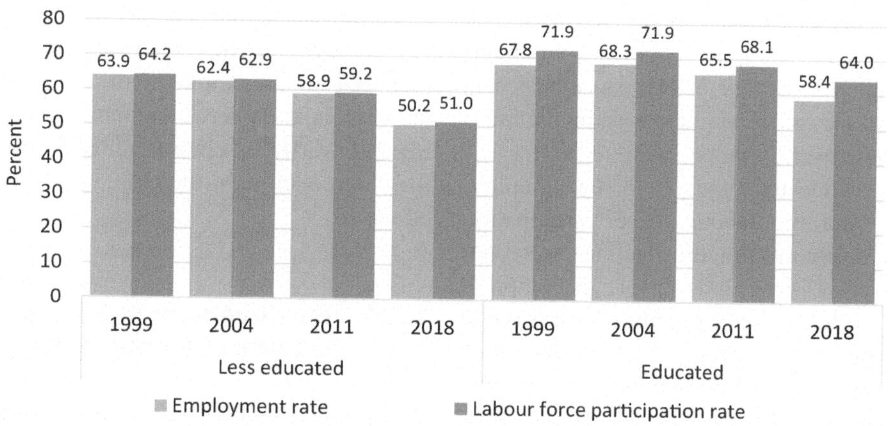

Fig. 2.2 Employment rate and labour force participation rate of the less educated and the educated. *Source* (data) Authors' estimates based on data in Appendix Table 2.23

the labour force participation rate followed the employment rate (Fig. 2.2 and Table 2.7). A similar adjustment is observed to have taken place during 2004–2011 when employment of the less educated had also declined in absolute terms (by 20 million) and yet the number in unemployment had actually declined (by 0.1 million).

All this stands in sharp contrast with what used to happen in the past when, for the less educated, it was employment growth that adjusted to labour force growth in a context where the scope for work sharing (*part-time* employment) was large. That was why, for this category of workers, employment growth was always positive (since labour force growth was always positive), open unemployment was always insignificant and underemployment was always significant. In the period since 1999, it is labour force growth that seems to have been adjusting to employment growth in a context where the scope for work sharing has been dwindling (so that underemployment has been declining). The effect on unemployment remains the same as before: open unemployment of the less educated remains insignificant. But an increasingly larger proportion of

Table 2.7 Growth (% p.a.) of employment, labour force and population

	Less educated			Educated		
	1999–2004	2004–11	2011–18	1999–2004	2004–11	2011–18
Total employment	0.6	−1.1	−2.7	4.6	4.2	2.7
Total labour force	0.6	−1.1	−2.5	4.5	4.0	3.5
Non-student population	1.1	−0.3	−0.4	4.5	4.9	4.3

Note Less educated—persons with 0–5 years of education; educated—persons with more than 5 years of education
Source Authors' estimates based on data in Appendix Table 2.23

the less educated non-student population now stays out of the labour force. So, a new question now arises: how do the less educated survive when they lose employment and move out of the labour force? Loss of employment means not just loss of income but also a simultaneous increase in the dependency ratio (the average number of persons that an employed less educated person must support) for those remaining in employment. The likely consequence is stalled decline (or even increase) in the incidence of poverty.

For the educated, too, both the employment rate and the labour force participation rate were declining, but the declines were not synchronised. During 2004–2011, the employment rate declined by 3 percentage points while the participation rate declined by 4 percentage points so that the unemployment rate actually declined. During 2011–2018, the employment rate declined by 7 percentage points while the participation rate declined by 4 percentage points so that the unemployment rate increased quite sharply. Thus, in the case of the educated, the labour force participation rate either did not adjust or adjusted only partially to the employment rate; when confronted with non-availability of jobs, the educated mostly remained in the labour force and joined the ranks of the unemployed. As a rule, therefore, declining employment rate was associated with both declining labour force participation rate and rising unemployment rate. Between 2011 and 2018, employment of the educated increased by 44 million but this fell seriously short of the increase in educated labour force (60 million) so that the number in unemployment increased by 16 million at the same time. And the increase in educated labour force itself fell seriously short of the increase in educated non-student population (112 million).

We can sum up the findings as follows. For the less educated, declines in employment rate tend to induce matching declines in labour force participation rate. For the educated, declines in employment rate tend to induce declines in labour force participation rate together with increases in unemployment rate. The sharp increase in the overall unemployment rate during 2011–2018 essentially reflected the effect of the sharp decline in the employment rate of the educated.

2.2.4 Improvement in Average Quality of Employment

While the overall employment conditions were worsening, the average quality of employment was improving throughout 1999–2018. This can be discerned from the changing weights of the different types of employment (that are found to exist in India) in total employment. Regular-formal employment is salaried employment that also offers entitlement to some form of social security benefit. Regular-informal employment is salaried employment that offers no entitlement to any kind of social security benefit. Casual employment is employment on a daily basis for a daily wage. Self-employment is engagement in work in own enterprise that generates output and income for those engaged.

In terms of quality, these four types of employment fall into a neat hierarchical order: regular-formal employment is best-quality, regular-informal employment is second-best-quality,

Table 2.8 Structure of total employment by type

	Percentage distribution			
	1999	2005	2012	2018
Type of employment				
Regular-formal		7.0	7.7	9.7
Regular-informal		9.2	11.3	14.1
Regular	15.0	16.2	19.0	23.8
Self	52.3	54.1	51.8	52.0
Casual	32.7	29.7	29.2	24.2
All types	*100*	*100*	*100*	*100*

Source Authors' estimates based on data in Appendix Table 2.25

self-employment is third-best-quality, and casual employment is worst-quality.[13] The change in the average quality of employment during a period can thus be read from the changes in the shares of these different types of employment in total employment (Table 2.8).

The data in Table 2.8 clearly suggest that the average quality of employment was improving throughout the period 1999–2018. The improvement derived basically from a growing regularisation of wage employment in the economy; the share of regular employment, formal and informal, in total employment was increasing while the share of casual wage employment was declining. The share of self-employment in total employment did not show a clear trend.[14]

The steady improvement in the average quality of employment can be read more readily from the changing values of a summary statistic—the Employment Quality Index (EQI); the higher the value of this Index, the higher is the average

Table 2.9 Employment Quality Index (EQI)

	1999	2004	2011	2018
EQI		1.935	1.975	2.093
EQI*	1.897	1.946	1.993	2.115

Note EQI is simply the weighted average of the quality-ranks assigned to different types of employment and is estimated as (percentage share of regular-formal employment × 4 + percentage share of regular-informal employment × 3 + percentage share of self-employment × 2 + percentage share of casual employment × 1)/100. The larger the value of EQI, the higher is the average quality of employment. For 1999, comparable data on formal employment is not available. So, we define EQI* as (percentage share of regular employment × 3.5 + percentage share of self-employment × 2 + percentage share of casual employment × 1)/100 so that comparable estimates can be derived for all four years

Source Authors' estimates based on data in Table 2.8

quality of employment.[15] These data, presented in Table 2.9, show not just that the average quality of employment was improving throughout 1999–2018, but, rather strikingly, also that the improvement was substantially larger during 2011–2018 than during 2004–2011.

All this appears rather puzzling at first sight. How could the average quality of employment have been improving when the overall employment conditions were worsening? And how could the average quality of employment have improved the most when the overall employment conditions also worsened the most? The answer lies in the fact that the very factors that worsened the overall employment conditions also improved the average quality of employment. The simultaneous processes of accelerating decline of employment of the less educated and decelerating growth of employment of the educated worsened the overall employment conditions but also improved the average quality of employment. As we have already seen, these trends had become much sharper during 2011–2018 than they had been during 1999–2011. Hence the period

[13] See Ghose (2016), pp. 26–29 for discussion and evidence.

[14] Several observers have noted that the average quality of regular-formal employment itself has been on the decline in the 2000s. Despite this, however, regular-formal employment still remains the best-quality employment so that the trend in the average quality of employment can still be read from the changes in the shares of the different types of employment in total employment.

[15] See Ghose (2016), Box 2.3, p. 31 and Ghose (2019), Box 3.4, p. 67 for discussions.

Table 2.10 Share (%) of the less educated in employment of different types

	1999	2004	2011	2018
Regular-formal employment		13.1	8.3	6.8
Regular-informal employment		41.9	34.2	28.5
Regular	28.9	29.5	23.7	19.7
Self-employment	66.7	62.3	53.6	46.1
Casual wage employment	84.4	81.1	73.2	64.1
Total employment	66.8	62.5	53.6	44.2

Source Authors' estimates based on data in Appendix Table 2.24

2011–2018 also witnessed the largest improvement in the average quality of employment.

There has always been a systematic relationship between the type of employment and the educational status of the employed. Regular wage employment, formal and informal, has been and is held overwhelmingly by the educated while casual wage employment has been and is held overwhelmingly by the less educated (Table 2.10). Viewed in another way, a large section of the less educated workers has been and still is in casual wage employment while a large section of the educated workers has been and still is in regular wage employment (Table 2.11). As a rule, therefore, growth of jobs for the educated means growth of regular wage employment, formal and informal, while decline of jobs for the less educated means decline of casual wage employment. Thus, simultaneous processes of declining employment of the less educated and increasing employment of the educated worsen aggregate employment conditions but also improve the average quality of employment at the same time.[16]

2.2.5 The Evolution of Employment Conditions, 1999–2018: A Summary View

Progressive exclusion of the less educated from employment and decelerating employment growth of the educated were the two defining trends for the period since the beginning of the millennium. Employment of the less educated was declining at an increasing rate while employment of the educated was increasing at a declining rate. The employment rate (with respect to non-student population) was declining for both but it was declining faster for the less educated than for the educated. For the less educated, however, declining employment rate engendered declining labour force participation rate rather than rising unemployment rate; when confronted with employment loss, the less educated became "discouraged workers" and moved out of the labour force. For the educated, on the other hand, declining employment rate was associated with both declining labour force participation rate and rising unemployment rate. When confronted with non-availability of jobs, only some of the educated moved out of the labour force while others joined the ranks of the unemployed. The large rise in unemployment between 2011 and 2018 was, in essence, a large rise in unemployment of the educated.

Ironically, while the overall employment conditions were worsening, the average quality of employment was actually improving. The reason that these two trends came to co-exist is that they were generated by exactly the same factors: progressive exclusion of the less educated and decelerating employment growth of the educated. It is striking but not surprising that the average quality of employment improved the most when the overall employment conditions worsened the most.

[16] Note the implication that any observed improvement in the average quality of employment does not automatically mean improvement in overall employment conditions.

Table 2.11 Distribution of employed persons by type of employment (percentages)

	Less educated				Educated			
	1999	2004	2011	2018	1999	2004	2011	2018
Regular-formal		1.5	1.2	1.5		16.3	15.2	16.1
Regular-informal		6.2	7.2	9.1		14.3	16.0	18.1
Regular	6.5	7.7	8.4	10.6	32.2	30.6	31.2	34.2
Self	52.1	53.9	51.7	54.2	52.4	54.4	51.9	50.2
Casual	41.4	38.4	39.9	35.2	15.4	15.0	16.9	15.6
	Not literate				Graduates and above			
	1999	2004	2011	2018	1999	2004	2011	2018
Regular-formal		0.8	0.7	0.9		40.7	40.9	40.8
Regular-informal		4.1	5.0	6.8		16.4	18.8	20.9
Regular	4.2	4.9	5.7	7.7	59.8	57.1	59.7	61.7
Self	50.7	53.4	52.1	55.0	38.8	41.5	38.1	36.2
Casual	45.1	41.7	42.2	37.3	1.4	1.4	2.2	2.1

Source Authors' estimates based on data in Appendix Table 2.24

2.3 Economic Growth and Employment

Why were the employment conditions steadily worsening over a fairly long period of high economic growth? To answer this question, we need to analyse the pattern of employment growth across economic sectors and the associated movement of workers across sectors to see how these relate to the employment trends discussed above. For, it is through an examination of the relationship between the pattern of employment growth across sectors and the pattern of output growth across sectors that we can develop an understanding of why growth performed so poorly in terms of employment generation.

2.3.1 Employment in Sectors and Inter-sector Movement of Workers

Underlying the decelerating growth of employment in the economy were two striking trends in sector-level employment (Table 2.12). First, employment in agriculture has been declining at an increasing rate since 2004 (employment growth was close to zero during 1999–2004). Second, employment in non-agriculture has been increasing at a decreasing rate since 1999. These two trends seem to bear remarkable resemblance to two other trends we noted earlier: that employment of the less educated has been declining at an increasing rate since 2004 and that employment of the educated has been increasing at a declining rate since 1999. The resemblance is not accidental; the two sets of trends were in fact closely related, as we shall see below.

The decline of employment in agriculture in the period since 2004 derived basically from decline in casual employment, which has been declining at an accelerating rate since 1999. Self-employment recorded decelerating growth but not persistent decline. And regular employment has never been significant in agriculture so that the observed trends do not mean much.

In non-agriculture, the deceleration in employment growth during 2004–2011 was due solely to deceleration in the growth of self-employment. Casual employment recorded rapid growth in non-agriculture during this period; indeed, the increase in non-agriculture outweighed the decline in agriculture so that casual employment in the economy recorded a positive growth. The sharp deceleration in employment growth in non-agriculture in the next period (2011–2018) resulted from sharp deceleration

Table 2.12 Employment by sector

	Change (in millions)			Growth (per cent per annum)		
	1999–2004	2004–2011	2011–2018	1999–2004	2004–2011	2011–2018
Total (UPSS) employment	39.9	33.2	1.7	2.0	1.1	0.05
Regular	11.1	17.6	22.7	3.6	3.3	3.4
Self	28.8	7.7	1.6	2.7	0.5	0.1
Casual	0.0	7.9	–22.6	0.0	0.9	–2.6
Agriculture	3.1	–18.0	–20.9	0.3	–1.3	–1.4
Regular	0.0	–0.7	–0.7	0.0	–2.9	–3.6
Self	10.2	–3.0	4.3	1.5	–0.3	0.4
Casual	–7.1	–14.3	–24.5	–1.6	–2.5	–5.7
Non-agriculture	36.8	51.1	22.6	4.4	3.5	1.3
Regular	11.1	18.3	23.4	3.8	3.6	3.6
Self	18.6	10.7	–2.7	5.0	1.7	–0.4
Casual	7.1	22.1	1.9	4.1	6.6	0.4

Source Authors' estimates based on data in Appendix Table 2.27

in both self-employment growth and casual employment growth.

Within non-agriculture, the two sectors in which self-employment has traditionally been of much importance are manufacturing and services. In 2004, for example, self-employment accounted for 50% of total employment in each of these two sectors and self-employment in the two sectors together accounted for 95% of total self-employment in non-agriculture.[17] Thus, the decelerating growth of self-employment in non-agriculture during 1999–2018 essentially reflected the decelerating growth of self-employment in these two sectors (Table 2.13). Noticeably, the deceleration was similar in the two sectors during 1999–2011 but was sharper in manufacturing than in services during 2011–2018.[18] Indeed, self-employment in manufacturing declined in absolute terms in the later period.

Table 2.13 Trends in self-employment and casual employment in non-agriculture

	Change (in millions)		
	1999–2004	2004–2011	2011–2018
Self-employment	**18.6 (5.0)**	**10.7 (1.7)**	**–2.7 (–0.4)**
Manufacturing	5.5 (4.8)	3.1 (1.6)	–5.1 (–2.7)
Services	11.5 (4.8)	6.4 (1.6)	1.8 (0.4)
Casual employment	**7.3 (4.2)**	**22.2 (6.6)**	**1.9 (0.4)**
Construction	6.9 (8.8)	20.7 (10.7)	6.0 (2.0)

Note Figures in parentheses are average annual growth rates (percentages)
Source Authors' estimates based on data in Appendix Table 2.25

The only sector of non-agriculture in which casual employment has traditionally been of importance is construction.[19] In 2004, casual employment accounted for 78% of total employment in construction, and the sector accounted

[17] Bulk (97%) of the self-employment in non-agriculture was of course in unorganised or informal part, i.e., in informal manufacturing and informal services. See Appendix Table 2.25.

[18] Arguably, the adverse effect of demonetisation was stronger on informal manufacturing than on informal services.

[19] Again, bulk of the casual employment was in unorganised or informal construction. See Appendix Table 2.25.

Table 2.14 Growth of self-employment

	Change (in millions)			Growth (per cent per annum)		
	1999–2004	2004–2011	2011–2018	1999–2004	2004–2011	2011–2018
Economy						
Self employed	28.8	7.7	1.6	2.7	0.5	0.1
Own account workers	23.4	9.6	16.2	3.7	0.9	1.5
Employers	2.4	0.5	3.6	10.1	1.1	6.3
Unpaid family workers	3.0	–2.4	–18.2	0.8	–0.4	–3.7
Agriculture						
Self employed	10.2	–3.0	4.3	1.5	–0.3	0.4
Own account workers	8.0	0.1	15.0	2.3	0.0	2.6
Employers	0.7	0.0	0.8	5.1	0.0	3.2
Unpaid family workers	1.5	–3.1	–11.5	0.5	–0.7	–2.9
Non-agriculture						
Self employed	18.6	10.7	–2.7	5.0	1.7	–0.4
Own account workers	15.4	9.5	1.2	5.4	1.9	0.2
Employers	1.7	0.5	2.8	17.2	2.2	8.6
Unpaid family workers	1.5	0.7	–6.7	1.9	0.6	–6.6

Note Figures in parentheses are average annual growth rates (percentages)
Source Authors' estimates based on data in Appendix Tables 2.27 and 2.28

for 51% of total casual employment in non-agriculture. Naturally enough, the time-trend in casual employment in non-agriculture basically reflected the time-trend in casual employment in construction. Thus, between 2004 and 2011, casual employment increased by 22 million in non-agriculture when it increased by 21 million in construction alone (Table 2.13). Between 2011 and 2018, casual employment in construction increased by 6 million while this increased by just about 2 million in non-agriculture (implying a decline of 4 million in the rest of non-agriculture). As a matter of fact, casual employment in construction was growing at a rapid and accelerating rate during 1999–2011 but then the growth decelerated very sharply during 2011–2018. Here we need to note that expanding government programmes—rural employment guarantee schemes, rural housing schemes and rural roads schemes—had contributed much to the growth of casual employment in construction during 1999–2011.[20] These programmes did not expand much between 2011 and 2018.

One remarkable aspect of the decelerating growth of self-employment in the economy (in agriculture as well as in non-agriculture) is that this was due entirely to decelerating growth of unpaid family work (Table 2.14). In the aggregate economy, the number of unpaid

[20] This can be seen from the fact that employment growth in construction was much faster in rural areas (10.3% per annum during 1999–2004 and 12.3% per annum during 2004–2011) than in urban areas (4.2% per annum during 1999–2004 and 4.8% per annum during 2004–2011). Employment in construction in rural areas as percentage of employment in construction in the economy was 56.8 in 1999, 63.5 in 2004 and 73.8 in 2011. During 2011–2018, employment growth in construction slowed down in both rural and urban areas (it was 2.4% per annum in rural areas and 0.7% per annum in urban areas) though it continued to be higher in rural areas. The rural share increased to 76% in 2018.

Table 2.15 Growth of employment of the educated in the economy

	Change (in millions)			Rate of growth (% per annum)		
	1999–04	2004–11	2011–18	1999–04	2004–11	2011–18
Regular employment	6.9	19.4	21.7	3.2	5.0	4.2
Self-employment	22.4	20.2	18.6	4.9	3.0	2.3
Casual employment	2.7	13.2	4.0	2.6	6.8	1.5
Total employment	32.0	52.8	44.3	4.6	4.2	2.8

Source Authors; estimates based on data in Appendix Table 2.24

family workers recorded little growth during 1999–2004 and was declining at an increasing rate after 2004. Thus, self-employment was increasingly becoming own account employment; the share of own account workers in all self-employed was 59% in 1999, 62% 2004, 64% in 2011 and 70% in 2018. Between 2011 and 2018, when the number of unpaid family workers declined by as much as 18 million, the number of own account workers increased by 16 million and the number of employers increased by 4 million.[21] These facts are of significance for two reasons. First, of the self-employed, the unpaid family workers had the lowest level of education. The decline of unpaid family work was thus consistent with the decline in the employment of the less educated. Second, the unpaid family workers, when they could no longer find work in family enterprises, were quite unlikely to look for wage employment outside home. And this surely is one reason why, for the less educated, declining employment meant declining labour force.

The broad picture is now clear. In agriculture, the employment of less educated workers—unpaid family workers and casual workers—was declining quite rapidly. While the unpaid family workers are likely to have moved out of the labour force following loss of employment, the casual workers, having lost employment in agriculture, would have looked for jobs in non-agriculture. Between 1999 and 2011, construction was generating jobs at a very rapid pace and most of these jobs were casual. So, many of the casual workers who lost jobs in agriculture could and did move into casual wage employment in construction. After 2011, job growth in construction collapsed so that most of the casual workers who lost employment in agriculture had nowhere to go and were forced out of the labour force. Growth of self-employment—particularly of unpaid family work—in non-agriculture was also decelerating rapidly between 1999–2018; it was negative and large during 2011–2018. Thus, throughout the period 1999–2018, employment opportunities for less educated workers were rapidly dwindling in both agriculture and non-agriculture.

But, even for the educated, employment growth was decelerating in both agriculture and non-agriculture (Table 2.15). While regular jobs in non-agriculture were growing steadily and most of these jobs went to the educated, this did not ensure steady growth of employment of the educated in non-agriculture. For, a majority of the educated workers had in fact been in self-employment throughout the period (see Table 2.9) and the growth of self-employment was decelerating rapidly in both agriculture and non-agriculture. So, even for the educated, employment growth in the economy was decelerating throughout 1999–2018.

[21] The trends, of course, would have looked different had we left the data for 2004 unmodified (see Appendix Table 2.27). Then, the number of unpaid family workers would have shown an inexplicable increase of 23.8 million between 1999 and 2004 and an equally inexplicable decline of 23.2 million between 2004 and 2011 (notice, too, how remarkably similar the two numbers are) with corresponding consequences for trends in self-employment and in total employment in the economy.

Table 2.16 Growth of output (real value added) and employment

	Output growth			Employment growth			Ratio: productivity growth to output growth		
	1999–2004	2004–2011	2011–2018	1999–2004	2004–2011	2011–2018	1999–2004	2004–2011	2011–2018
Economy	6.0	8.1	6.8	2.0	1.1	0.1	0.667	0.864	0.985
Agriculture	1.4	3.6	3.2	0.3	−1.3	−1.5	0.786	1.306	1.469
Non-agriculture	7.1	9.0	7.5	4.4	3.6	1.3	0.380	0.611	0.827
Manufacturing	5.7	8.7	7.6	4.7	2.1	−1.0	0.193	0.770	1.132
Construction	8.2	8.2	4.0	7.8	9.9	1.9	0.049	−0.146	0.525
Mining and utilities	4.2	5.0	6.0	2.5	3.0	6.6	0.405	0.400	−0.100
Services	7.6	9.6	8.1	3.6	2.5	2.0	0.526	0.760	0.753

Note In estimating output growth, the old National Accounts data series (base: 2004–05) have been used for the period 2000–12 and the new data series (base: 2011–12) have been used for the period 2012–18. Employment elasticity is defined as the ratio of rate of growth of employment to rate of growth of output

Source Output growth—authors' estimates based on data in Appendix Table 2.29 and 2.30; employment growth—authors' estimates based on data in Appendix Table 2.25

2.3.2 Economic Growth and Employment

That 1999–2018 was a period of high economic growth (at 6% or more per annum) is evident from the data in Table 2.16. It is also clear that while growth was undoubtedly services-led throughout this period, it still was quite broad-based. The individual sectors, with very few exceptions (agriculture during 1999–2004 and construction during 2011–2018), recorded high growth. We must remember, of course, that there are problems of comparability of the growth rates for the period 1999–2011 with those for the period 2011–2018 since, for the later period, we have to use a new series of national accounts data (with a new base year and changed methodology of estimation). Some have argued that the data for the period after 2011 are flawed so that there is serious overestimation of growth rates.[22] The economy, moreover, suffered two shocks, delivered by the sudden demonetisation of 2016 and the introduction of the Goods and Services Tax in 2017, which are widely thought to have had serious adverse effects on national output.[23] And yet, the observed growth rates of output for the period 2011–2018 do not seem to suggest that the shocks mattered at all. This is no place for undertaking a thorough evaluation of national accounts statistics in light of these arguments, however, and we can do precious little beyond noting them as caveats.

This period of high growth, as it happens, was also a period of sharply declining employment intensity of growth. The data in Table 2.16 show that labour productivity growth was accelerating throughout the period and the ratio of labour productivity growth to output growth was rapidly rising so that the employment elasticity was

[22] See Subramanian (2019a, 2019b), and Morris and Kumari (2019).

[23] See Lahiri (2020) for discussion of the effects of de-monetisation.

rapidly declining.[24] Remarkably, this tendency is observed not just in the aggregate economy but also in both agriculture and non-agriculture. Indeed, it is observed in the major sectors of non-agriculture as well.[25]

The accelerating growth of labour productivity in agriculture is most likely to have been the result of increasing mechanisation of agricultural operations.[26] Accelerating growth of labour productivity in the broadly defined non-agricultural sectors (particularly in manufacturing and services), on the other hand, could conceivably have resulted from any or all of three possible developments: technological advances and associated increases in capital intensity in the constituent (narrowly defined) sub-sectors, changes in the structure of output involving increases in the shares of more technology-and-capital-intensive products, and increase in the share of large enterprises (which generally employ more technology-and-capital-intensive methods of production than small enterprises) in the sector's output. Empirical investigation into the extent and relative significance of these developments within the broad sectors of India's economy is beyond the scope of this paper. What we can say with some confidence is that the rapid and accelerating growth of output per worker that we observe to have occurred in the broad non-agricultural sectors of the economy during the period under study constitutes evidence of rapid technological advances and rising capital intensity—"skill-biased" technological change for short—resulting from some combination of the three possible developments listed above.

But this is only one part of the story. By itself, "skill-biased" technological change can engender education-biased employment growth but not progressive exclusion of the less educated from employment nor decelerating growth of employment of the educated. This is where the phenomenon of rising ratio of labour productivity growth to output growth comes into play. For, this shows that the demand growth in the economy (as reflected in output growth) persistently lagged behind the expansion of production potential generated by the "skill-biased" technological change (as reflected in labour productivity growth). And it is this persistent failure of effective demand to keep pace with productivity growth that explains the decelerating employment growth, which incorporated both the negative growth of employment of the less educated and the positive but decelerating growth of employment of the educated.

It is a remarkable fact that these employment trends are observed not just at the level of the economy but also in each of the broad sectors (Table 2.17). Employment of the less educated declined in agriculture during 1999–2004, in agriculture and services during 2004–2011 and in all sectors except "mining and utilities" (which employed very few less educated workers) during 2011–2018.[27] And the growth of employment of the educated was steadily decelerating throughout 1999–2018 in all sectors except "mining and utilities", which employed few educated workers.[28] The rising trend in the ratio of labour productivity growth to output growth, we may recall, is observed not just in the aggregate economy but also in the broad sectors. The demand growth persistently lagged behind the potential supply growth even in the individual sectors.

[24] Employment elasticity, defined as the ratio of employment growth to output growth, equals [1—ratio of labour productivity growth to output growth]. Hence, rising ratio of labour productivity growth to output growth means declining employment elasticity.

[25] The exceptions are to be found in construction during 1999–2011, in "mining and utilities" during 1999–2018, and in services during 2011–2018.

[26] By 2015/16, percentages of agricultural operations mechanized were: soil working & seed bed preparation: 40; seeding and planting: 29; plant protection: 34; irrigation: 37; harvesting and threshing: 60–70 (for wheat and rice). See National Bank for Agriculture and Rural Development (NABARD), *Sectoral Paper on Farm Mechanization*, Mumbai, 2018.

[27] In 2011, for example, "mining and utilities" employed less than 1 million less educated workers when non-agriculture employed 99 million.

[28] In 2011, for example, "mining and utilities" employed just 3 million educated workers when non-agriculture employed 143 million.

Table 2.17 Employment growth for the less educated and the educated (per cent per annum)

	Less educated			Educated		
	1999–2004	2004–2011	2011–2018	1999–2004	2004–2011	2011–2018
Economy	0.6	−1.1	−2.7	4.6	4.2	2.7
Agriculture	−0.7	−2.9	−3.1	3.7	2.8	1.6
Non-agriculture	3.6	2.0	−1.9	5.1	4.9	3.2
Manufacturing	3.8	0.2	−5.3	5.9	4.0	2.0
Construction	7.1	9.0	−0.6	9.5	11.6	5.8
Mining and utilities	2.3	1.2	4.2	2.8	4.8	7.6
Services	2.2	−0.8	−1.1	4.4	4.2	3.4

Source Authors' estimates based on data in Appendix Table 2.26

2.4 Concluding Observations

In the early 2000s, India's economy reached a turning point very different from the Lewis turning point. Agriculture not just stopped accommodating new workers but was increasingly rendering many of the already employed workers—mostly less educated—redundant. All this was not because non-agriculture was generating employment at a rapid rate and pulling labour out of agriculture. As a matter of fact, non-agriculture was generating employment at an increasingly slower pace and was also generating it basically for the educated. Under these conditions, overall employment conditions were steadily deteriorating with the extent of deterioration increasing over time. Progressive exclusion of the less educated from employment and decelerating employment growth of the educated were the underlying trends that showed up in declining employment rate, rising unemployment rate and, ironically enough, steady improvement in the average quality of employment.

As it happens, the employment conditions were worsening in a period of high economic growth. The proximate reason is that "skill-biased" technological change in production (reflected in accelerating labour productivity growth) was out of sync with demand growth (reflected in output growth) in the economy. This is what is indicated by the fact that the ratio of labour productivity growth to output growth was steadily rising throughout 1999–2018.

Table 2.18 Trends in income distribution

	Share (%) of national income			
	1999	2004	2011	2015
Share (%) of adult population				
Richest 10%	39.5	44.3	54.1	56.1
Middle 40%	39.8	36.8	30.5	29.2
Poorest 50%	20.7	18.8	15.4	14.7

Source World Inequality Database (available online: www.wid.world)

But why was technical progress in production so "skill-biased" in an economy with abundance of low-skilled workers and scarcity of skilled workers? And why was it unaccompanied by commensurate expansion of effective demand? These are the questions that we need to answer if we are to gain a full understanding of why rapid economic growth was accompanied by worsening employment conditions. Unfortunately, the kind of empirical analysis required to answer the questions cannot be undertaken here and has to be left for future research. Here we can offer some reflections on the characteristics of the growth process that produced the observed employment outcomes.

Existing research has highlighted the fact that the incremental incomes generated by India's services-led growth have been going principally to the richest 10% of the adult population (Table 2.18).[29] How is this to be explained?

---

[29] The empirical evidence is analysed in Chancel and Piketty (2019).

Growth of the lead-sector—services—involved growth of skill-intensive services such as information technology and enabled services, financial services and business services. Such growth naturally generated jobs and incomes for the already rich and educated—a thin top layer of the population. The consequent growth of demand stimulated growth of other sectors and subsectors of the economy, which also were skill-intensive and whose growth also had similar employment and distributional consequences. For, demand grew not so much for manufactures and services already being produced but for newer, high-end manufactures (e.g., consumer electronics, white goods, automobiles, etc.) and services (e.g., IT services, e-commerce, shopping malls for retail trade, professional services, private education and health services, and a variety of social and personal services associated with luxury consumption) whose production is more technology-and-skill-intensive. In successive rounds, therefore, the rich beneficiaries of growth generated demand for goods and services intensive in factors of production held by the rich. The overall outcome has been a rapidly growing concentration of jobs and incomes in a narrow segment of the population on the one hand and growth of capital-and-skill-intensive manufacturing and services on the other.

The problem with this kind of growth is that the growth of demand for high-end manufactures and services decelerates rather quickly as the consumption of the rather small (and non-expanding) class of beneficiaries of growth inevitably approaches saturation levels.[30] The capacity to produce persistently grows faster than the demand. This shows up in rising ratio of labour productivity growth to output growth.

When its benefits accrue to a thin layer of already rich population, even rapid economic growth is accompanied by worsening employment conditions and hence fails to bring commensurate development in its wake. Moreover, the rapid growth itself cannot be sustained for long. For growth to improve employment conditions and to be sustainable, the class of beneficiaries has to be continuously expanding and this happens when, in successive rounds, the beneficiaries of growth generate demand for goods and services intensive in factors of production held by the non-beneficiaries. India's growth has not been of this kind.

Appendix 1: Changes in Employment Conditions Between 2017 and 2018

Overall employment conditions appear to have improved between 2017 and 2018 (Table 2.19). Total employment in the economy increased by 11.6 million, which appears impressive, given that it had actually declined by 9.9 million between 2011 and 2017.[31] The employment rate (defined with reference to non-student population), which had declined from 61.6% in 2011 to 54.0% in 2017, increased to 54.5% in 2018. The rate of unemployment in 2018—at 5.7%—was marginally lower than that in 2017 (6.1%) (Table 2.19).

However, a close scrutiny shows the improvement to have been rather insignificant. Of the incremental employment of 11.6 million, 10.2 million (88%) was in agriculture, a sector in which employment had declined by 31 million during 2011–2017 (i.e., by 5 million per year on average). In non-agriculture, employment growth continued to decelerate; it increased by just 1.4 million while it had increased by an average of 3.5 million per year during 2011–2017. The growth of employment in

[30] This would mean either rising saving or rising consumption of goods and services produced outside India or perhaps both. Demand growth could in principle have been maintained by net export growth but India's imports persistently exceeded its exports throughout the period.

[31] *Full-time* employment, which had increased by 12.7 million (or by 2.1 million per year on average) increased by 7.6 million between 2017 and 2018. And *part-time* employment, which had declined by 22.6 million (or by 3.8 million per year on average) increased by 4 million between 2017 and 2018.

Table 2.19 Changes in employment (numbers in millions)

	Total		Less educated		Educated	
	2011–2017	2017–2018	2011–2017	2017–2018	2011–2017	2017–2018
Economy	–9.9	11.6	–46.8	4.5	36.9	7.1
Agriculture	–31.1	10.2	–34.4	5.5	3.3	4.7
Non-agriculture	21.2	1.4	–12.4	–1.0	33.6	2.4
Manufacturing	–1.7	–2.7	–7.3	–1.8	5.6	–0.9
Construction	3.3	3.7	–3.4	2.1	6.7	1.6
Mining and utilities	1.7	0.0	0.5	–0.2	1.2	0.2
Services	17.9	0.4	–2.2	–1.1	20.1	1.5

Source Authors' estimates based on data in Appendix Table 2.26

construction between 2017 and 2018—by about 4 million (compared to 0.6 million per year on average during 2011–2017)—was impressive, but employment in manufacturing showed a large decline—by about 3 million (compared to a decline of about 2 million during the six-year period 2011–2017). Employment in services showed a near-zero growth (compared to a growth of 3 million per year on average between 2011 and 2017). Thus, between 2017 and 2018, employment increased basically in agriculture and in construction; employment in manufacturing recorded a large decline while employment in services showed near-zero growth.

Employment of the less educated, which had declined by 47 million (by 7.8 million per year on average) during 2011–2017, showed an increase of 4.5 million between 2017 and 2018. But the increase was confined to agriculture (5.5 million) and construction (2.1 million), the two sectors in which employment of the less educated had declined very substantially during 2011–2017. And, in manufacturing and services, employment of this category of workers declined much more rapidly than it had done during 2011–2017.

Employment of the educated increased by 7 million (compared to 6 million per year on average during 2011–2017), but nearly 5 million (66%) of the incremental employment was in agriculture (just 9% of the incremental employment of the educated was in agriculture during 2011–2017). Of the incremental employment of 2.4 million in non-agriculture, 1.6 million (67%) was in construction (compared to 20% during 2011–2017). And employment of the educated declined in manufacturing (where it had increased during 2011–2017) and recorded very slow growth in services (where it had recorded rapid growth during 2011–2017).

Thus, between 2017 and 2018, employment growth occurred basically in agriculture and in construction for both the less educated and the educated. The incremental employment in the two sectors was of two different types. The kind of employment that grew in agriculture was self-employment (Table 2.20), which increased by 10 million (after having declined

Table 2.20 Changes in employment (numbers in millions)

	Economy		Agriculture		Non-agriculture	
	2011–2017	2017–2018	2011–2017	2017–2018	2011–2017	2017–2018
Total	–9.9	11.6	–31.1	10.2	21.2	1.4
Regular	23.8	–1.1	0.2	–0.9	23.6	–0.2
Self	–8.4	10.0	–5.7	10.0	–2.7	0.0
Casual	–25.3	2.7	–25.6	1.1	0.3	1.6

Source Authors' estimates based on data in Appendix Table 2.25

Table 2.21 Changes in casual employment (numbers in millions)

	2011–2017	2017–2018
Non-agriculture	0.3	1.6
Construction	2.4	3.6

Source Authors' estimates based on data in Appendix Table 2.25

by about 9 million during 2011–2017). The type of employment that grew in construction, on the other hand, was casual employment (Table 2.21). Thus, for both the less educated and the educated, it was primarily self-employment in agriculture and secondarily casual employment in construction that recorded growth.

The growth of casual employment in construction is easily explained: the sector recorded much higher output growth between 2017 and 2018 than during 2011–2017 (Table 2.22). However, employment growth was actually faster than output growth so that labour productivity declined. The likely explanation is that the government's special employment schemes also expanded between 2017 and 2018, which contributed significantly to growth of casual employment in construction.[32] In agriculture, however, output growth was actually slower between 2017 and 2018 than it had been during 2011–2017. Yet, the growth of employment (essentially of self-employment) between 2017 and 2018 was very rapid (5.5%) so that labour productivity also declined rapidly (by 3.1%). It is not easy to see how and why this kind of employment growth might have occurred. A possible explanation is reverse migration of workers (whose families had remained engaged in agriculture) from manufacturing and services to agriculture. The fact that it was only self-employment that grew is supportive of this view. But the fact that such reverse migration had not happened during 2011–2017 when many less educated workers had lost jobs in non-agriculture then becomes inexplicable.

While this puzzle must remain unresolved here, it is abundantly clear that the employment growth between 2017 and 2018 was associated with serious decline in labour productivity. As such, it does not indicate significant improvement in employment conditions.

Table 2.22 Output (gross value added) growth (per cent per annum)

	Gross value added		Employment	
	2011–2017	2017–2018	2011–2017	2017–2018
Total	6.8	6.0	–0.4	2.6
Agriculture	2.8	2.4	–2.6	5.5
Non-agriculture	7.6	6.7	1.4	0.5
Manufacturing	7.7	5.7	–0.5	–4.6
Construction	3.8	6.1	1.1	7.1
Mining and utilities	6.2	0.2	7.8	0.0
Services	8.4	7.7	2.2	0.3

Source Authors' estimates based on data in Appendix Tables 2.26, 2.29 and 2.30

[32] Person days of employment generated under the MGNREGS, for example, increased from 23.4 billion during 2017–2018 to 26.8 billion during 2018–2019.

Appendix 2

See Tables 2.23, 2.24, 2.25, 2.26, 2.27, 2.28, 2.29 and 2.30.

Table 2.23 Population, non-student population, labour force and employment (age: 15 years or more, in millions)

	Population					Non-student population				
	1999	2004	2011	2017	2018	1999	2004	2011	2017	2018
Not literate	271.1	266.7	245.6	245.9	246.3	270.8	266.2	245.5	245.6	245.9
Below primary	64.3	70.7	78.2	49.6	47.3	63.2	69.4	77.2	49.3	47.0
Primary	71.2	91.2	97.7	107.7	114.5	66.5	86.9	92.5	103.3	110.2
Middle	95.7	116.2	139.6	192.8	194.8	78.6	97.4	114.4	162.9	165.6
Secondary	67.3	75.0	120.2	132.5	134.9	52.6	58.2	87.3	94.3	94.8
Higher secondary	35.1	53.0	88.1	116.6	122.2	25.1	38.6	58.7	75.7	80.6
Tertiary	34.4	43.7	73.4	106.4	108.0	31.4	39.7	65.6	95.4	97.1
Total	**639.1**	**716.5**	**842.8**	**951.5**	**968.0**	**588.2**	**656.4**	**741.2**	**826.5**	**841.2**
	Labour force (UPSS)					**Employment (UPSS)**				
Not literate	169.0	159.4	136.0	112.7	112.5	168.6	158.9	135.5	111.4	111.5
Below primary	43.2	46.9	49.1	28.3	27.8	42.8	46.4	48.7	27.9	27.2
Primary	45.4	59.6	60.9	60.2	65.2	44.7	58.7	60.2	58.3	63.6
Middle	55.1	68.5	77.2	99.8	102.5	53.2	66.7	75.6	94.2	97.5
Secondary	37.2	40.1	56.7	58.8	58.6	35.1	38.1	55.2	55.4	55.4
Higher secondary	18.5	29.3	40.1	50.1	52.7	17.0	27.1	38.0	44.1	47.1
Tertiary	24.2	29.9	46.8	66.0	66.8	22.0	27.4	43.3	55.3	55.9
Total	**392.6**	**433.7**	**466.8**	**475.9**	**486.1**	**383.4**	**423.3**	**456.5**	**446.6**	**458.2**
	Labour force (UPS)					**Employment (UPS)**				
Not literate	152.1	142.5	119.8	107.8	105.9	151.6	141.7	119.2	106.4	104.6
Below primary	40.9	44.6	46.1	27.6	27.1	40.4	44.0	45.5	27.1	26.5
Primary	42.8	55.9	56.8	58.8	63.1	42.0	54.8	55.9	56.8	61.4
Middle	52.4	64.5	72.6	98.1	100.4	50.2	62.1	70.8	92.4	95.3
Secondary	35.5	37.8	53.7	57.9	57.1	33.3	35.5	52.0	54.3	53.7
Higher secondary	17.7	27.7	38.1	48.7	51.1	16.1	25.2	35.7	42.6	45.3
Tertiary	23.9	29.8	46.1	65.5	66.1	21.4	26.6	42.3	54.5	54.9
Total	**365.3**	**402.8**	**433.2**	**464.4**	**470.8**	**355.0**	**389.9**	**421.4**	**434.1**	**441.7**

Note We have available: actual census data on population (rural male, rural female, urban male, urban female) on 1 March of the census years – 1991, 2001 and 2011 – by 5-year age-group; official projections of population (rural male, rural female, urban male, urban female) on 1 July of the years 2011 and 2017; and official projections of population (male, female) on 1 March of 2016 and 2021 by 5-year age-group. We have used linear interpolation to derive estimates of population (rural male, rural female, urban male, urban female) aged 15 years or more on 1 July of the years 1999, 2004, 2011, 2017 and 2018. These population estimates have then been used together with the relevant ratios (non-student population-to-population, labour force-to-population, and workforce-to-population) for rural male, rural female, urban male and urban female, derived from the NSSO surveys, to arrive at estimates of non-student population, labour force and employment

Table 2.24 Level of education and type of employment (UPSS) (numbers in millions)

	1999				2004			
	A	**B**	**C**	**D**	**A**	**B**	**C**	**D**
Illiterate	2.1	4.9	85.4	76.2	1.2	6.6	85.2	66.7
Below primary	1.4	2.7	22.9	15.8	1.0	3.6	24.7	17.1
Primary	1.9	3.7	25.1	14.0	1.7	6.2	32.7	18.0
Middle	3.4	6.1	30.6	13.2	3.4	8.2	38.2	16.6
Secondary	6.2	5.2	18.8	4.9	4.6	5.6	22.6	5.0
Higher secondary	4.5	2.5	8.8	1.2	6.6	4.4	14.1	1.8
Graduates and above	9.9	3.2	8.5	0.3	11.2	4.5	11.4	0.4
All	**29.4**	**28.3**	**200.1**	**125.6**	**29.7**	**39.1**	**228.9**	**125.6**
	2011				**2017**			
	A	B	C	D	A	B	C	D
Illiterate	1.0	6.8	70.8	57.2	1.0	8.0	60.9	41.5
Below primary	0.7	4.1	24.8	19.3	0.4	3.2	14.8	9.6
Primary	1.2	6.7	31.1	21.2	1.3	7.5	31.1	18.3
Middle	2.6	10.4	41.6	20.9	4.4	15.2	51.0	23.4
Secondary	4.5	8.4	32.2	10.0	5.1	10.7	29.7	9.8
Higher secondary	7.4	6.9	19.7	4.0	8.9	9.3	21.6	4.5
Graduates and above	17.6	8.1	16.4	0.9	23.5	11.7	19.1	1.1
All	**35.0**	**51.4**	**236.6**	**133.5**	**44.6**	**65.6**	**228.2**	**108.2**
	2018							
	A	B	C	D				
Illiterate	1.0	7.6	61.3	41.6				
Below primary	0.5	2.6	14.6	9.5				
Primary	1.5	8.3	33.8	20.0				
Middle	4.7	15.3	53.5	24.0				
Secondary	4.9	9.7	31.0	9.8				
Higher secondary	8.9	9.6	23.8	4.8				
Graduates and above	22.8	11.7	20.2	1.2				
All	**44.3**	**64.8**	**238.2**	**110.9**				

Note A—regular-formal employment, B—regular-informal employment, C—self-employment, D—casual employment

Source The distributions of the employed by level of education are derived from the NSSO surveys; estimates of employed population are from Appendix Table 2.23

Table 2.25 Employment (UPSS) by type (numbers in millions)

	Economy					Agriculture				
	1999	2004	2011	2017	2018	1999	2004	2011	2017	2018
Regular formal	29.4	29.7	35.0	44.6	44.3	1.9	1.2	1.3	1.2	0.2
Regular informal	28.3	39.1	51.4	65.6	64.8	1.9	2.6	1.8	2.1	2.2
Regular	57.7	68.8	86.4	110.2	109.1	3.8	3.8	3.1	3.3	2.4
Self employed	200.1	228.9	236.6	228.2	238.2	132.5	142.7	139.7	134.0	144.0
Casual workers	125.6	125.6	133.5	108.2	110.9	93.7	86.6	72.3	46.7	47.8
Total	**383.4**	**423.3**	**456.5**	**446.6**	**458.2**	**230.0**	**233.1**	**215.1**	**184.0**	**194.2**
	Manufacturing					**Construction**				
Regular formal	5.5	5.7	6.4	9.0	9.2	0.3	0.3	0.6	0.9	0.8
Regular informal	7.9	11.1	15.4	15.6	15.1	0.7	0.9	2.1	2.2	2.2
Regular	13.4	16.8	21.8	24.6	24.3	1.0	1.2	2.7	3.1	3.0
Self employed	20.6	26.1	29.2	24.8	24.1	3.1	4.6	5.2	5.7	5.9
Casual workers	7.5	9.1	9.2	9.1	7.4	13.1	20.0	40.7	43.1	46.7
Total	**41.5**	**52.0**	**60.2**	**58.5**	**55.8**	**17.2**	**25.8**	**48.6**	**51.9**	**55.6**
	Services					**Non-agriculture**				
Regular formal	20.6	21.7	25.8	32.3	32.3	27.5	28.5	33.7	43.4	44.1
Regular informal	17.5	24.3	32.1	44.2	44.1	26.4	36.5	49.6	63.5	62.6
Regular	38.1	46.0	57.9	76.5	76.4	53.9	65.0	83.3	106.9	106.7
Self employed	43.9	55.4	61.8	63.1	63.6	67.6	86.2	96.9	94.2	94.2
Casual workers	9.8	8.2	9.9	7.9	7.9	31.9	39.0	61.2	61.5	63.1
Total	**91.8**	**109.6**	**129.6**	**147.5**	**147.9**	**153.4**	**190.2**	**241.4**	**262.6**	**264.0**
	Non-agriculture, organised					**Non-agriculture, unorganised**				
Regular formal	26.5	27.8	33.1	41.8	41.7	1.0	0.7	0.6	1.6	2.8
Regular informal	8.1	12.7	23.1	26.6	27.4	18.3	23.8	26.5	36.9	34.8
Regular	34.6	40.5	56.2	68.4	69.1	19.2	24.5	27.1	38.5	37.6
Self employed	1.8	2.6	2.7	3.1	2.7	65.8	83.6	94.2	91.1	91.8
Casual workers	5.9	10.3	19.2	14.6	13.1	26.0	28.7	42.0	46.9	49.7
Total	**42.3**	**53.4**	**78.1**	**86.1**	**84.9**	**111.1**	**136.8**	**163.3**	**176.5**	**179.1**

Note Casual workers in the organised sector include those employed in public works and special employment schemes
Source Distributions of the employed by employment status/sector are derived from the NSSO surveys; estimates of employed population are from Appendix Table 2.23

Table 2.26 Employment by level of education (in millions)

	Economy					Agriculture				
	1999	2004	2011	2017	2018	1999	2004	2011	2017	2018
Illiterate	168.6	158.9	135.5	111.4	111.3	130.9	117.4	89.8	70.9	72.9
Below primary	42.8	46.4	48.7	27.9	27.2	26.3	27.5	26.4	14.1	14.0
Primary	44.7	58.7	60.2	58.3	63.6	25.3	31.2	29.0	25.8	29.4
Middle	53.2	66.7	75.6	94.2	97.5	26.2	31.7	33.0	37.5	38.4
Secondary	35.1	38.1	55.2	55.4	55.6	13.1	14.5	21.0	18.1	19.4
Higher Secondary	17.0	27.1	38.0	44.1	47.1	5.3	7.2	10.8	11.5	12.9
Graduates and above	22.0	27.4	43.3	55.3	55.9	2.9	3.6	5.1	6.1	7.2
Total	**383.4**	**423.3**	**456.5**	**446.6**	**458.2**	**230.0**	**233.1**	**215.1**	**184.0**	**194.2**
	Non-agriculture					**Manufacturing**				
Illiterate	37.6	39.4	45.1	40.5	38.4	12.0	13.1	12.6	9.3	7.8
Below primary	16.5	18.9	22.2	13.7	13.2	5.3	5.9	6.7	3.5	3.2
Primary	19.4	27.5	31.2	32.6	34.2	6.2	9.3	9.5	8.7	8.7
Middle	27.0	35.0	42.6	56.7	59.1	7.5	10.7	12.0	14.4	14.3
Secondary	22.1	23.6	34.1	37.2	36.2	5.2	5.9	8.2	9.0	8.3
Higher Secondary	11.7	20.0	27.2	32.6	34.2	2.5	4.0	6.2	7.3	7.8
Graduates and above	19.1	23.8	38.4	49.3	48.7	2.8	3.2	5.0	6.3	5.7
Total	**153.4**	**190.2**	**241.4**	**262.6**	**264.0**	**41.5**	**52.1**	**60.2**	**58.5**	**55.8**
	Construction					**Services**				
Illiterate	7.1	9.2	16.9	14.5	15.4	17.5	18.0	16.2	15.9	14.6
Below primary	2.4	3.4	6.3	4.0	4.0	8.4	9.2	8.9	6.1	5.6
Primary	2.6	4.5	8.1	9.4	10.1	10.2	13.2	13.1	14.0	14.7
Middle	2.8	4.8	9.0	12.9	14.4	16.1	19.0	21.0	28.6	29.5
Secondary	1.3	1.7	4.8	5.9	6.2	15.0	15.5	20.7	21.5	21.0
Higher Secondary	0.5	1.0	2.3	3.4	3.5	8.5	14.5	18.3	21.1	22.1
Graduates and above	0.5	0.6	1.2	1.8	1.7	15.5	19.6	31.4	40.3	40.4
Total	**17.2**	**25.2**	**48.6**	**51.9**	**55.3**	**91.2**	**109.0**	**129.6**	**147.5**	**147.9**

Source Distributions of the employed in different sectors by level of education are derived from the NSSO surveys; estimates of employment in sectors are from Appendix Table 2.25

Table 2.27 Modification of data on self-employment in 2004 (numbers in millions)

	1999	2004 (u)	2004 (m)	2011	2017	2018
Self employed	**200.1**	**249.7**	**228.9**	**236.6**	**228.2**	**238.2**
Employers	3.9	6.3	6.3	6.8	9.1	10.4
Own account workers	118.1	141.5	141.5	151.1	161.7	167.3
Unpaid family workers	78.1	101.9	81.1	78.7	57.4	60.5
Rural male	28.4	33.5	28.4	28.4	21.6	21.8
Rural female	40.6	55.4	39.7	38.7	27.3	31.7
Urban male	5.6	7.5	7.5	7.2	5.1	4.3
Urban female	3.5	5.5	5.5	4.4	3.4	2.7

Note 2004 (u) – unmodified data for 2004; 2004 (m) – modified data for 2004. Only the figures for rural male unpaid family workers and rural female unpaid family workers are modified. The modified figures are simple averages of the figures for 1999 and 2011. If we leave the data for 2004 unmodified, we observe the number of unpaid family workers to increase by 23.8 million between 1999 and 2004 and to decline by 23.2 million between 2004 and 2011
Source Distributions of the employed by employment status are derived from NSSO surveys; estimates of employment are from Appendix Table 2.23

Table 2.28 Self-employment in agriculture and non-agriculture (numbers in millions)

	1999	2004 (u)	2004 (m)	2011	2017	2018
Agriculture						
Self employed	132.5	155.7	142.7	139.7	134.0	144.0
Employers	2.5	3.2	3.2	3.2	3.2	3.9
Own account workers	67.3	75.3	75.3	75.4	84.7	90.5
Unpaid family workers	62.7	77.2	64.2	61.1	46.1	49.6
Non-agriculture						
Self employed	67.6	94.0	86.2	96.9	94.2	94.2
Employers	1.4	3.1	3.1	3.6	5.9	6.5
Own account workers	50.8	66.2	66.2	75.7	77.0	76.9
Unpaid family workers	15.4	24.7	16.9	17.6	11.3	10.8

Source Same as in Appendix Table 2.5

Table 2.29 A: gross value added (Rupees crores in constant 2004–05 prices)

	1999	**2000**	**2001**	**2002**	**2003**	**2004**
Agriculture	529,850	528,016	559,809	518,956	568,642	565,426
Manufacturing	335,944	361,613	369,997	394,980	419,715	453,225
Construction	149,540	158,701	164,685	177,674	198,266	228,855
Mining and utilities	119,816	122,386	124,251	132,974	137,546	147,703
Services	1,060,054	1,189,802	1,268,909	1,355,102	1,463,967	1,576,255
2005	**2006**	**2007**	**2008**	**2009**	**2010**	**2011**
594,487	619,190	655,080	655,689	660,987	713,477	739,495
499,020	570,458	626,073	656,302	730,435	801,476	823,023
258,129	284,806	314,595	332,329	354,436	390,692	412,412
153,264	165,940	175,427	181,105	192,077	201,711	207,063
1,748,173	1,923,970	2,121,561	2,333,251	2,578,165	2,829,650	3,061,589

Source Central Statistical Office, National Accounts Statistics (available online)

Table 2.30 B: gross value added (Rupees crores in constant 2011–2012 prices)

	2011	2012	2013	2014	2015	2016	2017	2018
Agriculture	1,501,947	1,524,288	1,609,198	1,605,715	1,616,146	1,726,004	1,828,329	1,872,339
Manufacturing	1,409,986	1,486,873	1,560,709	1,683,938	1,903,850	2,054,764	2,190,791	2,316,643
Construction	777,335	780,050	800,771	835,229	865,335	916,445	962,009	1,020,314
Mining and utilities	447,703	454,244	462,708	502,732	542,132	595,744	640,600	641,629
Services	3,969,975	4,300,820	4,630,263	5,084,519	5,564,407	6,035,327	6,452,684	6,952,203
GVA	8,106,946	8,546,275	9,063,649	9,712,133	10,491,870	11,328,284	12,074,413	12,803,128

Source Central Statistical Office, National Accounts Statistics (available online)

References

Chancel L, Piketty T (2019) Indian income inequality, 1922–2015: from British raj to billionaire raj? Rev Income Wealth 65(51):533–562

Ghose AK (2013) The enigma of women in the labour force. Indian J Labour Econ 56(4):635–642

Ghose AK (2016) India employment report 2016: challenges and the imperatives of manufacturing-led growth. Oxford University Press and Institute for Human Development, New Delhi

Ghose AK (2019) Employment in India. New Delhi: Oxford University Press (Oxford India Short Introductions)

Kannan KP, Raveendran G (2019) From Jobless to job-loss growth: gainers and losers during 2012–2018. Econ Pol Wkly 54(44):38–44

Kapsos S, Silberman A, Bourmpoula E (2014) Why is female labour force participation declining so sharply in India? Research Paper No. 10, International Labour Office, Geneva

Lahiri A (2020) The Great Indian demonetization. J Econ Perspect 34(1):55–74

Mehrotra S, Parida JK (2019) India's employment crisis: rising education levels and falling non-agricultural job growth. Working Paper 4, Centre for Sustainable Employment (Azim Premji University), Bengaluru

Morris S, Kumari T (2019) Overestimation in the growth rates of national income in recent years?—An analysis based on extending GDP04–05 through other indicators of output", Working Paper 2019–01–01, Indian Institute of Management, Ahmedabad

Nath P, Basole A (2020) Did employment rise or fall in India between 2011 and 2017? Estimating absolute change in the workforce", Working Paper 29, Centre for Sustainable Employment (Azim Premji University), Bengaluru

Subramanian A (2019a) India's GDP Mis-estimation: Likelihood, Magnitudes, Mechanisms, and Implications. Faculty Working Paper 354, Center for International Development (Harvard University), USA

Subramanian A (2019b) Validating India's GDP growth estimates. Faculty Working Paper 357, Center for International Development (Harvard University), USA

3 Potential of the Manufacturing Sector to Absorb Rural Labour: A Case Study of Two Towns in Bihar

Tanuka Endow and Sunil Kumar Mishra

3.1 Introduction

There has been much discussion around India's demographic dividend resulting from a faster growth of the working-age population vis-à-vis the total population. Some estimates indicate that with this dividend, the dependency ratio[1] is slated to decline from 59.6% in 2012 to 50.8% by 2026 (Krishnamurty 2015). The youth population (15–24 years) is expected to increase from 233 million in 2011 to 251 million in 2021 and then go on to decline to 229 million in 2036.[2] While the potential for harnessing a youthful population for the economic growth process in the country has been highlighted in the policy discourse, it has also been realized that without proper deployment of potential workforce, not only will it be a missed opportunity for India but it may also give rise to adverse consequences for the society. The recent pronouncements of 'Make in India' by the new government and the skill development initiatives undertaken by both the previous and the present government are all efforts towards productively absorbing the labour force available in large numbers in the country. The manufacturing sector, which has a weight of 75.5% in the Index for Industrial Production in India, is therefore very much at the centre stage of deliberations for Indian policymakers at present.

Industrialization, with a focus on manufacturing, has traditionally been the route to economic growth for many developed countries in

[1] The estimates are from Krishnamurty (2015). The dependency ratio is defined as the proportion of 0–14 years plus 60 years and over population to population aged 15–59 years.

The paper is based on a study supported by Indian Council of Social Science Research and was carried out by the Institute for Human Development, New Delhi. The draft report was co-authored by the authors and Dr Abhay Kumar. The authors are grateful to Professor Alakh N. Sharma, Prof.Ajit Ghose, Prof. Dev Nathan, and Prof. Sandip Sarkar from the IHD for their comments. They also acknowledge the contributions of Vikas Dubey of IHD for data processing and B.K.N.Singh, IHD, for the field survey. The chapter draws on two papers: 'Potential of the Manufacturing Sector to Absorb Rural Labour: A Case Study of Two Towns in Bihar' (Volume 35, Issue No. 3&4, pp. 305–322) and 'Rural–Urban Linkages in Bihar Manufacturing Sector and Jobs for Unskilled Labour', the latter based on a primary survey in six towns in Bihar in 2015, published by the authors in the Economic and Political Weekly (Vol. 56, Issue No. 45–46, pp. 51–59). The requisite permissions have been obtained.

T. Endow (✉) · S. K. Mishra
Institute for Human Development, New Delhi, India
e-mail: tanuka.endow@ihdindia.org

[2] The Report of the Technical Group on Population Proiections, Ministry of Health and Family Welfare, Government of India, July,2020 accessed at https://main.mohfw.gov.in/sites/default/files/Population%20Projection%20Report%202011-2036%20-%20upload_compressed_0.pdf.

A. Alakshendra et al. (eds.), *Development, Inclusion and Sustainability*,
Sustainable Development Goals Series, https://doi.org/10.1007/978-981-97-6863-9_3

the West as well as for some countries in East Asia. However, it is the service sector which has emerged as the driver of growth among the South Asian countries including India over the last 25 years (Ghani and Kharas 2010). Not only have services been the largest contributor to GDP levels but has also been the largest contributor to GDP growth in South Asia.

Yet, service sector successes have been based on workers with relatively high education levels (Dahlman 2010). India, which experienced the highest rate of growth in South Asia, benefited from export growth in information-enabled services 'based on a critical mass of highly educated workers'.[3] Given that usually service-led growth has been associated with educated and skilled workers, a very persuasive argument for choosing the manufacturing route for growth, particularly in the Indian context, is that only manufacturing is capable of productively absorbing the '*...large mass of surplus low-skilled labour in agriculture that must be moved to productive employment in non-agriculture*' (Ghose 2014).

South Korea has often been held up as an example of manufacturing-led growth. The experience of South Korea showed that the share of employment in manufacturing started very low at around 5% during the sixties and rose over time for more than two decades to almost 30% before starting to decline after a fairly high level of GDP has been reached (Amirapu and Subramanian 2015). It has been found for India that the dominant employment within different manufacturing sectors is low-medium-skilled employment, and that the manufacturing sectors such as 'textiles' and 'wood and wood products, furniture and fixtures' create low-medium-skilled employment, both directly as well as indirectly (Bhattacharya and Bhandari 2021). Thus, manufacturing has a high potential for absorbing low-skilled labour to promote inclusive development.

Thus India, which currently has a huge labour force with low levels of education, possibly cannot afford to ignore the manufacturing and industrialization route for development. It has also been felt that service-led growth is difficult to sustain in view of resultant trade deficits (Ghose 2014). This is notwithstanding the decline in employment intensity in manufacturing over time, since indirect employment in manufacturing is also generated via the services needed for manufacturing.

The sectors such as manufacturing, as well as sub-sectors within services to assess their potential as a vehicle for transforming the Indian economy have been assessed in the Economic Survey 2014–15.[4] It is pointed out that within manufacturing, it is the formal or the registered manufacturing sector that has some of the requisite characteristics. But such manufacturing processes require skilled workers and thus do not use India's comparative advantage in low-cost, unskilled labour, opting for the costlier skilling alternative. The construction sector, on the other hand, uses such labour. However, this sector would not be suitable as a transformative sector due to its lack of tradability and productivity issues. Overall, no clear winner emerged among various strategies but it was observed that in case the manufacturing option is selected, then education and skilling would be a pre-requisite.

Whichever route is adopted for the Indian economy for moving ahead, the surplus labour, due to new entrants as well as release of workers from agriculture sector, would have to be productively absorbed. In India, rural population forms the bulk of the total population and the majority of the workforce is still in the rural areas. According to the 2011 census, the percentage share of rural population was as high as 69%, and in absolute terms, the number of people living in rural areas was 833 million. As of 2011–12, out of a total workforce of 474.2

[3] Dahlman in Ghani (ed) (2010).

[4] Five characteristics have been used to assess the transformational potential of a sector: high levels of productivity, rapid rate of growth in productivity, ability to attract resources, alignment with country's underlying resources and tradability of the sector (Government of India 2015).

million employed in the country, around 49% (231.9 million) was employed in agriculture (Mehrotra et al. 2014).

But in the second half of the 2000s, a structural shift has been observed (ibid). While the total workforce increased from 374 million in 1993–94 to around 474.2 million in 2011–12, the workforce in agriculture underwent a decline from 241.5 million to 231.9 million in the corresponding period. Following an initial rise after 1993–94 in agricultural workforce, there was a massive decline after 2004–05, which was absorbed to a great extent by non-agricultural activities (ibid).

One way of absorbing labour that is unemployed or underemployed in rural areas is to provide them productive employment opportunities in urban areas. The urban centres can work as a pull-factor for the surrounding rural areas through many rural–urban linkages, providing scope for absorbing workers from rural areas. They can also provide growth impulses to surrounding rural hinterland through material linkages (backward and forward), especially in manufacturing activities. The process of urbanization has traditionally been closely associated with industrialization. An urban centre in India is *defined,* among other criteria, in terms of the share of male population engaged in non-agricultural activities.[5]

In the context of the above discussion, a recent study in two towns of different sizes in Bihar, one Class-I[6] and the other a Class-II town, [7] looked at the sources of urban output and growth as well as rural–urban linkages there in terms of employment and input–output with the help of detailed enterprise surveys. Ironically, it was found that although the manufacturing sector shows the strongest rural–urban linkages compared to other sectors such as retail or trade and other services, making it the sector with the maximum potential to create employment for absorbing the surplus labour in the rural areas, this sector has been languishing in the sample towns and is constrained by severe infrastructure bottlenecks, among other constraints. The findings are discussed in greater detail below.

3.2 The Context

There has been increasing urbanization in India over time, albeit at a slow pace. Starting from 1991, the census figures indicate urbanization of 25.5% which rose to 27.2% in 2001 and further to 31.2% in 2011.[8] It is projected that around half of India's population will be urban by the year 2050.[9] The urban landscape, however, is quite diverse, ranging from small urban settlements to mega-cities. The towns and cities have been classified according to different criteria from time to time. At present, the census has six size classes for urban centres: (i) Class I, population 1 lakh and above, (ii), Class II, population 50,000–99,999, (iii) Class III, population 20,000–49,999, (iv) Class IV, population 10,000–19,999, (v) Class V population 5000–9,999 and (vi) Class VI, population below 5000.

In another earlier classification used in NSS 1987–88 (43rd) and 1987–88 (50th) rounds, the following categories have been used[10] (Mahadevia and Sarkar 2012):

[5] A human settlement is called urban when it fulfils three criteria: (i) it has a minimum population of 5000, (ii) it has a population density of at least 400 per sq km, and (iii) it has 75% of the male population working in non-agricultural sector. Such activities, apart from manufacturing, can include construction, trade and service-related activities.

[6] A Class-I town has a population 1 lakh and above.

[7] A Class-II town has a population of 50,000–99,999.

[8] In the Indian context, a human settlement is called urban when it has a minimum population of 5000, has a population density of at least 400 per sq km, and has 75% of the male population working in non-agricultural sector.

[9] 'Sustainable Development Goals: Strengthening rural–urban linkages is the key for India', by Shenggen Fan & PK Joshi accessed at https://www.financialexpress.com/opinion/sustainable-development-goals-strengthening-rural-urban-linkages-is-the-key-for-india/663049/.

[10] In subsequent rounds, these categories have been re-organized.

Table A1 Urban population percentage shares for size-class of towns in Bihar, 2001 and 2011

	Class I 1 lakh and above	Class II 50,000–99,999	Class III 20,000–49,999	Class IV 10,000–19,999	Class V 5,000–9,999	Class VI Less than 5,000	Total
2001	59.3	13.7	23.6	3.2	0.2		100.0
2011	57.5	15.6	21.6	2.8	2.2	0.3	100.0

Source Calculations based on data available from Census 2001 and 2011
Note The percentage shares vary slightly from those indicated by Mahadevia and Sarkar (2012), which are likely to be due to variations in estimates after provisional population figures are finalized

- Small Towns—Size class 4 (C4), population up to 50,000.
- Medium Towns—Size class 3 (C3) population 50,000 to 200,000.
- Large cities—Size class 2 (C2), population 200,000–1 million.
- Metropolitan cities—Size class 1 (C1), population more than 1 million.

The Small and Medium Towns (SMTs) have aroused interest from time to time as an alternative option for economic and social development, as a counterpoint to the concentration of high-end livelihood opportunities, economic and social infrastructure development, etc., in large cities and metropolises. It has been felt that the SMTs could be a vehicle for extending urban services to rural areas via rural–urban linkages and generally lead to some sort of decentralization in the process of development away from the larger cities (Minocha and Yadav 1989 in Mahadevia and Sarkar 2012). The SMTs could also give a boost to local enterprises and help in wealth accumulation locally. However, developing countries have not been very successful in reshaping city-centric urban development and making it more widespread and taking it to the SMTs (Hardoy and Satterthwaite 1986 cited in Mahadevia and Sarkar 2012).

In the above context, the 2001 census data show that the relative concentration of urban population in the state of Bihar in SMTs at 63.71% was much higher compared to the all-India average of 49.87% (Table 1.5, p 17 in Mahadevia and Sarkar 2012). Both 2001 and 2011 census data are available for size class of Class I–Class VI towns (see Table A1 in Annexure). Comparing the data for the 2 years, it is seen that, even after the lapse of a decade, there is not a big difference in percentage shares of population in different towns categorized by size class. The share of Small towns remains more or less the same after a decade at around 27%. While the share for Medium towns is not clear from Table A1, it can be seen that there has been a slight increase in the population share of one component of Medium towns, i.e. in Class II towns, from 13.7 to 15.6%. Thus, the share of population living in Small and Medium towns would be higher than 42.6%, once we take into account the share of people residing in towns with population between 1 and 2 lakh. Thus, even in 2011, the share of urban population living in SMTs is likely to be nearly half the urban population in the state, underscoring the need for a close look at the urban development in such towns.

3.2.1 The Survey

It has been contended that SMTs can play an important role in spreading urban development via rural–urban linkages. Bihar is a state characterized by low industrialization and urbanization. In 2011, the urbanization was only 11.3% vis-a-vis 31.1% urbanization at the national level. Yet it is the state with the highest population density in the country at 1106 people per sq. km. National population density increased from 324 in year 2001 to 382 in 2011, while that for Bihar increased from 880 in 2001 to 1106 in 2011 (Census, 2001 and Census 2011). The high fragmentation of land, large number of people depending on agriculture for livelihood and the poor economic base in the state, have led to migration as a survival strategy, and remittances today comprise a major source of earnings for the people there.[11]

An attempt was made to understand the urban development in two towns in Bihar with the help of enterprise surveys. The towns were selected on the basis of their size, location, main economy, level of economic development, proximity to rural centres and large urban centres. Madhubani, a Class II town with a population of 75,736 (2011) is located in northern Bihar and belongs to one of the poorest regions of the state. Biharsharif, a Class-I town, has a population of 2.97 lakhs (2011). It is not a Medium Town, but its population is much closer to the lower end limit for a large city, and thereby, to that of a Medium town, than to the higher end limit, which is a million. It is a prosperous urban centre, close to Patna, and is the only town in Bihar with manufacturing activities as a dominant activity, i.e. manufacturing employs more workers there compared to trade and transport and services. The two towns belong to districts with very different urbanization rates. Madhubani town is the district headquarters of Madhubani district, which has an urbanization of 3.6%, while Biharsharif belongs to Nalanda district, which has an urbanization of 15.9%.

The enterprise survey was conducted in 2014 in the factories, shops and establishments in the sample towns and covered identification particulars of the firms, details for enterprise, workers, related information about the own account enterprises, factory workers and challenges and problems faced by all types of enterprises.

Using lists of all the factories and shops/establishments from the concerned departments of the state government, smaller numbers of units were selected on the basis of stratified random sampling. The stratification was done on the basis of broad classification of activities. Among each group, proportionate sample has been drawn. Using the above methodology, an enterprise survey of 251 units was conducted in the two sample towns; 139 units were covered in Biharsharif and 112 units covered in Madhubani.

3.3 Sources of Urban Output

The survey indicated a general industrial decline in the sample towns. In Biharsharif, there is fertile land, high agricultural productivity and plentiful crops such as rice, wheat, maize, potato, onion, tomato, cauliflower, cabbage and green chili. But there is a string of closed industrial ventures, where the major blame lay with an alarming power situation. Biharsharif earlier had a number of cold storage units in the surrounding areas to match its abundant agricultural produce. It was found during the survey that many of these cold storage units, some thread and paper-making factories, bidi-making factories and a shoe factory closed down in the last 20–30 years.

Madhubani, located in North Bihar, is less developed than Biharsharif and has much lower urbanization. It is a small town with nearby areas producing *makhana*,[12] fish, vegetables, etc. Madhubani painting and handicraft based on Sikka grass is also renowned. In Madhubani district, a thread factory and sugar factory closed down in the past and key informants reported that Madhubani has little industrial activity at present.

In terms of types of activity, the findings from the enterprise survey indicated that the main source of urban output is traded in both sample towns. Manufacturing and other services came second and third, respectively (Table 3.1). There were a handful of units (six) engaged in service and trade simultaneously. Manufacturing units included agro-processing units such as rice, oil and flour mills, as well as furniture units, sawmills, etc., and were very small scale in size. Trading was observed in food, textiles & garments, utensils, electrical items, jewellery, watches, shoes, etc. Other services comprise

[11] For details, see Rodgers (2013).

[12] Makhana, or fox-nuts, are usually grown in ponds and North Bihar is an important producer of makhana.

Table 3.1 Distribution of surveyed enterprises by type of activity (% in brackets)

Type of unit	Biharsharif	Madhubani	Total
Manufacturing	38 (27)	8 (7)	46 (18)
Trade	84 (60)	87 (78)	171 (68)
Other services	12 (9)	16 (14)	28 (11)
Service and trade	5 (4)	1 (1)	6 (3)
Total	139 (100)	112 (100)	251 (100)

Source Field survey for ICSSR supported study, IHD

Table 3.2 Distribution of Surveyed Enterprises by Type of Enterprise (% in brackets)

Type of enterprise	Biharsharif	Madhubani	Total
Own account enterprises	37 (27)	56 (50)	93 (37)
Non-directory enterprises	72 (52)	43 (38)	115 (46)
Directory enterprises	30 (21)	13 (12)	43 (17)
All enterprises	139 (100)	112 (100)	251 (100)

Source Field Survey for ICSSR supported Study, IHD

cyber cafes, hotels, restaurants, medical services, cinema halls, etc. Six units which combined services and trading, consisted of enterprises which sell motor parts, pump sets, cycles/motorcycles, cycle tyre & parts, motorcycle parts, etc., and provide repairing services for the same.

Out of the 251 units surveyed, the share of Non-directory Enterprises (NDEs) was highest (46%), followed by Own Account Enterprises (OAEs) (37%) and Directory Enterprises (DEs) with only 17% (Table 3.2).[13] Since the OAE units were of relatively smaller size and these do not employ hired workers, the impact of such units on the growth in other segments of the economy is limited. The smallness of size of most enterprises is manifested in the 76% share of enterprises with a turnover of less than Rs 20 lakh (Table 3.3), while around 82% of the surveyed enterprises were found to employ less than 5 workers (Table 3.4).

Thus the type of units which has potential to absorb workers from surrounding areas had low representation in the sample towns. Among the two towns, Biharsharif has more potential since it accounted for the majority of the DEs.

3.4 Rural–Urban Linkages

3.4.1 *Linkages* via *the Labour Market*

A total of 158 enterprises were found to hire workers. Analysis of enterprise related information indicates hiring of permanent employees and to a lesser extent, temporary employees. Around 16–17% enterprises reported hiring casual and contract workers. The hiring of unskilled workers was reported by 74% of the 158 enterprises, followed by hiring of skilled workers (50% of enterprises) and semi-skilled workers (21.4% of enterprises).[14]

Of the 158 units that reported using hired workers, more than half (55%) reported hiring workers from rural areas, usually along with workers from urban areas. Of the 87 units that reported hiring workers, a higher percentage of Directory Enterprises (72%) hired workers from rural areas vis-a-vis Non-directory Enterprises (48.7%). Similarly, a higher share of rural workers was reported for Biharsharif compared to Madhubani.

The rural–urban linkage is identified with more clarity when we observe that of the 610

[13] An enterprise, which is run usually without the help of any hired worker employed on a fairly regular basis, is defined as an Own Account Enterprise (OAE). The DEs are enterprises which employ 6 or more workers (household and hired workers taken together) of whom at least one hired worker is employed on a fairly regular basis. The NDEs refer to enterprises which employ less than 6 workers (household and hired workers taken together) of whom at least one hired worker is employed on a fairly regular basis.

[14] The hiring of different types of workers is not mutually exclusive.

Table 3.3 Distribution of surveyed enterprises by annual turnover (Rs) of enterprise (percentage)

Unit type	Up to Rs 2 lakh	Above Rs 2lakh and up to Rs 4 lakh	Above Rs 4 lakh and up to Rs 8 lakh	Above Rs 8 lakh and up to Rs 12 lakh	Above Rs 12 lakh and up to Rs 20 lakh	Above Rs 20 lakh and up to Rs 40 lakh	Above Rs 40 lakh and up to Rs 50 lakh	Above Rs 50 lakh	Total
Manufacturing	8.7	10.9	15.2	15.2	15.2	15.2	6.5	13.0	100.0
Trading	11.7	13.5	19.3	21.1	15.2	9.9	3.5	5.8	100.0
Services	21.4	14.3	10.7	14.3	3.6	14.3	7.1	14.3	100.0
Service and trade	0.0	0.0	66.7	16.7	0.0	16.7	0.0	0.0	100.0
Grand total	12.0	12.7	18.7	19.1	13.5	11.6	4.4	8.0	100.0

Source Field Survey for ICSSR supported Study, IHD

Table 3.4 Distribution of Enterprises by Number of Workers (number)

Type of enterprise	Total worker			
	1–5	6–9	>10	Total
Own Account Enterprises	93	0	0	93
Non-directory Enterprises	115	0	0	115
Directory Enterprises	0	24	19	43
Total (number) Percentage share (%)	208 (82.9)	24 (9.6)	19 (7.6)	251 (100.0)

Source Field survey for ICSSR supported study, IHD

adult workers hired for the survey enterprises,[15] as many as 443 workers, or 72.6% are from rural areas (Table 3.5). The DEs' share of rural workers is higher than NDEs, and Biharsharif data show nearly 80% workers coming from nearby villages, while for Madhubani, the corresponding share is lower.[16]

Among different types of activities, the manufacturing enterprises use maximum hired labour from rural areas (90% of workers), followed by service and trade (84.6% of workers).[17] The other two categories also have a high share of around 60%. Cold storage units,[18] sawmills and furniture-making units are some of the types of manufacturing units which hire many rural workers. Among the service and trade units, the automobile workshops, units engaged in cycle repairing and selling cycle parts, etc., hire more rural workers. In trading activity, the enterprises which hire more rural workers are grocery shops, general stores, garment stores, electronic shops, chemists' shops, etc. Other

[15] The analysis is based on the reports of 152 enterprises out of the 158 survey enterprises. Four enterprises hire only child workers and two enterprises could not provide disaggregated data for workers from rural and urban areas.

[16] It may be noted, however, that many of the daily commuters from nearby rural areas such as rickshaw-pullers, masons, etc., are not part of the enterprise survey, and were covered under qualitative interviews.

[17] These were, however, very few in number.

[18] Cold Storage units have been included as part of manufacturing rather than trade or services because of their scale of operation and dissimilarity with either trade or services.

Table 3.5 Hired workers among enterprises and in sample towns

		Hired adult worker (Current year)			Percentage of hired worker from rural areas (%)
		Rural area	Urban area	Total	
Type of establishment	NDE	126	92	218	57.8
	DE	317	75	392	80.9
Town	Biharsharif	360	91	451	79.8
	Madhubani	83	76	159	52.2
Type of activity	Manufacturing	207	23	230	90.0
	Trading	148	91	239	61.9
	Other services	77	51	128	60.2
	Service and trade	11	2	13	84.6
	Total	443	167	610	72.6

Source Field survey for ICSSR supported study, IHD

service enterprises hiring large number of rural workers constitute cinema halls, hotel/restaurants, courier services, automobile showrooms, tailors, etc.

Manufacturing units and service & trade units were among the largest hirers of rural labourers, and also more from remote areas. Nearly half of the hired workers reported travelling a distance of 4–6 km daily and spending around Rs 16–20 daily for reaching the workplace. The average distance travelled daily by them occurred most in 4–6 km range (49% of units which hire rural workers), slightly less in 7–9 km (22%) and much less in shorter distances of 1–3 km (14%) as well as for larger distances of more than 10 km (14%).

NDEs were found to employ more workers coming from closer distances while DE units hire workers from villages further away from town. Biharsharif, which has relatively more manufacturing units, has enterprises hiring labourers from a longer distance compared to Madhubani. Other services also showed a high share of workers commuting from remote areas.

Informal hiring practices were found to hold sway in the labour market as around half of the 158 enterprises reported that they hire workers through family/friends connections. Around 40% enterprises reported that workers come seeking employment. It is revealing that advertisement and employment exchange play hardly any role in hiring workers and social connections seem to matter the most. The enterprises reported that they find it more difficult to recruit skilled than unskilled workers.

3.4.2 Backward and Forward Linkages

Manufacturing enterprises exhibited more rural–urban linkage in terms of input–output, represented by forward and backward linkages.[19] But the linkages do not appear as strong as the labour market linkages. The wholesalers and retailers who procure their merchandise from various places are the major sources of raw material and intermediate products (Table 3.6). On an average, 57.4% is sourced from wholesalers, 5.6% from retailers and 8.4% from wholesalers and retailers both, making a total of 71.7%. Further disaggregation indicates that some manufacturing units, such as the flour mills, oil mill, two cold storage units and the *gul* mill, among others, directly procure raw materials from local farmers. But their share for procuring raw materials/intermediate products from wholesalers/retailers is still over 50%.

[19] Forward linkages refers to linkages to where the products of the enterprises are sold and backward linkages refer to sectors from where the raw materials/inputs of the enterprises are obtained.

Table 3.6 Source of raw material and intermediate product/merchandise (Percentage)

Unit type	Wholesaler	Retailer	Both wholesaler & retailer	Under putting out system	Distributor	Others	Total
Manufacturing and Cold Storage units*	34.8	10.9	15.2	4.3	0.0	34.8	100.0
Trading	67.3	2.3	5.3	0.0	21.1	4.1	100.0
Services	32.1	14.3	14.3	3.6	14.3	21.4	100.0
Service and Trade	66.7	16.7	16.7	0.0	0.0	0.0	100.0
Grand Total	57.4	5.6	8.4	1.2	15.9	11.6	100.0

*Cold Storage units are not included in manufacturing activities as per Annual Survey of Industries at present
Source Field Survey for ICSSR supported Study, IHD

Table 3.7 Destination of final product (Percentage)

Unit type	Wholesaler	Retailer	Both wholesaler & retailer	Directly to consumer	Person who supplied the inputs	Others	Total
Manufacturing and Cold storage units	4.3	4.3	4.3	78.3	2.2	6.5	100.0
Trading	2.9	7.6	5.8	83.0	0.0	0.6	100.0
Services	3.6	3.6	0.0	85.7	0.0	7.1	100.0
service and trade	0.0	0.0	0.0	100.0	0.0	0.0	100.0
Grand total	3.2	6.4	4.8	82.9	0.4	2.4	100.0

Source Field survey for ICSSR supported study, IHD

Two flour mills also have a putting out system. As for the service industries, there are film hall owners who source their raw material from distributors, doctors who treat patients who visit their clinics, etc. Rather, linkages were observed with areas far away from the sample towns. For instance, while cold storage units source potatoes and onions from local farmers, many of these showed links with Himachal Pradesh, Punjab and Delhi in North India. The entrepreneurs source fruits in such places and bring to Biharsharif, for cold storage, so that their capacity utilization is improved.[20]

[20] Subsequent visits to various small and medium towns of Bihar during 2015 reinforced the evidence for low local backward linkage for all types of enterprises, including manufacturing. For instance, a timber manufacturing enterprises in the small town of Hilsa near Patna was found to source the timber mostly from Kolkata at huge transport costs. Cement dealers in the same town were procuring sand from Gaya and gravel from Jharkhand.

In terms of forward linkage, the survey findings showed that 83% of the 251 survey enterprises sell final products directly to consumers and 88% of the enterprises sell locally (Table 3.7). Among manufacturing units, there are a handful of units selling to wholesalers/retailers, among which there are flour mills, ice cream factory and *gul* factory. Cold storages are an exception insofar as none of the units have reported selling directly to consumers. Wholesalers/retailers and people who supply the inputs have been reported as destination of the final products by cold storage units.

The local destination can have consumers both from the town itself as well as the rural inhabitants coming to town to buy these products. For instance, people come from rural areas to the town for buying garments, shoes, food, etc., as well as for medical treatment, education-related services, etc. With improved transport connectivity over time, such linkages of the rural areas with the nearby towns have increased. Many also watch movies in the cinema hall in the town, although this linkage is weakening

Table 3.8 Market for the final product produced/sold in the enterprise (Number)*

Unit type	Local	Within state	National	Export	Total
Manufacturing	43 (93.5)	3 (6.5)	0 (0.0)	0 (0.0)	46 (100.0)
Trading	153 (89.5)	12 (7.0)	3 (1.8)	3 (1.8)	171 (100.0)
Services	21 (75.0)	4 (14.3)	3 (10.7)	0 (0.0)	28 (100.0)
Service and trade	5 (83.3)	0 (0.0)	1 (16.7)	0 (0.0)	6 (100.0)
Grand total	222 (88.4)	19 (7.6)	7 (2.8)	3 (1.2)	251 (100.0)

Source Field Survey for ICSSR supported Study, IHD
* Figures in brackets indicate percentage shares of total

due to competition from television. There is traffic in the opposite direction as well, such as ice cream vendors, utensil vendors and the like peddle their goods in the nearby villages and in the town. Thus there seems to be some evidence of forward linkage in the sample towns.

For the manufacturing enterprises, the share of local destination is highest at 93.5% (Table 3.8). Out of a total of 46 units, only 2 sawmills and 1 furniture manufacturer sell their products outside the town, but still within the state of Bihar. The other enterprises have some representation of national as well as international destinations, albeit at very small percentages.

Analysing the destination of products by different types of enterprises, it is observed that it is only the Directory Enterprises, i.e. the bigger enterprises which sell some share of their products in far-away destinations (Table 3.9). Unlike the OAEs and NDEs, for which more than 90% sell their products in local destinations, nearly 28% of the DEs sell their products elsewhere within state boundaries. Most of the larger units sell within the district, Rural (around 60%) and urban (around 40%), and only an ice making unit was found to sell 10% of output in the urban area outside the district. It is the DEs which hire relatively more local labour from rural areas, but these are found to have relatively lower demand for their products from local areas, compared to other types of enterprises. Thus, the larger enterprises have relatively weaker forward linkages.

Table 3.9 Market for the final product produced/sold in the enterprise

Type of enterprise	Local	Within state	National	Export	Total
	Number				
Own account enterprises	89	2	0	2	93
Non-directory enterprises	106	5	4	0	115
Directory enterprises	27	12	3	1	43
Total	222	19	7	3	251
	Percentage				
Own account enterprises	95.7	2.2	0.0	2.2	100.0
Non-directory enterprises	92.2	4.3	3.5	0.0	100.0
Directory enterprises	62.8	27.9	7.0	2.3	100.0
Total	88.4	7.6	2.8	1.2	100.0

Source Field survey for ICSSR supported study, IHD

3.5 Source of Urban Growth

The assessment of urban growth was based on the perception of the respondents in the enterprises. Expansion in business was reported by a high share of 64% of the 224 units which have been operating for the last 5 years, but 20%

Table 3.10 Enterprise status over the last five years (number of Units)

Unit type	Expanding	Stagnant	Contracting	Total
Manufacturing	17	9	18	44
Trading	105	26	17	148
Services	17	1	8	26
Service and trade	5	0	1	6
Total (% in brackets)	144 (64)	36 (16)	44 (20)	224 (100)

Source Field Survey for ICSSR supported Study, IHD
Note Percentage in brackets

reported contraction and 16% reported stagnation (Table 3.10). The trading units provided the driving force for expansion, but among these, garments and shoes & leather businesses reported mixed progress due to sales dampened by competition. It is possible that the garments and shoes & leather businesses have faced competition from bigger shopping complexes that are coming up, especially in Biharsharif. This was also supported by qualitative data from interviews. Manufacturing activities showed contraction, with the exception of saw mills. Other service related units mostly exhibited growth, a notable exception being cinema halls, which faced competition from television, pirated CDs, etc. These were also affected by high cost.

The survey also included questions about the problems faced by the enterprises. The findings indicate that most enterprises face constraints related to electricity shortage, general infrastructure, lack of access to funds, corruption, lack of skilled workers, limited market, policy issues including red tape, taxation and lack of stability in government policy. Many respondents in Madhubani also mentioned a general dearth of economic development in the region, including poor availability of education and health facilities.

All types of enterprises identified electricity as the key bottleneck area. For Directory and Non-directory enterprises, poor public infrastructure was the next most important challenge. For Own Account Enterprises, though, lack of sufficient funds was the most important constraint after electricity. The entrepreneurs in DEs also emphasized training of agricultural development, skill development and general development of educational institutions, as enterprises find it difficult to recruit skilled workers.

3.6 Conclusions

The sample towns of Madhubani and Biharsharif have witnessed a gradual decline in industrial activity over the years. The survey findings indicated that trade is the main driver of urban growth in both towns. The share of Own Account Enterprises in total enterprises is quite high at 37% and it is particularly high for Madhubani at 50% of total enterprises. Therefore, in the context of the stated objective of the study, i.e. exploring the potential of absorption of rural labour, the high share of OAEs implies limited scope for absorption of rural labour.

With regard to the potential of manufacturing sector, in particular, to absorb rural labour productively, the survey found that the rural–urban linkage is highest for manufacturing activities both in terms of labour and input–output linkage. Yet, the manufacturing sector, notwithstanding its close links with the rural hinterland in both survey towns, and great potential for absorbing surplus labour, is clearly suffering from contraction, according to the enterprise survey in the two towns of Biharsharif and Madhubani. It was found that in both towns, there has been a lot of closure of industries over the last two or three decades. While several explanations were put forth by key informants in the two towns, some key factors stood out, endorsed by the respondents from enterprises (largely owners) included in the survey.

Electricity was the main bottleneck area indicated by respondents for all types of enterprises. For Directory Enterprises and Non-directory Enterprises, poor public infrastructure was the second major hurdle, while, for Own Account Enterprises, lack of sufficient funds was the second most cited factor. Some of the other important problem areas outlined by the enterprises were inflation, corruption/bribery, lack of skilled workers and lack of a stable tax policy. An overall deficit in terms of development, industrialization and urbanization leading to deficient demand was cited by several respondents from Madhubani.

Given the above findings for the two sample towns, it appears that if the manufacturing route for development is to be followed to harness the youth potential in a densely populated state such as Bihar while exploiting the rural–urban linkages, then the underlying infrastructure framework as well as enabling governance must be provided for enterprises to grow. In Biharsharif, agro-processing industries such as making juice/pulp, etc., as well as frozen and packaged fruits and vegetables hold promise with easy supply of raw material from rural areas. This town, being in the Buddhist circuit, is already seeing expansion in service areas such as hotels and restaurants and this aspect of tourism can be promoted. Madhubani would benefit from focus on *makhana* production, pisciculture and promotion of traditional Mithila handicrafts and paintings. Where demand is a limiting factor, the government may need to buttress the same by expanding public investment. Recent visits[21] to Bihar indicate somewhat improved electricity supply in many towns. With ease in one of the major bottlenecks that was indicated in the survey, it remains to be seen whether meaningful strides can be made by the state towards expanding industrialization and generating employment. Or would migration continue to hold sway as a major strategy for tackling poverty and unemployment in the state?

[21] The authors visited several towns, including Madhubani, in Bihar, during May 2015.

References

Amirapu Amrit, Arvind Subramanian (2015) Manufacturing or services? An Indian illustration of a development dilemma, Working Paper 409, Center for Global Development, Washington D.C.

Bhattacharya Tulika, Bornali Bhandari (2021) Where are the jobs? Estimating skill-based Employment Linkages across Sectors for the Indian Economy: An Input-Output Analysis', Working paper no. WP 120 National Council of Applied Economic Research/NCAER, New Delhi

Dahlman Carl J (2010) Education and growth of services, in Ejaz Ghani (ed), *The Service Revolution in South Asia*, pp 261–262. Oxford University Press, New Delhi

Ghani Ejaz, Homi Kharas (2010) The service revolution in South Asia: an overview, in Ejaz Ghani (ed), *The Service Revolution in South Asia*, Oxford University Press, New Delhi

Ghose Ajit (2014) Manufacturing-led growth in India, paper presented at 'Make in India' Seminar organized by Institute for Human Development in New Delhi

Government of India (2015) Economic Survey of India-Union Budget. www.indiabudget.nic.in/survey.asp

Krishnamurty J (2015) Tackling India's Shrinking Youth Bulge' in *Financial Express*

Mahadevia D, Sarkar S (2012) Handbook of urban inequalities. Oxford University Press, New Delhi

Mehrotra S, Parida J, Sinha S, Gandhi A (2014) Explaining employment trends in the Indian economy: 1993–94 to 2011–12. Econ Pol Week XLIX(32)

Minocha AC, Yadav HS (1989) Small and medium towns and their role in regional development. India Gyan Books Pvt. Ltd., New Delhi

Rodgers Gerry, Datta A, Rodgers J, Mishra SK, Sharma AN (2013) The challenge of inclusive development in Bihar, IHD and Manak, NewDelhi

Dynamics of Impoverishment Impact of Healthcare Payments in Nepal: An Analysis of Synthetic Panel Data Between 2004 and 2011

4

Shiva Raj Adhikari

Abstract

Poverty dynamics encompass the duration of experiencing poverty and elucidate the transitions into and out of poverty. Evidence concerning poverty dynamics is crucial for policymakers in formulating effective anti-poverty strategies. Additionally, healthcare payment significantly influences the dynamics of poverty. Utilizing panel data is pivotal for gaining a deeper comprehension of poverty dynamics. In the absence of such data, this study constructs a synthetic panel dataset by merging two rounds of cross-sectional surveys conducted in Nepal in 2003/04 and 2010/11 to evaluate poverty dynamics. It's important to note that the incidence of poverty estimated from the hybrid dataset may not be directly comparable to estimations derived from conventional approaches. The findings reveal that chronic poverty, without adjusting for healthcare payment, stood at approximately 21% for both 2003/04 and 2010/11. Furthermore, healthcare payment contributed to a 1% increase in chronic poverty. Movements into and out of poverty, such as transitions from non-poor to poor and vice versa, signify transient poverty. Healthcare payment also influences transient poverty, with payments increasing the transition from non-poor to poor while reducing the likelihood of households moving from poor to non-poor status. Chronic poverty is prevalent across all regions and among marginalized ethnic and Dalit (occupational caste) groups. Tailored anti-poverty policies are necessary to address both chronic and transient poverty. Policies aimed at stabilizing short-term income fluctuations, such as increasing access to credit facilities, services, remittances, and social safety net programs, may be more effective in addressing transient poverty. Conversely, policies focusing on structural or long-term interventions, such as developing basic infrastructure, enhancing social and political inclusion, redistributing assets, and promoting capital accumulation, are imperative for addressing chronic poverty.

Keywords

Poverty dynamics · Healthcare payment · Transitory poverty · Chronic poverty · Nepal

S. R. Adhikari (✉)
Central Department of Economics, Tribhuvan University, Kirtipur, Nepal
e-mail: shiva.adhikari@cdec.tu.edu.np; sssadhikari@yahoo.com

A. Alakshendra et al. (eds.), *Development, Inclusion and Sustainability*, Sustainable Development Goals Series, https://doi.org/10.1007/978-981-97-6863-9_4

4.1 Introduction[1]

In recent years, Nepal has experienced a notable decline in poverty. This is evident from a significant decrease in poverty incidence, dropping from 41.8% in 1996 to 30.9% in 2004 and further to 25.16% in 2011 (CBS 2005, 2011). However, analyzing poverty solely through single cross-sectional household data provides a static view of poverty, offering limited insight into its persistence and dynamics. Static analysis fails to adequately explain the fluctuations and movements within poverty trends.

Poverty trends merely depict changes in poverty incidence over time, potentially masking the underlying poverty dynamics. Poverty dynamics, on the other hand, delve into the duration of poverty experiences and elucidate the transitions in and out of poverty. Understanding poverty dynamics is crucial for policymakers to design effective anti-poverty policies as it captures the economic mobility of households and focuses on their inter-temporal changes in poverty status (Ward 2016).

Persistent poverty over an extended period is termed chronic poverty, while individuals experiencing fluctuations in and out of poverty are described as facing transient poverty. Deepening our comprehension of poverty dynamics is imperative. To this end, this paper examines poverty dynamics in Nepal by constructing a hybrid dataset from the cross-sectional data of Nepal Living Standard Surveys conducted in 2003/04 and 2010/11.

4.2 Healthcare Payments and Poverty Dynamics

Between 2004 and 2009, Nepal underwent significant changes in its healthcare system. Notably, public health expenditure saw a marked increase. Data sourced from the World Health Organization's Global Health Expenditure Database in 2015 revealed that government health expenditure accounted for only 2.2% of Gross Domestic Product (GDP), slightly surpassing the low-income country average of 1.9% for that year. By 2012, total health expenditure rose to 5.5% of the country's GDP, closely aligning with the average of 5% observed in other low-income countries.

At that time, public allocations for the health sector constituted approximately 10% of total government expenditure, exceeding the average of 8.1% for other low-income countries. This underscored the government's commitment to healthcare funding. Despite this, Nepal relied significantly on donor financing, which comprised nearly a fifth (19%) of total health sector expenditure in 2012, lower than the low-income country average of 28%. However, this reliance left the country susceptible to fluctuations in donor funding.

Despite these changes, out-of-pocket payments remained the primary source of healthcare financing, constituting nearly half of total financing in 2012. While this figure was in line with the low-income average, it was considerably high on a global scale, where the average was 21%. Moreover, it exceeded the 20% threshold recommended by the 2010 World Health Report to mitigate financial catastrophes and impoverishment resulting from healthcare expenses (World Health Organization 2010). Private health insurance had negligible presence in Nepal in 2012.

Health care is a basic necessity of life, as important as food, shelter, and clothing; however, the current practice of estimating poverty incidence in Nepal, similar to other countries, does not capture the basic healthcare cost. Not accounted for out-of-pocket payment on health care (OOPH) in the poverty estimation could give a misleading picture of poverty dynamics. Ignoring healthcare costs altogether can result in misclassifying which households or individuals are in the greatest need. The OOPH causes hardship to people if an adequate financial protection mechanism is absent such as

[1] Earlier version of this chapter has been published in Social Science Baha Occasional paper. Adhikari, S. R Sapkota V.P (2018). Dynamics of Impoverishment Impact of Healthcare Payments in Nepal. Social Science Baha, Kathmandu.

health insurance. OOPH is impoverishing when they push the household into poverty or further into poverty (Van Doorslaer et al. 2006). The assessment of impoverishment resulting from OOPH describes the difference between the pre-payment poverty headcount before OOPH and post-payment poverty headcount after OOPH (Wagstaff and Doorslaer 2003). OOPH leads to poverty; however, this poverty is not captured by the standard method of poverty measurement. Health care is a basic need, but it is difficult to capture the uncertain nature of healthcare needs through a constant poverty line (Wagstaff 2008). Poor health causes poverty through several pathways, including impaired labor participation, thereby disrupting people's income-generating capabilities (Deaton and Zaidi 2002). The proposed method in several studies to measure the impoverishment impact of OOP doesn't provide sufficient information for policymakers. For example, low-income countries have various poverty alleviation programs and how healthcare payments interact with these programs to change the poverty levels- might be interesting for policymakers. Current literature on measuring impoverishing impact provides information on whether the OOPH pushes the household into below the poverty line or not. However, OOPH may be a source of poverty spiral for the household (Adhikari et al. 2007). OOPH affects the household economy through at least two channels: (a) it increases the incidence of poverty because non-poor households are pushing below the poverty line; (b) it increases the intensity of poverty because poor households are pushing further poor. In other words, the households can move from poor to non-poor due to some anti-poverty programs, at this time OOPH can pull the households to remain the poor. OOPH not pushes the non-poor to poor and poor to further poor but also affects the movement of poverty—in and out of poverty. The method of estimating the improvising effects of OOPH provides the importance of healthcare financing while designing the poverty alleviation programs. An extensive search of literature confirms none of the studies estimates the impoverishing impact of OOPH in the case of poverty dynamics. Such measurement requires panel data; however, panel data are rarely available of nationally representative data. The paper creates the synthetic panel data from two cross-sectional data using the method suggested by Dang and Lanjouw (2013).

Health care is a fundamental necessity, akin to food, shelter, and clothing. However, the current method of estimating poverty incidence in Nepal, as in many other countries, fails to account for basic healthcare costs. Neglecting out-of-pocket payments for health care (OOPH) in poverty estimation can distort the understanding of poverty dynamics and misclassify households or individuals in dire need. The absence of adequate financial protection mechanisms, such as health insurance, exacerbates the hardships caused by OOPH, often driving households into poverty or deeper into it (Van Doorslaer et al. 2006).

Assessing the impoverishing effects of OOPH involves comparing pre-payment poverty headcounts with post-payment headcounts (Wagstaff and Doorslaer 2003). While OOPH undoubtedly contributes to poverty, conventional poverty measurement methods fail to capture this aspect due to the intricate and unpredictable nature of healthcare needs (Wagstaff 2008).

Poor health can lead to poverty through various avenues, including diminished labor participation, thereby reducing income-generating capacities (Deaton and Zaidi 2002). Policymakers need comprehensive information on how healthcare payments interact with poverty alleviation programs in low-income countries to effectively address poverty levels. While existing literature provides insights into whether OOPH pushes households below the poverty line, it overlooks the potential for OOPH to initiate a poverty spiral within households (Adhikari et al. 2007).

OOPH impacts household economies through multiple channels: it not only increases the incidence of poverty by pushing non-poor households below the poverty line but also intensifies poverty by driving poor households deeper into poverty. Furthermore, OOPH disrupts the movement of households in and out of poverty, even when they are benefiting from anti-poverty programs. Therefore, accurately estimating the impoverishing effects of OOPH is crucial for designing effective poverty alleviation programs that consider healthcare financing.

Despite the importance of this measurement, none of the existing studies on poverty dynamics estimate the impoverishing impact of OOPH, largely due to the scarcity of panel data. To address this gap, this paper utilizes a synthetic panel data approach, as suggested by Dang and Lanjouw (2013), by merging two cross-sectional datasets.

Over the past two decades, extensive empirical research has focused on understanding the dynamics of poverty (Addison et al. 2009). This body of work has enhanced our comprehension of poverty by quantifying its persistence and identifying the factors influencing individuals' ability to escape poverty, as well as the events that may precipitate poverty over their lifetimes. While such investigations were previously hindered by the absence of well-designed panel data in developing countries, Dang and Lanjouw (2013) developed a methodology for generating synthetic panels to analyze household-level poverty dynamics, requiring at least two years of cross-sectional survey rounds. Their methodology has been shown to outperform true panel data in terms of asymptotic properties and the robustness of estimates, provided certain assumptions are met.

In this study, we employ the same methodology to examine poverty dynamics related to gross and net out-of-pocket (OOP) payments for health care, utilizing the Nepal Living Standards Survey (NLSS) data for 2003/04 and 2010/11. Our aim is to contribute further to understanding how OOP payments impact transitory and chronic poverty.

Understanding poverty dynamics—such as the percentage of individuals remaining in poverty, moving out of poverty, or falling into poverty during a given period—is essential for realistic poverty reduction strategies. Chronic poverty refers to poverty persisting over a prolonged period, while transitory poverty involves movement in and out of poverty. Deepening our comprehension of poverty dynamics is imperative, and this paper seeks to complement existing literature on poverty measurement in Nepal.

Our contribution to this literature includes a discussion on the impoverishing impact of OOP payments, utilizing recently developed methods of poverty dynamics. We measure the impoverishment impact of OOP payments in terms of their incidence on poverty. Using a synthetic panel dataset constructed from cross-sectional NLSS data for 2003/04 and 2010/11, we estimate poverty dynamics both before and after adjusting for OOP payments. This allows us to quantify the difference in poverty dynamics with and without OOP payment adjustments.

4.3 Methods and Materials

4.3.1 Conceptual Framework of Poverty Dynamics

Poverty is not a static condition, but rather a dynamic phenomenon that varies across different population groups defined by their place of residence, as well as their socio-economic and demographic conditions. Analyzing a single cross-sectional household dataset provides only a snapshot of poverty, offering limited insight into its persistence. Fluctuations in poverty trends alone do not adequately capture its dynamics. Comparative static methods are commonly used to analyze poverty trends, but they fail to fully capture the economic mobility of households over time.

Nepal's economic landscape has undergone significant changes over the past three decades, positioning it among the most dynamic economies in recent times. Household consumption capabilities evolve over time, emphasizing the need for measuring poverty dynamics to comprehend the processes central to its persistence and reduction.

Dynamic analysis enables a deeper understanding of both permanent and transient poverty, which is crucial from a policy perspective. This paper adopts a novel approach to distinguish between permanent and transient poverty,

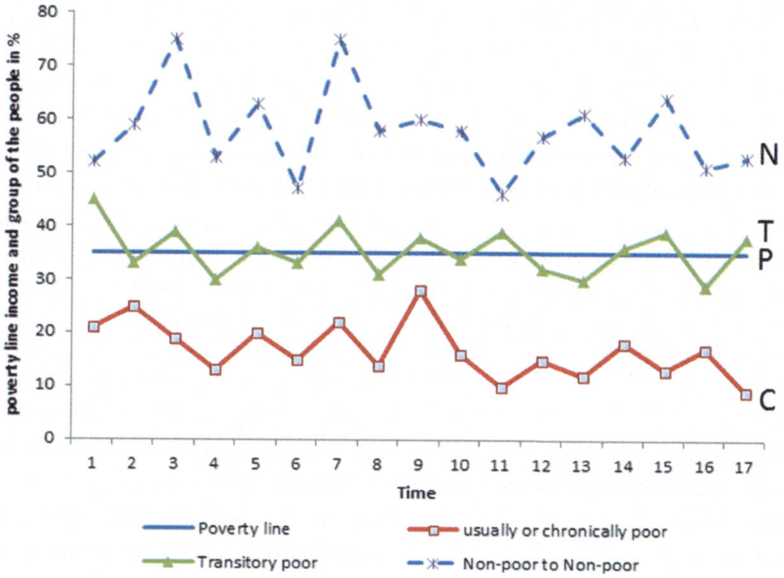

Fig. 4.1 Conceptual framework for poverty dynamics. *Source* Adhikari (2016)

aiming to complement existing poverty measurement efforts in Nepal. Drawing inspiration from recent literature on poverty, the study utilizes hybrid (or synthetic panel) data constructed from the Nepal Living Standards Surveys (NLSS II) 2003/04 and (NLSS III) 2010/11 to estimate poverty dynamics.

Chronic poverty pertains to long-term or persistent poverty. Vulnerability to poverty varies over time and among different demographic groups. While certain groups may consistently fall below the poverty threshold or poverty line, others may experience short but recurrent episodes of poverty. Some groups may never experience poverty at all. Broadly speaking, there are at least three main patterns of poverty (refer to Fig. 4.1): the chronically poor, the transient poor whose poverty status fluctuates, and those who never experience poverty.

Poverty dynamics encompass all these patterns, examining aggregate changes in poverty prevalence as well as changes specific to particular groups and regions. While the conceptual framework illustrates the long-term dynamics of poverty (see Fig. 4.1), the analysis of poverty dynamics in this study focuses on two specific periods: 2003/04 and 2010/11.

In Fig. 4.1, a linear horizontal line labeled P represents the poverty line. The fluctuating curve labeled T within the poverty line illustrates the movement of individuals in and out of poverty over time. This group experiencing fluctuation in poverty status is referred to as the transient poor. The transient poor can be further classified into two groups: those transitioning from non-poor to poor and those transitioning from poor to non-poor. Within a given period, individuals may move between these categories. Curve C, consistently below the poverty line, signifies chronic poverty. The fluctuating curve labeled N, consistently above the poverty line, represents individuals who never experience poverty. Consequently, the total population can be classified into four groups: chronically poor, non-poor to poor, poor to non-poor, and non-poor to non-poor.

Out-of-pocket health expenditure (OOPH) influences the incidence of these four categories in the following ways:

(a) OOPH can push non-poor individuals into poverty, as those who have never experienced poverty may fall below the poverty line due to health expenditure. This results in an increase in poverty incidence.

(b) There may be an increase in chronic poverty or an increase in the likelihood of individuals transitioning from being poor to remaining poor.
(c) OOPH reduces the capacity of individuals with the potential to move from poverty to non-poverty.
(d) OOPH can accelerate the transition of individuals from being non-poor to becoming poor. These scenarios provide insight into the extent of the impoverishing effect of OOPH.

4.3.2 Methodology for Construction Synthetic Panels

The analysis of poverty dynamics entails the identification of the same economic unit through time. It requires longitudinal data that tracks individuals or households over time. Hybrid data (synthetic panel) are used in the absence of longitudinal data to analyze the poverty dynamics. Recently, pseudo-panel approaches are popular to create hybrid or synthetic panel data using repeated cross-sectional surveys (Cruces et al. 2014). Synthetic panels based on cohorts have been widely used to track income and consumption outcomes over time (Deaton and Paxson 1994; Banks et al. 2001; Pencavel 2007; Dang et al. 2014; Dang and Lanjouw 2013). The methodology of this paper largely relies on recently developed procedures and methods of constructing hybrid dataset from cross-sectional surveys (Dang et al. 2014; Dang and Lanjouw 2013) and several studies which validate the method have generally yielded encouraging findings (Dang et al. 2014; Dang and Lanjouw 2013; Cruces et al. 2014).

It is assumed that y_{it} round t household capita consumption (where t=1, 2) of household i and Z_k is the poverty line (where k=1, 2). Let X_{it} be a vector of household characteristics observed in survey round $t(t = 1, 2)$ that are also observed in the other survey round for household $i, i = 1, \ldots, N$. These household characteristics include time-invariant variables such as household head's ethnicity, education, and sex of the household heads which remain the same across survey rounds.

The paper seeks to estimate the fraction of poor population in the first round of the survey who escaped poverty $P(y_{i1} \leq z_1 \text{ and } y_{i2} \geq z_2)$ or remain poor $P(y_{i1} \leq z_1 \text{ and } y_{i2} \leq z_2)$ in the second round of the survey, and the fraction of non-poor population in the first round of the survey who became poor $P(y_{i1} \geq z_1 \text{ and } y_{i2} \leq z_2)$ or remained non-poor $P(y_{i1} \geq z_1 \text{ and } y_{i2} \geq z_2)$ in the second round of the survey. z_1 *and* z_2 are poverty lines for NLSS II and III, respectively. Both surveys, NLSS II and NLSS III adopted random sampling procedures to select the households.

We can estimate the relationship between consumption and time-invariant characteristics in each round:

$$y_{it} = \beta_1' x_{it} + \varepsilon_{it} t = 1, 2 \tag{4.1}$$

ε_{it} is an error term.

Using observations from the NLSS III round, we can predict consumption in the NLSS II round utilizing the same observed vector of time-invariant or retrospective characteristics and the NLSS II round OLS estimates of parameters. As suggested by Dang and Lanjouw (2013) the mobility into and out of poverty in the context of synthetic panels can be defined as follows:

(a) Probability of being poor in 2003/04 remaining poor in 2010/11.

$$P(y_{i1} \leq z_1 \text{ and } y_{i2} \leq z_2) = \phi\left(\frac{z_1 - \beta_1' x_{i1}}{\sigma_{\varepsilon_1}}, \frac{z_2 - \beta_1' x_{i2}}{\sigma_{\varepsilon_2}}, \rho\right) \tag{4.2}$$

(b) Probability of being poor in 2003/04 becoming non-poor in 2010/11.

$$P(y_{i1} \leq z_1 \text{ and } y_{i2} \geq z_2) = \phi\left(\frac{z_1 - \beta_1' x_{i1}}{\sigma_{\varepsilon_1}}, -\frac{z_2 - \beta_1' x_{i2}}{\sigma_{\varepsilon_2}}, -\rho\right) \tag{4.3}$$

(c) Probability of being non-poor in 2003/04 becoming poor in 2010/11.

$$P(y_{i1} \geq z_1 \text{ and } y_{i2} \leq z_2) = \phi\left(-\frac{z_1 - \beta_1' x_{i1}}{\sigma_{\varepsilon_1}}, \frac{z_2 - \beta_1' x_{i2}}{\sigma_{\varepsilon_2}}, -\rho\right) \quad (4.4)$$

(d) Probability of being non-poor in 2003/04 becoming non-poor in 2010/11.

$$P(y_{i1} \geq z_1 \text{ and } y_{i2} \geq z_2) = \phi\left(-\frac{z_1 - \beta_1' x_{i1}}{\sigma_{\varepsilon_1}}, -\frac{z_2 - \beta_1' x_{i2}}{\sigma_{\varepsilon_2}}, \rho\right) \quad (4.5)$$

We need to repeat the procedure R times and calculate the average of the equations mentioned above to estimate movements in and out of poverty. This is necessary due to the randomly drawn distribution of estimated errors. Sampling errors were determined using a bootstrap method, and bootstrap standard errors were computed based on 1000 replications. Dang and Lanjouw (2013) proposed a method for generating synthetic panels from repeated cross-sectional surveys to obtain estimates of poverty dynamics over the given period.

Our endeavor involved measuring poverty dynamics at the household level, both including and excluding out-of-pocket health expenditure (OOPH). The disparity between poverty incidence with and without accounting for OOPH yields an estimate of the poverty impact of healthcare expenses.

4.3.3 Data Sources and Poverty Lines

The paper analyzed two rounds of cross-sectional household Surveys: NLSS II in 2003/04 and NLSS III in 2010/11. Both surveys were conducted by the Central Bureau of Statistics (CBS) with the support of the World Bank and were nationally representative, with the household as the sampling unit. NLSS II covered 3912 households from 326 Primary Sampling Units (PSUs) across the country, while NLSS III covered 5988 households from 500 PSUs. Both surveys adhered to the Living Standards Measurement Survey methodology developed and endorsed by the World Bank, providing detailed consumption-related information.

Table 4.1 Poverty lines for 2003/04

Analytical domain	Food (in Rs)	Non-food (in Rs)	Total (in Rs)
Kathmandu	6722.0	4334.8	11,056.8
Other urban	4919.2	2981.9	7901.1
Rural Western Hill	5613.0	3288.5	8901.5
Rural Eastern Hill	5311.2	2758.5	8069.7
Rural Western Terai	4308.4	3110.0	7418.4
Rural Eastern Terai	4323.2	1755.6	6078.8
Nepal	4966.4	2729.4	7695.8

Source World Bank (2006)

Out-of-pocket health expenditure (OOPH), including costs related to consultation, diagnosis, and medicine, was collected from both NLSS II and NLSS III survey data, as recommended by the literature. Official poverty lines were utilized for both surveys in estimating poverty dynamics (refer to Tables 4.1 and 4.2). The official poverty lines used were Nepalese rupees (Rs.) 7696 for NLSS II and Rs. 19,261 for NLSS III. Average poverty lines (z_1; z_2) are used while measuring poverty dynamics for the national level.

4.4 Estimation of Poverty Dynamics

4.4.1 Impact of OOPH on National Poverty[2]

The results of poverty dynamics in Nepal between the periods of 2003/04 and 2010/11 are presented in Table 4.3 (gross poverty column), revealing that nearly twenty-one percent of the

[2] Poverty estimation of this has been published in Economic Review published by Nepal Rashtra bank, Nepal.

Table 4.2 Poverty lines for 2010/11

Analytical domain	Food (in Rs)	Non-food (in Rs)	Overall (in Rs)
Mountains	13,295	6564	19,859
Kathmandu Urban	14,610	26,323	40,933
Urban Hill	11,805	7772	19,577
Urban Terai	11,743	9390	21,133
Rural Hills Eastern	12,297	4254	16,551
Rural Hills Central	12,240	6448	18,688
Rural Hills Western	12,537	5891	18,428
Rural Hills Mid and Far Western	11,772	4583	16,355
Rural Terai Eastern	11,333	5524	16,857
Rural Terai Central	11,257	6283	17,540
Rural Terai Western	10,600	5398	15,998
Rural Terai Mid and Far Western	10,998	6321	17,319
Nepal	11,929	7332	19,261

Source CBS (2011)

population remained in poverty during both periods. Similarly, 58 percent of the population belonged to the non-poor to non-poor category (Adhikari 2016). The transition from poverty to non-poverty was more prevalent than the transition from non-poverty to poverty between the two periods, indicating an overall reduction in poverty levels by 2010/11. However, six percent of the non-poor population fell below the poverty line due to factors such as poor agricultural harvests, unrelated to healthcare costs.

Upon excluding out-of-pocket health expenditure (OOPH) from individual consumption, poverty incidences differ from gross poverty incidences, highlighting the impact of OOPH on poverty levels. The impacts of OOPH on poverty dynamics manifest in two ways: it increases chronic poverty and enhances vulnerability to poverty. The incidence of chronic poverty among those classified as poor in both 2003/04 and 2010/11 (poor to poor) increased by more than 1% due to OOPH. Additionally, the movement of individuals from poverty to non-poverty decreased by 0.8%, signifying that OOPH pulls individuals back below the poverty line after they have escaped it. Conversely, the movement of non-poor individuals to poverty increased by 0.7%, indicating that OOPH accelerates the rate at which individuals transition from non-poor to poor. The probability of remaining non-poor between 2003/04 and 2010/11 decreased by 1.1% as a result of OOPH.

The last column of Table 4.3 illustrates the impact of OOPH on poverty incidence. A positive sign signifies an increase in poverty incidence due to OOPH, while a negative sign indicates a reduction in the household's capacity to remain non-poor. In other words, reduced capacity to maintain non-poverty status correlates with an increase in poverty incidence.

4.4.2 Poverty Mobility for Population Sub-Groups

Tables 4.4, 4.5, and 4.6 provide insights into and compare the impact on poverty dynamics. Table 4.4 indicates that the percentage of individuals

Table 4.3 Poverty dynamics gross and net of OOP and medical expenditure (2004 and 2011)

	Gross		Net OOPH		Impact of OOPH (difference between gross and net incidences of poverty)
	Incidence of poverty	SE	Incidence of poverty	SE	
Poor to Poor	0.209	0.009	0.222	0.009	0.013
Poor to non-Poor	0.144	0.014	0.136	0.014	–0.008
Non-poor to Poor	0.064	0.009	0.071	0.009	0.007
Non-poor to Non-poor	0.582	0.014	0.571	0.014	–0.011

remaining non-poor to non-poor is lowest in mountainous regions (53%). Approximately one-fifth of the population falls into chronic poverty across all ecological belts. In mountainous regions, the percentage of individuals transitioning from non-poor to poor (13%) is higher than those transitioning from poor to non-poor (10.7%), suggesting a higher likelihood of poverty increase. Chronic (or permanent) poverty rates are higher in the Terai and Eastern regions compared to other regions. On average, nearly 20% of the population is classified as chronically poor, while the remaining 20% are classified as transitory poor, indicating that over 40% of the population is vulnerable to poverty.

Table 4.5 depicts poverty incidence after excluding out-of-pocket health expenditure (OOPH) from individual consumption data. Comparing the incidences presented in Tables 4.4 and 4.5 illustrates the impact of OOPH. Consistent with previous interpretations, OOPH increases poverty incidence and vulnerability to poverty while simultaneously reducing the capacity to remain non-poor.

The central and western development regions exhibit a higher impact of OOPH on increasing chronic poverty compared to other regions. The Terai region is more affected by OOPH compared to mountainous areas. However, a higher proportion of poor individuals reside in mountainous regions compared to hills and the Terai. The highest poverty incidence is observed in the mid-western region; nevertheless, chronic poverty in the Western region increases by 1.5 percentage points due to OOPH (Table 4.6). Chronic poverty predominantly affects Dalits, Disadvantaged non-Dalit Terai castes, and Disadvantaged Janajatis. Approximately 11% of

Table 4.4 Overall poverty dynamics for population Sub-groups

SN	Population Sub-groups	Poor to poor	Poor to Non-poor	Non-poor to Poor	Non-poor to Non-poor
Ecological Belts in % (Standard Error)					
1	Mountain	23.3 (0.014)	10.7 (0.016)	13.0 (0.015)	53.0 (0.018)
2	Hill	18.8 (0.008)	16.1 (0.015)	04.8 (0.008)	60.4 (0.015)
3	Terai	23.4 (0.011)	12.9 (0.014)	07.6 (0.010)	56.1 (0.015)
Development regions (Standard Error)					
1	Eastern	24.9 (0.011)	12.5 (0.015)	07.1 (0.011)	55.5 (0.016)
2	Central	16.2 (0.008)	19.0 (0.015)	03.1 (0.008)	61.7 (0.015)
3	Western	19.6 (0.011)	14.4 (0.016)	05.5 (0.010)	60.5 (0.016)
4	Mid-western	26.0 (0.012)	09.9 (0.015)	11.7 (0.013)	52.4 (0.016)
5	Far-western	27.4 (0.014)	06.1 (0.015)	14.0 (0.014)	52.5 (0.016)
Place of Residence in %(Standard Error)					
1	Urban	12.9 (0.007)	22.3 (0.015)	01.6 (0.006)	63.2 (0.014)
2	Rural	26.5 (0.014)	09.5 (1.014)	09.7 (0.012)	54.3 (0.015)
Caste (Standard Error)					
1	Dalit	27.6 (0.012)	09.1 (0.015)	10.9 (0.013)	52.5 (0.015)
2	Disadvantaged Janajatis	26.8 (0.012)	09.6 (0.014)	09.9 (0.012)	53.7 (0.015)
3	Disadvantaged Non dalit Terai caste	27.0 (0.013)	11.5 (0.015)	07.6 (0.012)	53.9 (0.016)
4	Religious Minorities	22.8 (0.014)	10.9 (0.018)	10.4 (0.015)	55.9 (0.018)
5	Relatively advantaged Janajatis	12.7 (0.007)	21.2 (0.016)	01.8 (0.006)	64.2 (0.016)
6	Upper caste groups	17.5 (0.009)	17.3 (0.015)	04.4 (0.008)	60.7 (0.015)

Source Authors

Table 4.5 Poverty dynamics for population sub-groups after netting OOPH

	Incidences			
	P2P	P2NP	NP2P	NP2NP
Belt				
Mountain	0.243 (0.014)	0.101 (0.016)	0.138 (0.016)	0.519 (0.017)
Hill	0.2 (0.01)	0.153 (0.015)	0.053 (0.009)	0.015)
Terai	0.247 (0.011)	0.12 (0.014)	0.084 (0.014)	–0.015
Caste				
1. Dalit	0.278 (0.013)	0.089 (0.015)	0.112 (0.013)	0.521 (0.015)
2. Disadvantaged Janajatis	0.272 (0.012)	0.093 (0.015)	0.103 (0.012)	0.532 (0.015)
3. Disadvantaged Non dalit Terai caste	0.274 (0.13)	0.112 (0.015)	0.08 (0.013)	0.534 (0.016)
4. Religious Minorities	0.231 (0.015)	0.106 (0.018)	0.107 (0.015)	0.555 (0.018)
5. Relatively advantaged Janajatis	0.133 (0.008)	0.207 (0.017)	0.02 (0.006)	0.639 (0.016)
6. Upper caste groups	0.18 (0.009)	0.17 (0.015)	0.046 (0.009)	0.604 (0.015)
Development region				
Eastern	0.26 (0.012)	0.116 (0.015)	0.079 (0.012)	0.544 (0.016)
Central	0.177 (0.01)	0.179 (0.015)	0.037 (0.009)	0.607 (0.015)
Western	0.21 (0.012)	0.136 (0.06)	0.063 (0.011)	0.591 (0.016)
Mid-western	0.269 (0.012)	0.095 (0.015)	0.123 (0.014)	0.513 (0.015)
Far-western	0.279 (0.014)	0.059 (0.015)	0.145 (0.015)	0.516 (0.016)
Place of residence				
Urban	0.135 (0.008)	0.218 (0.015)	0.018 (0.006)	0.629 (0.014)
Rural	0.268 (0.012)	0.093 (0.014)	0.1 (0.012)	0.539 (0.015)

Source Authors

Dalits transition from non-poor to poor, while only 9% move from poor to non-poor. OOPH has heightened vulnerability among Dalits and Disadvantaged non-Dalit Terai castes.

4.5 Discussions and Concluding Remarks

The cornerstone for poverty reduction in the country lies in providing accurate and up-to-date data for measuring poverty. While a substantial body of cross-sectional poverty studies exists in the literature, analyzing poverty trends solely based on cross-sectional data may not suffice to design effective anti-poverty policies. Poverty is inherently dynamic, with analysis of poverty dynamics highlighting both the duration of poverty and movements in and out of poverty. The duration and movement of poverty are also influenced by healthcare expenditures. Panel data plays a pivotal role in enhancing our understanding of poverty dynamics. Given the absence of actual panel data for Nepal, this paper employs a recently developed statistical method by Dang et al. (2014) and Dang and Lanjouw (2013) to construct synthetic panel data using two rounds of cross-sectional data from 2003/04 and 2010/11 to measure poverty dynamics. Under reasonably standard assumptions, time-invariant household characteristics in the two cross-sections serve as the panel connectors utilized to construct synthetic panel households. The adoption of measuring poverty dynamics has gained widespread acceptance in recent years as it addresses the processes central to both the persistence of poverty and its alleviation. Dynamic analysis aids in discerning between permanent and transient poverty, a crucial consideration from a policy standpoint (Ward 2016).

Table 4.6 Impact of OOPH on poverty dynamics for population sub-groups

	Incidences			
	P2P	P2NP	NP2P	NP2NP
Belt				
Mountain	0.01	–0.006	0.008	–0.011
Hill	0.012	–0.008	0.005	–0.01
Terai	0.013	–0.009	0.008	–0.013
Caste				
1. Dalit	0.002	–0.002	0.003	–0.004
2. Disadvantaged Janajatis	0.004	–0.003	0.004	–0.005
3. Disadvantaged Non-Dalit Terai caste	0.004	–0.003	0.004	–0.005
4. Religious Minorities	0.003	–0.003	0.003	–0.004
5. Relatively advantaged Janajatis	0.006	–0.005	0.002	–0.003
6. Upper caste groups	0.005	–0.003	0.002	–0.003
Development Region				
Eastern	0.011	–0.009	0.008	–0.011
Central	0.015	–0.011	0.006	–0.01
Western	0.014	–0.008	0.008	–0.014
Mid-western	0.009	–0.004	0.006	–0.011
Far-western	0.005	–0.002	0.005	–0.009
Place of Residence				
Urban	0.006	–0.005	0.002	–0.003
Rural	0.003	–0.002	0.003	–0.004

Source Authors

The results reveal that nearly 40 percent of the population in Nepal is vulnerable to poverty, a percentage that increases by over 1 percent due to out-of-pocket health expenditure (OOPH). Of this population, almost 21 percent are trapped in chronic poverty. There has been little to no decline in poverty among Dalit households, with chronic poverty predominantly affecting Dalits and certain ethnic groups. OOPH has exacerbated vulnerability among Dalits and disadvantaged non-Dalit Terai castes. The central and western development regions experience a higher impact of OOPH on increasing chronic poverty compared to other regions, with the Terai region being particularly affected compared to mountainous areas. OOPH amplifies poverty incidence and vulnerability while simultaneously reducing the likelihood of maintaining non-poverty status over time. However, the impact of OOPH on chronic poverty is contingent upon healthcare service utilization, as the analysis does not encompass chronic poor individuals who do not seek healthcare services.

In Nepal, poverty is exacerbated by a pervasive sense of uncertainty surrounding livelihoods and life, which threatens a broader segment of the population beyond those officially classified as poor. Poverty is a highly intricate phenomenon, characterized by a multitude of interconnected economic, political, and social deprivations. The benefits of development often fail to reach various segments of society, leaving some marginalized and untouched by progress. Uneven development persists in Nepal due to a myriad of factors, including social structures, limited access to services, shortcomings in the planning process, and deficiencies in government mechanisms, among others. Consequently, chronic poverty remains unacceptably high in Nepal.

Chronic poverty typically stems from structural characteristics of households and their environments, as individuals experience deprivation over prolonged periods, often spanning their entire lives. Addressing such entrenched poverty requires policies aimed at tackling structural issues, necessitating long-term and resource-intensive interventions such as infrastructure development, asset redistribution, enhancing economic, social, and political inclusion, and fostering higher rates of capital accumulation, among others.

Transient poverty, on the other hand, is often the result of temporary setbacks that reverse when better times ensue. Policies targeting transient poverty should focus on providing short-term stability through measures like increased access to credit facilities, enhanced service access, remittance support, and social safety net programs to stabilize income fluctuations. Distinguishing between transient and chronic poverty is crucial for assessing progress toward overarching development goals.

The current reduction in poverty in Nepal is largely attributed to the phenomenon of out-migration, leading to a significant rise in remittances. Remittances can play a vital role in alleviating transient poverty by serving as a form of insurance to shield families from adverse risks and shocks back home.

Acknowledgements I would like to express my sincere thanks to Mr. Vishnu Sapkota for his support for data analysis.

References

Addison T, Hulme D, Kanbur R (2009) Poverty dynamics: interdisciplinary perspectives: interdisciplinary perspectives, Oxford University Press

Adhikari SR et al (2009) Paying for hospital-based care of Kala-Azar in Nepal: assessing catastrophe. Impover Econ Conseq Health Policy Plan 24(2):129–139

Adhikari SR (2016) Poverty dynamics in Nepal between 2004 and 2011: an analysis of hybrid dataset. NRB Econ Rev 28(1):29–40

Adhikari SR, Sapkota VP (2018) Dynamics of impoverishment impact of healthcare payments in Nepal. Social Science Baha, Kathmandu

Banks J, Blundell R, Brugiavini A (2001) Risk pooling, precautionary saving, and consumption growth. Rev Econ Stud 68(4):757–779

Central Bureau of Statistics (CBS) (1996) Nepal living standard survey 1995/96: statistical report. Volume I & II. Central Bureau of Statistics, National Planning Commission Secretariat, Kathmandu, Nepal

Central Bureau of Statistics (CBS) (2004) Nepal Living Standard Survey 2003/04: Statistical Report. Volume I & II. Central Bureau of Statistics, National Planning Commission Secretariat, Kathmandu, Nepal

Central Bureau of Statistics (CBS) (2005) Poverty Trends in Nepal (1995–96 and 2003–04). Central Bureau of Statistics, National Planning Commission Secretariat, Kathmandu, Nepal

Central Bureau of Statistics (CBS) (2011) Nepal Living Standard Survey 2010/11: Statistical Report. Volume I & II. Central Bureau of Statistics, National Planning Commission Secretariat, Kathmandu, Nepal

Central Bureau of Statistics (CBS) (2011) Poverty in Nepal. Central Bureau of Statistics, National Planning Commission Secretariat, Kathmandu, Nepal

Chaubey PK (1995) Poverty measurement: issues, approaches, and indices. India, New Age International (P) Limited publications, New Delhi

Clark D, Hulme D (2005) Towards a unified framework for understanding the depth, breadth, and duration of poverty global poverty. Working paper GPRG-WPS-020 Research Group and Institute for Development Policy and Management University of Manchester, UK

Cruces G, Lanjouw P, Lucchetti L, Perova E, Vakis R, Viollaz M (2014) Estimating poverty transitions repeated cross-sections: a three-country validation exercise. J Econ Inequ. https://doi.org/10.1007/s10888-014-9284-9

Dang Hai-Anh, Peter Lanjouw (2013) Measuring poverty dynamics with synthetic panels based on cross-sections. World Bank Policy Research Working Paper No. 6504. World Bank

Dang H-A, Lanjouw P, Luoto J, McKenzie D (2014) Using repeated cross-sections to explore movements in and out of poverty. J Dev Econ 107:112–128

Deaton A, Paxson C (1994) Inter-temporal choice and inequality. J Polit Econ 102(3):437–467

Ferreira Francisco HG (2015) Measuring Global poverty past, present, and future. DECRG Policy Research Talk: November 30

Hasegawa H, Ueda K (2007) Measuring chronic and transient components of poverty: a Bayesian approach. Emp Econ 33:469–490

Haughton J, Khandker SR (2009) Handbook on poverty and inequality. The International Bank for Reconstruction and Development The World Bank Washington, DC

Jean-Yves D, Araar JA, Giles J (2010) Chronic and transient poverty: measurement and estimation, with evidence from China. J Devel Econ 91:266–277

National Planning Commission (NPC) (2015) Sustainable development goals 2016–2030 National (preliminary) report. The national planning commission, Government of Nepal, Kathmandu

O'Donnell OA, Wagstaff A (2008) Analyzing health equity using household survey data: a guide to techniques and their implementation: World Bank Publications

Pencavel J (2007) A life-cycle perspective on changes in earnings inequality among married men and women. Rev Econ Stat 88(2):232–242

Van Doorslaer E, O'Donnell O, Rannan-Eliya RP, Somanathan A, Adhikari SR, Garg CC, Ibragimova S (2006) Effect of payments for health care on poverty estimates in 11 countries in Asia: an analysis of household survey data. The Lancet 368(9544):1357–1364

Ward PS (2016) Transient poverty, poverty dynamics, and vulnerability to poverty: an empirical analysis using a balanced panel from rural China. World Dev 78:541–553

World Health Organisation (2015) Global Health Expenditure Database. http://apps.who.int/nha/database/Key_Indicators/Index/en

World Bank (2006) Nepal resilience amidst conflict an assessment of poverty in Nepal, 1995–96 and 2003–04" Report No. 34834-NP Poverty Reduction and Economic Management Sector Unit, South Asia Region, World Bank

Development and Exclusion: Intergenerational Stickiness in India

5

Rajarshi Majumder and Jhilam Ray

Abstract

The concept of *development* has matured from being indicative of aggregative progress to being sensitive to inequality and exclusion within the whole, giving rise to the coinage *Inclusive Development.* While India is doing quite well as a macroeconomic entity, increasing inequality and social tension puts up questions regarding nature of its developmental process and whether it has been sufficiently inclusive. Issues like labour market discrimination, (in)effectiveness of policies like reservation in education and employment, and growing inequality in the country have been flagged by several researchers in recent years. The present paper explores the role of intergenerational stickiness in perpetuating such disparity across social groups in India. Results suggest that though mobility rates are converging across social groups in recent decade, inequalities in basic indicators of human capital and livelihood outcomes are still quite high. It is suggested that the State should focus on providing quality education and skill for all so that livelihood opportunities become universal and inclusive, and the growth process becomes sustainable.

5.1 Introduction

The concept of *development* has matured from being indicative of aggregative progress to being sensitive to inequality and exclusion within the whole, giving rise to the coinage *Inclusive Development*. This notion speaks of bridging the gap between ethnic/social groups within a nation in domains of livelihood, social status, political empowerment and cultural freedom, among others. Social scientists, especially economists, have focussed primarily on livelihood and economic condition while speaking of inter-group equality and inclusive development. If however, a country or society is marked by severe initial disparity among social groups, convergence and catching up would depend on the temporal movement of them over the economic ladder. In this context, intergenerational mobility acts as

This paper draws from the authors' previous and continuing works on Mobility in India. Some of these have been published earlier. See Ray and Majumder (2010, 2013, 2015), and Majumder (2013) for previous works. The pre-print version of this paper is available at the pre-print repository at MPRA (https://mpra.ub.uni-muenchen.de/71182/1/MPRA_paper_71182.pdf). MPRA retains the right to use the pre-print version.

R. Majumder (✉)
Department of Economics, University of Burdwan, Burdwan, West Bengal, India
e-mail: rmajumder@eco.buruniv.ac.in

J. Ray
Department of Economics, Jadavpur University, Kolkata, West Bengal, India

A. Alakshendra et al. (eds.), *Development, Inclusion and Sustainability*, Sustainable Development Goals Series, https://doi.org/10.1007/978-981-97-6863-9_5

a mechanism through which social fluidity and equality may be achieved. Conversely, stickiness across generations acts as a barrier to inclusive development by keeping intact the traditional socioeconomic hierarchy characterised by a wide chasm between social groups at the top and those at the bottom. Historically, large developing countries were characterised by such unequal social hierarchy and restricted mobility, leading to persistence of inequality and vicious cycle of poverty among a large proportion of population. In India too, historically some groups were confined to lower strata of society through economic and social discrimination leading to socioeconomic stagnation. Over the last two decades, with its exceptional growth story and developmental agenda, India has attracted the attention of the world. While India is doing quite well as a macroeconomic entity, increasing inequality and social tension puts up questions regarding nature of its developmental process and whether it has been sufficiently inclusive. Issues like labour market discrimination, (in)effectiveness of policies like reservation in education and employment, and growing inequality in the country have been flagged by several researchers in recent years. The present paper explores the role of intergenerational stickiness in perpetuating such disparity across social groups in India. We argue that economic status is intricately linked to what a person does for livelihood, i.e. her occupation, and what remuneration she receives for it, i.e. her wages. In the present world system, occupation and wages are also critically determined by the human capital quotient of the individual, marked generally by her educational level. Therefore, the socioeconomic structure of a country and its temporal movement would be shaped by intergenerational mobility in education, occupation and income for different social groups. Higher (upward) mobility for the lagging classes would lead to catching up and convergence, while lower mobility for them would lead to widening gaps. Our objective is to examine whether persistence of economic inequality across social groups in India is associated with high parental impact and low intergenerational mobility for the historically lagging and excluded social groups.

Mobility may result from two sources—structural changes in the society or greater social fluidity. In case of former, structural changes in the society like economic policy shifts, jump in growth trajectory etc. may lead to transformation across the board resulting in increased share of specific education/occupation/income groups in the society. This will appear as intergenerational mobility in the dataset and is called Structural Mobility. When estimated separately for social groups, it also shows how structural factors have benefitted different social groups in the country. Alternatively, even if there is no significant change in the occupational pattern across the board, people may interchange positions due to social fluidity as children of parents in X category moves into category Y and vice versa. This part of the mobility is thus called Exchange mobility and gives a measure of social flux in the country. Since these have completely different policy implications, we decompose the observed mobility into structural and exchange mobility to understand the pattern and source of mobility across generations and social strata.

In addition, recent methodologies suggest that transition matrices (of two time periods) should be standardised before calculating stickiness/mobility figures to enable comparability and decomposability into structural and fluidity components (Altham and Ferrie 2007). These have been sometimes called effects of prevalence and association by researchers (see Bourdieu et al. 2006 for example). Equalising marginal frequencies of the two transition matrices provides us precise and comparable measures of stickiness for the two time periods, and also allows us to examine whether stickiness has changed significantly over time. This technique has been used in this paper.

While estimates of degree of stickiness/mobility are important, they offer only a partial understanding of the socioeconomic process. Unless we look at possible linkages between mobility and both causal and impact variables, it would not be possible to comprehend how stickiness acts as a barrier to development or how it can be scaled down. Surprisingly, researchers have still not touched this issue adequately. This paper looks at possible linkages of stickiness/mobility with several micro- and macro-economic indicators to bridge this gap in existing research.

Internationally there is a substantial literature on intergenerational mobility, mostly from developed countries. These include, but not limited to, studies on educational mobility by Halsey et al. (1980), Gang and Zimmerman (2000), Checchi and Flabbi (2007); those on occupational mobility by Behrman et al. (2001); and those on intergenerational income stickiness by Becker and Tomes (1979), Solon (1992), Minicozzi (2003) [see Solon (1999), Mazumder (2001), and Majumder (2013) for a good review]. Though any simplification is difficult, it is observed that mobility is higher in developed countries compared to underdeveloped ones.

Surprisingly, this area has remained underfocussed in Indian economic research, one of the major reasons being the absence of pan-generation data on education/occupation/income and allied factors. Some of the previous studies include an early work by Driver (1962), and those by Bhowmik (1992), Kumar et al. (2002a, 2002b), Munshi and Rosenzweig (2006), Barooah et al. (2007), Deshpande and Palshikar (2008), Majumder (2010, 2013), Ray and Majumder (2010, 2013, 2014, 2015), Motiram and Singh (2012), Hnatkovska et al. (2013), Azam and Bhat (2014)]. The present paper is related to these works and also to those on disparity and discrimination [Atkinson (1998), Takahiro (2007), Madheswaran and Attewell (2007), Majumder (2010), Mukherjee and Majumder (2011)].

The present paper is different from the existing works both in terms of methodology, tools used, coverage and inclusion of structural and exchange mobility in the analysis and adds to existing literature on five grounds. First, the paper uses both the Transition Matrix approach as well as the Regression/Odds Ratio approach to estimate mobility across generations—thus combining methods practised by sociologists and economists. Even within the Transition matrix, we estimate and present both the absolute and standardised stickiness/mobility figures and a measure of direction of change in stickiness. Second, to the best of our knowledge, this is the only paper that provides estimates of *upward mobility*—where children have moved into superior category compared to their parents. In our opinion this is a crucial contribution since the direction of mobility is more important from a developmental perspective than simply the quantum of mobility (e.g. a society in distress where large number of current generation workers move into occupations that are inferior to their parents will exhibit high mobility). Third, by estimating both detailed occupational mobility and broad-group mobility and by comparing between them we have provided a snapshot of vertical vis-a-vis horizontal occupational mobility in the country. Fourth, unlike other recent works which explore occupational mobility across generations in India our study facilitates comparison across time and uses a more realistic occupational classification. Last, but not the least, this paper provides interesting linkages between stickiness/mobility and selected socioeconomic variables to understand both the impact of stickiness on individuals and how it can be scaled down. Thus, our paper improves upon recent works by using a better methodological/conceptual framework and temporally comparable database as also expanding the coverage and framework of examination from merely descriptive to prescriptive.

5.2 Data and Methodology

5.2.1 Database

The study has used the Periodic Labourforce Survey (PLFS) data collected by National Statistical Office (NSO) pertaining to the year 2021–22. As a contrast we have also presented results for the NSSO 50th round data for the year 1993–94, the beginning of the structural adjustment process in India. Our study therefore provides a comparative view of mobility and stickiness over almost three decades of market led economic growth in the country. Family records have been matched with personal records to obtain multi-generational data on education, occupation, and income. Since our database is at household level, we have used only those parent–child pairs where both father and child live in the same household. Also, to allow for completion of formal education in Indian system, we have selected only those persons with age greater than 20 years as belonging to '*children*' group. We have also used the occupation of father as the parental occupation since female work participation is quite low in India (one-fourth of the corresponding male figures), especially for the previous generation.

We understand that restricting the sample to co-resident father–child creates possibility of sample selection bias but there are conflicting evidences regarding the significance of the selection bias when we ignore the split-off sons [Thomas et al. (2001) find a significant difference in case of Indonesia, while Alderman et al. (2001) report that the bias is non-significant in case of Bolivia, Kenya and South Africa]. We contend that split-off decisions are themselves random and not solely dependent on the characteristics of the children since robustness tests found that demographic/labour-market characteristics of our working sample are not significantly different statistically from the full sample.

5.2.2 Technical Methodology

We are interested in examining how children's education/occupation/income is related to parental occupation. More specifically, we want to quantify the degree of intergenerational stickiness and upward mobility. Stickiness can be measured by the percentage of children remaining stuck in his/her father's education/occupation/income class. In literature, this is done by following the ***Transition/Mobility Matrix*** approach. It can also be measured by the parental impact on child's education/occupation/income using the ***Regression Approach***. We have applied both the techniques in this paper. The transition matrix approach has been extended to identify upward and downward mobility, structural and exchange mobility and Altham d-Statistics (for details on this measure see Altham 1970; Altham and Ferrie 2007). For the sake of brevity, details of the methodology are not spelt out here. Interested readers can see Ray and Majumder (2010, 2013, 2016) and Majumder (2013).

5.2.3 Educational and Occupational Classification

We have classified education attainment into seven educational levels—illiterates, barely literate but without primary schooling, primary school passed, middle school passed, secondary school passed, higher secondary passed and graduates and beyond. We have also used completed years of schooling as a quantitative variable in the regression-based approach.

We have used the Indian NCO-1968 classification in our study and workers have been divided into ten occupational classes. Arranged in descending order of hierarchy, social prestige and average income, these are technical and scientific personnel, professionals, administrative, clerical, sales, service, farmers, production-related, transport and labourers not elsewhere classified. Occupational structure and mobility are discussed in terms of this structure. At the second level, we have clubbed similar occupations to form three broad groups—Grade-I (white-collar jobs—technical and scientific personnel, professionals and administrative); Grade-II (pink-collar jobs—clerical, sales and service); and Grade-III (blue-collar jobs—farmers, production-related workers, transport workers and labourers not elsewhere classified). This hierarchical structure has also been used in our study. It is however important to note that in the Indian context production-related jobs include construction jobs also. Both the 1993 and 2021 databases have been suitably adjusted using concordance tables to reflect the NCO1968 groups.

We understand that, while NSSO assigns a *single and unique* occupation to normally working population based on time involvement, in developing countries like India, especially in the rural areas, people are more often than not engaged in multiple occupations either at the same time or seasonally over the course of the year. Problem does not arise as long as one occupation truly dominates the rest. But in cases where several occupations are quite close in terms of time involvement, the NSSO database may not truly reveal the actual situation. This is a caveat we must keep in mind.

5.2.4 Deriving Lifetime Wage Income and Income Classification

In studying intergenerational income mobility, our basic objective is to examine whether current generation of workers is earning more than their parents, after controlling for factors like age, experience, etc. Ideally, we should consider permanent incomes of children and parents, which is very difficult to observe and not possible with the data available to researchers. The solution suggested in literature is to compute some form of *synthetic variable* that proxies for the permanent wage income of parents and children, filtering out the effects of age, experience, stage of life cycle, etc. A double isolation method is used here where both father and child's *Isolated Wage Income* is derived after controlling for age, experience and occupation [see Ray and Majumder (2013) for details].

Instead of computing synthetic wages as above, we may group the parents and children into quintile classes according to their respective actual/current wage incomes and use the transition matrix between children's quintile group membership and that of parents to estimate *Relative Income Mobility*—a comparison between relative position of a child within his peers with relative position of the parent among *their* peers. This method has also been used here.

5.3 Social Hierarchy in India

To better comprehend the results, we must understand that traditionally Indian society can be segmented into four broad groups—the upper castes (mostly Hindus), the adivasis (or aboriginals), the dalits and the shudras or mahadalits. Historical socioeconomic structure gave birth to severe social disparity with the Hindu upper castes at the pinnacle and dalits, mahadalits and adivasis at the bottom (Ambedkar 1936, 1987; Akerlof 1976; Lal 1988; Scoville 1991; Thorat and Newman 2009). Caste system and associated discrimination strengthened during colonial rule due to codification and recording of caste in the decennial population census, and co-option of members from the upper castes into the ruling coalition. As a result, administrative hierarchy mirrored existing social stratification. After independence, India tried to reverse the exploitation by giving express recognition to lagging castes and tribes in the First Schedule of its constitution in 1950. These groups came to be known as Scheduled Castes and Scheduled Tribes and various positive discriminatory measures through reservation and quota in elementary education and government jobs were provided to bring them up to a level playing field. In 1993, the *shudras* or *mahadalits* were also brought under the purview of reservation by designating them as Other Backward Castes or OBCs. Such efforts notwithstanding, discrimination still exists in India in the form of 'hidden apartheid'. Members of upper castes, knowingly or unknowingly as part of social custom, discriminate against and exploit the lagging castes. Since caste is considered hereditary, advantages and disadvantages transmit from one generation to the next, projecting the upper castes to a virtuous self-propelling upward spiral while trapping the backward castes in a vicious low level cycle. Thus, the social hierarchy at present ranges from the General/Advanced Castes at the top; followed by the Other Backward Castes, Scheduled Castes and Scheduled Tribes at the bottom. This is mirrored in the education, occupation and income level of the groups as well. Throughout this paper we must keep this hierarchy at the back of our mind. Also, in this paper, we have contrasted the situation of the two most lagging social classes—the SCs and the STs—with the General Caste for examining the issue of exclusion and disparity. It is also to be noted that the figures for the General Caste are not strictly comparable across 1993 and 2021 because of the OBCs being included within the general caste in 1993. For some later tables, we have clubbed the OBCs with the General Caste to compute comparable figures for 2021.

Table 5.1 Educational indicators in India

Year	Education						Formal Schooling					
	Literacy (%)				Gap from Gen		High School Passed (%)				Gap from Gen	
	ST	SC	OBC	Gen	ST	SC	ST	SC	OBC	Gen	ST	SC
1983	23.3	27.9	na	45.3	−22.0	−17.4	1.8	2.3	na	8.0	−22.0	−17.4
1987	25.8	29.9	na	48.9	−23.1	−19.0	2.6	2.9	na	9.7	−23.1	−19.0
1993	32.7	36.9	na	55.2	−22.5	−18.3	4.0	4.7	na	13.3	−22.5	−18.3
1999	40.6	44.7	52.4	66.9	−26.3	−22.2	6.2	6.8	10.3	21.7	−26.3	−22.2
2004	46.7	51.1	57.9	71.4	−24.7	−20.3	6.4	8.6	12.7	25.5	−24.7	−20.3
2011	62.4	63.2	98.6	80.1	−17.7	−16.9	15.7	16.8	20.7	35.8	−20.1	−19.0
2021	70.8	72.8	77.5	85.7	−14.9	−12.9	22.6	26.6	33.3	45.1	−22.5	−18.5
	Employment						Poverty Situation—HCR					
	Regular Salaried (%)				Gap from Gen		Incidence of Poverty (%)				Gap from Gen	
	ST	SC	OBC	Gen	ST	SC	ST	SC	OBC	Gen	ST	SC
1983	8.3	9.3	na	21.0	−12.7	−11.7	63.6	58.5	na	40.2	23.4	18.3
1987	8.8	12.5	na	16.9	−8.1	−4.4	57.5	52.6	na	35.7	21.8	16.9
1993	6.8	10.4	na	17.5	−10.7	−7.1	69.5	64.4	na	45.3	24.2	19.1
1999	6.8	10.4	11.6	21.6	−14.8	−11.2	49.9	43.6	33.5	23.8	26.1	19.8
2004	7.6	14.2	13.5	23.4	−15.8	−9.2	60.3	54.8	41.3	27.1	33.2	27.7
2011	10.2	16.0	16.3	27.4	−17.2	−11.4	45.5	38.4	27.0	17.8	27.7	22.6
2021	13.3	19.7	20.7	31.2	−17.9	−11.5	16.6	14.0	9.9	7.4	9.2	6.4

Source Authors' Calculations based on NSSO (Various Years) and PLFS (2021)

We start with an empirical background of the condition of the two marginal social classes (the SCs and the STs) vis-a-vis the General Caste in terms of education, employment and poverty over the last three decades. It appears that the SC/STs have lower educational attainment levels, poorer employment situation and higher incidence of poverty compared to the General Caste (Table 5.1). The disparity has been persistent over the last three decades, even increasing in the 1990s and early years of this century for the STs. Also remarkable is that for all three dimensions figures for the marginal social classes in 2011–12 were close to what the general caste had during 1993–94, and in 2021–22 was what the general caste had during 2004–05, indicating that they are lagging by almost 20 years from the advanced class!

A quick look at the distribution of population over educational groups and monthly consumption deciles for the year 2021–22 shows that SC/STs being over-represented in the lower levels and under-represented in the upper levels (Fig. 5.2). This indicates that the development process in India has not been as inclusive as it should have been in context of the historical social disparity and recent high economic growth. It is our argument that the answer lies in high stickiness and low upward mobility, especially of the marginalised social classes (read SCs and STs). We examine the empirical evidences in the subsequent sections (Fig. 5.1).

5.4 Stickiness and Mobility: Regression Approach

5.4.1 Educational Stickiness

We want to examine how family background and parental educational level influences children's education. To do so, we have considered son's status as the dependent variable while status of father (and mother in some cases) is taken as explanatory variable. Several other

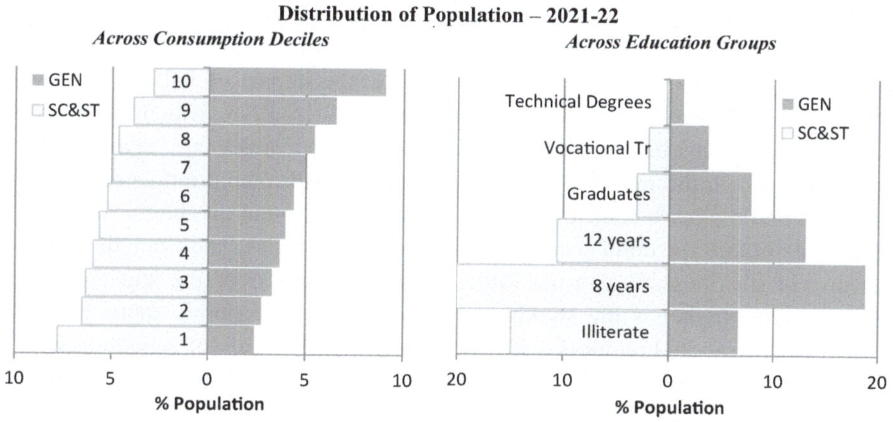

Fig. 5.1 Distribution of population—2021–22. *Source* Authors' Calculations based on NSSO (2021)

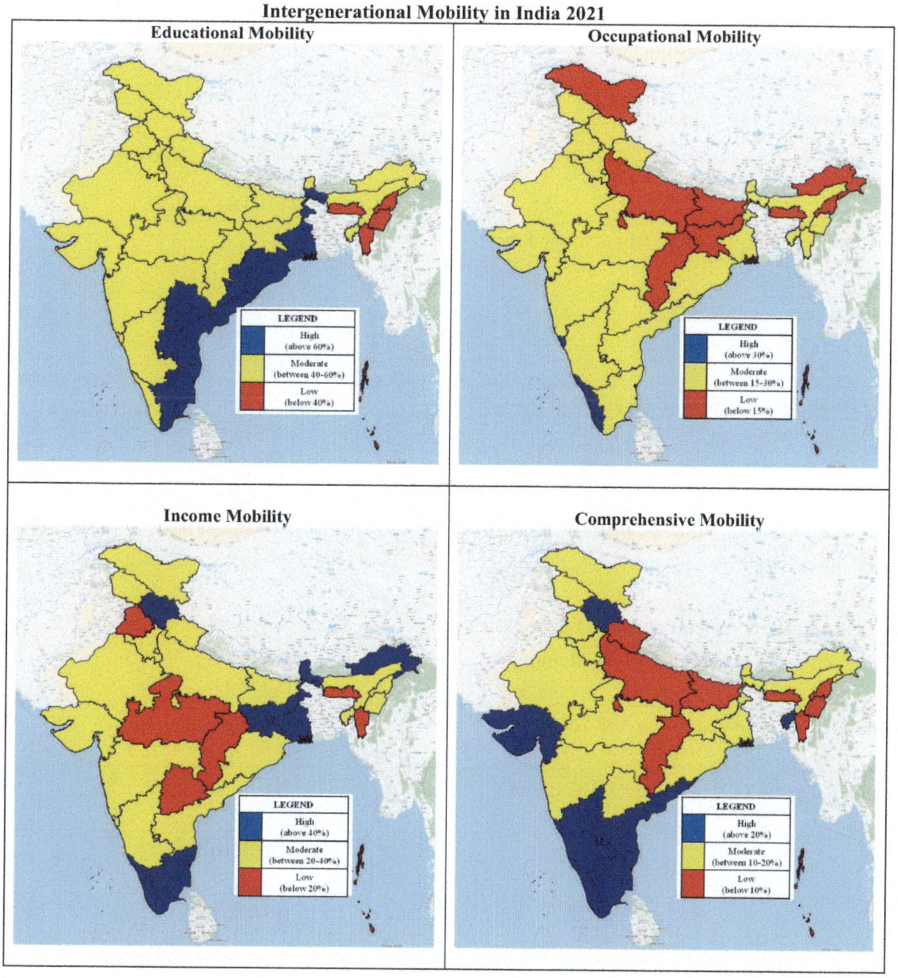

Fig. 5.2 Intergenerational mobility in India 2021

socioeconomic indicators (explained later) are also considered. Since we hypothesise that the social background in terms of caste status also has a major role on child's achievement, a caste dummy and its interaction effect with father's status are also included in the model. The coefficients of father's and mother's education status measures parental impact on child's education and are measures of stickiness. Higher the value of these β-coefficients, higher is the stickiness, and less is the intergenerational mobility.

Table 5.2 gives us the results of regression analysis for the year 1993 and 2021. The value of β_F, i.e. the impact of father's education on child was 0.429 in 1993, which declined to 0.215 in 2021. Both values are statistically significant indicating substantial influence of father's educational background on child's education and existence of stickiness. The strength of such stickiness has decreased over the three decades. Among other factors, economic status of the household affects child's years of schooling with base level

Table 5.2 Regression-based estimates of stickiness of child's education

Dependent variable: son's completed years of schooling		
Independent variables ↓	1993	2021
(Constant)	3.657** (2870.3)	6.943** (2568.3)
Father's completed years of schooling	0.429** (2338.2)	0.360** (2725.4)
Mother's completed years of schooling	0.309** (1365.8)	0.220** (1587.4)
Poverty dummy (control: non-poor)	–1.274** (1185.4)	–1.182** (991.1)
Father's occupation (Control Group: Blue-Collar Jobs)		
Pink collar	0.628** (456.8)	0.342** (214.5)
White collar	0.216** (135.2)	0.664** (568.5)
Social group (control group: general or hindu upper caste)		
ST dummy	–1.512** (675.6)	–0.976** (483.5)
SC dummy	–1.023** (656.7)	–0.258** (154.4)
OBC dummy		–0.0635** (42.9)
Interaction (control group: general or hindu upper caste)		
Father's completed years of schooling * ST dummy	0.121** (175.6)	0.062** (107.3)
Father's completed years of schooling * SC dummy	0.071** (152.7)	0.057** (145.8)
Father's completed years of schooling * OBC dummy	Na	–0.022** (125.7)
F Value	$42.6X10^5$ **	$28.4X10^5$ **
Adj R Sq	0.363	0.346

Note Figures in parenthesis are t-values; * and ** denotes significance at 10 per cent and 5 per cent levels, respectively

Source Author's calculations

nearly 20–30 per cent lower for households below poverty line compared to non-poor families. This negative impact of poverty on education has, however, decreased over time indicating perhaps increased accessibility of public education in recent times. Base level schooling is significantly lower for the marginal classes compared to the advanced class. Schooling is also significantly higher for children whose parents are in higher occupational groups (white- and pink-collar jobs) compared to the lowest occupational group (blue-collar jobs). Also, the role of white-collar occupation of father in raising the education level of children is increasing over time.

However our main focus is examining difference in intergenerational stickiness across social groups. The interaction coefficients of social group dummies with father's education provide the difference of impact of father's education across social groups. These interaction coefficients are positive and significant for SCs and STs but negative for OBCs indicating that the parental influence is significantly higher for the first two social classes compared to the advanced group but lower for the OBCS. This implies that stickiness is higher and mobility is lower for the SCs and STs but the reverse for the OBCs. One reason behind lower stickiness for the OBCs may be the reservation in educational institutions for OBCs since the 1990s. Also to be noted is that this additional stickiness for the excluded classes has been declining over time, indicating that mobility rates are coming closer across social classes in recent times. This is surely a welcome trend, but unless mobility rates of excluded classes are substantially higher than the advanced classes, convergence of educational achievement levels will take a very long time since the former have started from a much lower level.

5.4.2 Occupational Stickiness

While multinomial logit regression models are best suited to estimate probabilities of child entering into different occupation groups and expected changes in probability for changes in explanatory variables, we have used a simpler approach here. We have provided scores to the occupational groups on a scale of 1–10, depending on social hierarchy and status. Thus, the technical occupations are given a value of 10, followed by professional, administrative, clerical, sales, services, production, transport and farming in that order. Workers in non-specified activities are given a score of 1. Intergenerational stickiness can then be estimated by considering child's occupation score as a dependent variable and occupation score of father as one of several explanatory variables. Other causal variables generally used in such studies and also included here are age of child, completed years of schooling of child and completed years of schooling of father. We have also used interaction between father's occupation and social class dummy to understand the difference in stickiness across social groups.

Table 5.3 Regression-based estimates of stickiness of child's occupation score

Dependent variable: son's occupational score		
Independent variables ↓	1993	2021
(Constant)	2.030** (1514.3)	0.778** (493.3)
Age of child	0.015** (344.0)	0.127** (361.6)
Child's completed years of schooling	0.070** (1137.1)	0.056** (721.2)
Father's completed years of schooling	0.024** (289.1)	0.007** (107.3)
Father's occupation score	0.359** (2002.8)	0.266** (452.6)
Interaction (control group: general or Hindu upper caste)		
Father's occupation score * ST dummy	0.008** (41.6)	–0.077** (287.2)
Father's occupation score * SC dummy	0.011** (71.6)	–0.076** (377.7)
Father's occupation score * OBC dummy	na	–0.020** (134.2)
F value	$1.6X10^5$ **	$1.7X10^5$ **
Adj R Sq	0.224	0.307

Note Figures in parenthesis are t-values; * and ** denotes significance at 10 per cent and 5 per cent levels, respectively
Source Author's calculations

Results indicate that, with increase in completed years of schooling and age, there is a substantial increase in occupational score, indicating that probability of being in higher level occupations increases with rise in education and age (Table 5.3). Occupational score increases with increase in completed years of schooling of father also. However, the highest impact is of Parental Occupation Score. This indicates substantial stickiness between father's occupation group and child's occupation group, confirming lack of occupational mobility across generations in India. The stickiness has decreased over the entire study period. To determine whether parental influence is different across social groups we look at the coefficients of interaction terms. It is observed that for the marginal social classes the coefficients were significantly positive in 1993, indicating higher stickiness compared to the general caste. But in three decades, these coefficients have turned significantly negative, indicating that impact of parental occupational score is lower for them compared to the advanced class in 2021, in contrast to earlier results where it was observed to have increased during 1993–2004 period (Ray and Majumder 2010). Thus, occupational diversification is more common among the marginalised social classes in recent times compared to the general caste.

Table 5.4 Results of income stickiness estimation—regression output

Dependent variable: (ln) child's wage		
Independent variables ↓	1993	2021
(Constant)	4.338 (773.2)	3.859(2976.1)
Education	0.051** (278.5)	0.040** (1725.3)
(ln) Father's Wage	0.418** (583.3)	0.265** (1186.4)
Social Class (Control Group: General or Hindu Upper caste)		
ST_dummy	–2.725** (134.3)	–0.325** (113.2)
SC_dummy	–1.482** (144.9)	–0.311** (151.9)
OBC_dummy		–0.010* (5.5)
Interaction (Control Group: General or Hindu Upper caste)		
Wage_father*ST dummy	0.288** (120.8)	–0.102** (195.6)
Wage_father*SC dummy	0.152** (129.6)	–0.080** (221.2)
Wage_father*OBC dummy		–0.017** (57.9)
F Value	$3.56X10^5$**	$1.28X10^5$**
Adj R-square	0.4	0.32

Source Author's calculations
Note Figures in parenthesis are t-ratios; ** denotes significance at 1 per cent level

5.4.3 Income Stickiness

In the Regression or elasticity-based approach, estimating stickiness between child's wage income and that of the parent requires computing a wage function with (log of) child's income as the dependent variable and (log of) parent's income as the independent, along with other causal variables like parental education, location, regional and social class dummies, etc. As stated earlier, we have used *Isolated Wage Income* of Child and *Isolated Wage Income* of Father for the estimation. A higher value for the coefficient of father's wage income implies stronger parental effect on the children, higher intergenerational stickiness and, therefore, less mobility.

Results show that education has a positive effect on wage income, and marginal social classes earn less than the general castes on average (Table 5.4). There is substantial stickiness and impact of father's wage income on child's wage income is high. In 1993, all the marginal social classes had higher stickiness compared to the general castes. There has been a decline in stickiness over time for the marginalised classes, as a result of which, in 2021, all these groups had lower stickiness and higher mobility than the general castes.

It is this rising income mobility, among the marginalised classes in particular, that has been termed by contemporary researchers as convergence among social classes and rising up the ladder of the marginalised groups. We however beg to differ on two counts. The first point to note is that the marginalised classes started with a low base as evident from Table 5.1, and so naturally they have more scope for mobility compared to the general caste who already had reached a higher echelon and hence there was a limited scope for further movement. Second, regression-based measure does not indicate whether the mobility is dominated by upward or downward movements. It may well happen that higher mobility is because of downward shifts and is an indicator of deteriorating living conditions rather than improving. There may be lots of movements no doubt, but we have to examine whether improvements are dominating or deteriorations! Otherwise, increased social fluidity may often be misjudged as development and progress while all we have is pauperisation of a large section of the masses. The answer to this will come up in the next section when we use the transaction matrix approach to derive upward mobility.

5.5 Intergenerational Mobility: Transition Matrix Approach

We are more interested in examining to what extent children's status (educational/occupational/income) has ***improved*** over parental standards. More specifically, we want to quantify the degree of intergenerational upward mobility. This would be given by the percentage of children moving to a higher educational/occupational/income class as compared to their parents.

5.5.1 Educational Mobility

Substantial upward mobility is observed in terms of educational attainment levels and this has increased over time even after controlling for changes in marginal distribution (Table 5.5). This is true for all the social groups indicating a decline in educational stickiness across the social spectrum during these three decades. What is also noteworthy is that the marginal social

Table 5.5 Intergenerational mobility in India: summary measures

	Scheduled tribes		Scheduled castes		Others (General + OBC)		Aggregate	
	1993	2021	1993	2021	1993	2021	1993	2021
	Intergenerational stickiness							
Education	32.4	13.4	29.7	11.7	27.1	11.5	27.9	11.8
Occupation group	76.6	39.2	66.9	29.0	67.0	30.4	67.9	28.5
Occupation grade	84.5	84.2	82.5	74.6	70.2	67.9	75.8	71.4
Income quintile	66.4	34.9	58.0	34.5	53.6	35.6	56.1	34.8
	Intergenerational upward mobility							
Education	59.6	81.4	60.4	81.4	61.1	78.2	60.9	79.6
Occupation group	7.4	23.8	13.6	30.8	15.7	32.5	15.1	35.9
Occupation grade	4.5	11.1	8.3	16.4	11.6	21.2	9.5	19.2
Income quintile	18.8	36.9	19.3	38.8	19.6	35.4	19.4	36.5

Source Author's calculation based on NSSO (1993), PLFS (2021)
Note Stickiness and Upward Mobility are derived from the standardised transition matrix after converting 1993 matrix to 2021 marginal frequencies; For Income quintiles standardisation is not required; Occupation Group refers to tenfold NCO 1968 classification; Occupation Grade refers to threefold classification as explained in text

classes have higher upward mobility than the general/advanced class in 2021, which is surely a welcome trend.

5.5.2 Occupational Mobility

Occupational mobility figures using detail occupational classification show that, while, in 1993, close to 68 per cent of working adults were continuing in their father's footsteps, the figure has come down to roughly 29 per cent in 2021. Upward mobility has increased from 15 per cent to more than 35 per cent in 2021. Similar trend is observed for all the social groups indicating a decline in stickiness and a rise in mobility across the social spectrum during these three decades. However, the gap between the STs and advanced groups has remained mostly stagnant during this period, indicating that they have not been able to enter the mainstream of the economy.

Mobility among occupational groups that are pretty close in terms of social hierarchy and possible income often gives a false sense of fluidity. To explore that we have also looked at mobility across broad occupational classifications (White Collar, Pink Collar and Blue Collar, discussed earlier). Results indicate that, while only about one-tenth of working adults had better occupational grades compared to their fathers in 1993, the figure increased to one-fifth in 2021. Upward mobility of advanced group is higher than the SCs and STs in both time points and there seems to be no convergence. What is more interesting is that grade level upward occupational mobility is lower than the earlier estimated detailed occupational level mobility, indicating that most of the mobility across generations are lateral movements among related occupations while vertical movements are low. More than 84 per cent of STs and about 75 per cent of the SCs are stuck in their parental occupational grade. Therefore it would not be too wide to comment that the occupational transformation supposedly evident in India in recent times is more apparent than real at the micro level and does not appear to be improving the condition of the households en masse. This stickiness is a substantial barrier to inclusive and sustainable development.

5.5.3 Income Mobility

As mentioned earlier, we have also tried to estimate relative wage income mobility—indicating how current generation fares among their peers compared to the position of their father among their peers. For that, we have computed quintile membership of fathers and sons with respect to actual wage income. Cross tabulation of quintile groups of father and child gives us the Transitional Mobility Matrix for Income. It is observed that this (relative) upward income mobility was 19.4 per cent in 1993, but has increased to about 37 per cent in 2021. Upward income mobility figures are almost similar across social groups. However, in this case too, stickiness is markedly higher among STs compared to the other groups.

5.5.4 Age Cohorts

Another interesting feature emerges if we look at the age-cohorts (Table 5.6). For education and occupation, mobility is higher among the younger age cohort compared to the 40+age cohort. This gap is higher for the STs and SCs, indicating that they are late starters and improvement in educational and occupational levels among them have taken place mainly in the post-1990 period making the younger age cohort more upwardly mobile than the older ones. For income level however, the opposite is true. This is but natural as income increases with age and experience and so the younger age cohorts are still to enjoy the phase of increased income compared to the older cohort.

Thus, the greater social fluidity and convergence, as indicated by the regression approach discussed earlier and as claimed by contemporary researchers is restricted to educational achievement where close to four-fifth of males have better education compared to their fathers. For occupation and income however, mobility is caused as much by *downward* movement as upward and majority of adult male workers have lower or same occupation and income compared to their parents. The sense of development and progress is therefore only half-baked where

Table 5.6 Upward intergenerational mobility of different age-cohorts in India—2021 (%)

Social groups	Education		Occupation Group		Occupation Grade		Income	
	20–40 age group	40+age group	20–40 age group	40+age group	20–40 age group	40+age group	20–40 age group	40+age group
Scheduled tribes	81.9	76.2	20.6	12.1	11.2	9.1	37.0	34.6
Scheduled castes	82.3	70.0	26.9	21.7	16.3	20.0	38.4	59.0
Others (General+OBC)	78.2	73.5	33.3	22.5	21.5	21.4	34.5	53.8
Aggregate	80.3	73.3	28.7	20.9	19.1	19.9	35.8	53.4

Source Author's calculations
Note Same as Table 5.8

educational mobility is not being translated to occupational and income mobility.

5.6 Structural and Exchange Mobility

As mentioned earlier, Intergenerational mobility measured by mobility tables/transitional matrix are result of two different types of flows or movement—structural movement and exchange movement. *Structural mobility* refers to changes in the positions of individuals as a result of differences in proportions of members in hierarchical groups between two generations. It results from overall shifts in the economy status or increase in the opportunity available to all and is thus a signal of structural changes that has affected the society in general over time. *Exchange mobility* denotes changes in position of individuals consisting of substitution as a result of vacating of position in specific groups by those who do not inherit their father's position. Exchange mobility is defined as that portion of total change in status between two generations that is independent of structural change. High exchange mobility is a sign of social fluidity but keeps the aggregate socioeconomic situation almost stagnant. On the other hand, Structural Upward Mobility indicates net increase in number of persons with higher educational/occupational/income levels compared to their parents and therefore high level of this type of mobility indicates improvement of the society and would be desirable. Let us now explore the results briefly.

It is observed that most of the upward educational mobility observed in India came from structural mobility and over the period its contribution is increasing whereas the contribution of exchange mobility is low and is declining (Table 5.7). This perhaps is a result of substantial

Table 5.7 Contribution of structural mobility to aggregate mobility in India (%)

Social group	Educational mobility		Occupational mobility		Income mobility	
	1993	2021	1993	2021	1993	2021
Scheduled tribes	88.2	88.0	33.6	40.5	20.0	13.4
Scheduled castes	89.9	84.4	17.8	13.2	15.0	18.5
Others (General+OBC)	85.3	76.7	30.5	6.6	10.0	9.9
Aggregate	86.2	80.5	26.8	0.4	15.2	12.0

Source Author's calculations

expansion of educational infrastructure in recent decades through the Sarva Sikhsa Mission.

On the contrary occupational mobility is fuelled more by exchange and less by structural mobility. Moreover, the share of structural mobility in this case is decreasing over time, except for the STs.

Most of the upward income mobility also is due to exchange mobility and lesser can be attributed to structural mobility. While in 1993 about 85 per cent of the upward income mobility was due to exchange, by 2021 the contribution further increased to 88 per cent. It thus appears that the income levels did not scale up and the movement was more a social churning rather than structural improvement of the economy. Only for the SCs there is a marginal increase in share of structural mobility in aggregate upward mobility.

5.7 Interlinking the Three Mobility Dimensions

The previous sections presented us with details regarding intergenerational mobility in India across three spectrums—Education, Occupation and Income. It has been observed that moderate to high educational mobility has resulted in only marginal occupational shifts and income mobility has been generally low. In addition, the conditions of backward classes have been quite different from that of the advanced classes. Let us now try to examine whether mobility in the three spheres are interlinked and what macro variables may have had impact on them. This can be examined from several angles. First we may check whether individuals experiencing one type of mobility have the other two types of mobility as well. This is essentially a micro interlinkage and works at the household level through reciprocity and complementarity of factors. Second, we may examine whether regions having higher mobility in one sphere also have higher mobility in terms of the other two. This is a macro view where regional macro characteristics like state's effort at educational expansion or economic factors like per capita net state domestic product (PCNSDP) and its growth may have impact on the mobility situation. Let us now examine these issues one by one using the 2021 dataset.

5.7.1 Micro-Issue: Interlinkage at Individual Level

The simplest way to examine the interlinkage at the household level would be to cross-tabulate individuals according to their mobility groups across education–occupation, occupation–income and education–income pairs and look at chi-square and likelihood ratio tests of association. These chi-squares and likelihood ratios were observed to be significant, prompting us to look deeper. It is observed that conditional probability of upward occupational mobility is highest for the group with upward educational mobility (Table 5.8). The same is true for education–income pair and occupation–income pair as

Table 5.8 Conditional probability of upward mobility—2021

Conditional upon		Upward occupational mobility					Upward income mobility				
		ST	SC	OBC	GEN	ALL	ST	SC	OBC	GEN	ALL
Educational mobility	Downward	0.8	0.6	0.6	0.7	0.6	1.4	2.8	1.6	1.7	1.9
	Static	1.0	1.6	1.6	2.8	1.8	10.9	5.2	6.2	5.7	6.3
	Upward	4.5	7.0	7.0	8.0	7.0	19.5	24.2	24.0	24.2	22.7
Occupational mobility	Downward						15.1	14.1	12.6	10.3	12.6
	Static						13.5	13.7	11.4	8.4	11.4
	Upward						16.1	19.4	22.3	27.6	22.3

Source Author's calculation

well. Thus, upward mobility in one dimension is linked to upward mobility in other two dimensions as well.

If we now look at social groups, it is observed that the conditional probabilities are lowest for the STs and highest for the General or advanced class—indicating that the interlinked probability of upward mobility is highest for this group, reflecting perhaps the continuing hegemony of upper castes in the society and economy.

The proportion of population with matching mobility groups across pairs of dimensions would also provide us a measure of how one mobility type overlaps with another and is useful in this respect (Table 5.9). It is observed that the overlap between educational mobility and occupational mobility is strongest, followed by that between education and income, while that between occupation and income is weakest. This partly reflects our earlier inferences where we commented that occupational mobility has weak reflection on income mobility. The issue of occupational shifts under distress to low-income jobs in recent times is also a possibility reaffirmed by such results.

If the trios of mobility dimensions are considered in unison, it is observed that just about 18 per cent of current generation wage earners have upward educational, occupation and income mobility simultaneously and are the charmed group in this atmosphere of general gloom in the labour market. This proportion is marginally higher for the advanced class and lowest for the STs—bringing to the fore once again the disparity among social classes and persistence of poverty among excluded social groups, especially the STs.

Table 5.9 Proportion of population with matched mobility groups—2021

	ST	SC	OBC	GEN	ALL
Education–Occupation	29.0	34.3	37.7	38.0	36.3
Occupation–Income	13.1	14.0	17.1	25.3	18.2
Education–Income	31.9	24.1	27.5	28.0	27.6
Education–Occupation–Income	16.0	17.5	17.8	18.8	17.8

Source Authors' calculation

5.7.2 Macro Issue: Interlinkage at Regional Level

There are vast regional disparities in terms of intergenerational educational and occupational mobility in India. Such disparities are evident at the aggregate levels of mobility, differential mobilities for the excluded and advanced classes, and their time trends. If we concentrate on the 2021 data to look at the interlinkage between income mobility on one hand and educational and occupational mobility on other, several issues emerge (Table 5.10 and Fig. 5.2).

It seems that Andhra Pradesh, Gujarat, Karnataka, Kerala, Tamil Nadu and Tripura are enjoying high educational, occupational and income mobility simultaneously. These states are therefore successful in transforming livelihood through educational expansion and occupational shift. At the other end of the spectrum lies Bihar, Manipur, Meghalaya, and Uttar Pradesh—where we see overlap of low educational mobility, low occupational mobility and low-income mobility. Together these four states account for about one-fourth of India's population and thereby underline the strong interlinkage between the three spheres of mobility examined by us. There are several layers in between also. Chhattisgarh, Madhya Pradesh, Odisha and Telangana have high educational mobility but could not transform this rising human capital potential to occupational or income mobility perhaps because of natural resource dependent economic base of these states. Maharashtra had high educational and occupational mobility but could not achieve much income mobility—perhaps because the state had already achieved a high-income base. Jharkhand, Rajasthan and Bengal did have educational mobility or income mobility though experiencing low occupational mobility indicating perhaps a disconnect between education, occupation and income in these states. Arunachal Pradesh, Assam and Nagaland could

Table 5.10 Mobility matrix at regional level 2021

		Occupational mobility		Income mobility	
		High	Low	High	Low
Educational mobility	**High**	Andhra Pradesh, Gujarat, Karnataka, Kerala, Maharashtra, Tamil Nadu, Tripura	Chhattisgarh, Jharkhand, Madhya Pradesh, Odisha, Rajasthan, Telangana, West Bengal	Andhra Pradesh, Gujarat, Jharkhand, Karnataka, Kerala, Rajasthan, Tamil Nadu, Tripura, West Bengal	Chhattisgarh, Madhya Pradesh, Maharashtra, Odisha, Telangana
	Low	Goa, Haryana, Himachal Pradesh, J&K, Mizoram, Punjab, Uttarakhand	Arunachal Pradesh, Assam, Bihar, Manipur, Meghalaya, Nagaland, Uttar Pradesh	Arunachal Pradesh, Assam, Himachal Pradesh, J&K, Nagaland, Tamil Nadu	Bihar, Goa, Haryana, Manipur, Meghalaya, Mizoram, Punjab, Uttar Pradesh, Utta-rakhand
Occupational Mobility	High			Andhra Pradesh, Gujarat, Himachal Pradesh, J&K, Karnataka, Kerala, Tamil Nadu, Tripura	Goa, Haryana, Maha-rashtra, Mizoram, Punjab, Uttarakhand
	Low			Arunachal Pradesh, Assam, Jharkhand, Nagaland, Rajasthan, West Bengal	Bihar, Chhattisgarh, Madhya Pradesh, Manipur, Meghalaya, Odisha, Telangana, Uttar Pradesh
		High	**Low**	**High**	**Low**
		Occupational Mobility		**Income** Mobility	

Source Authors' calculation

not achieve educational or occupational mobility but could achieve relatively higher income mobility—signalling perhaps faster growth of returns from existing forest and plantation based activities and government services in these states. Finally, in states like Goa, Haryana, Mizoram, Punjab and Uttarakhand, we perceive high occupational shifts but low educational and income mobility indicating perhaps distress driven occupational changes in these states. Rank correlation between states in terms of mobility rates also confirm that mobility across these three dimensions are interlinked (Table 5.11). The interdependence is particularly strong for the STs, signalling the need for a rounded policy intervention for this group, where benefits (or lack thereof) from one sphere will percolate to the other spheres as well.

Table 5.11 Association between regional mobility dimensions 2021

Rank correlation between mobility measures	Social group				
	ST	SC	OBC	GEN	ALL
Education–occupation	0.30**	0.02	0.16*	0.23*	0.25*
Occupation–income	0.33**	0.32**	0.10	0.14	0.31**
Education–income	0.35**	0.05	0.16	0.16	0.32**

Source Author's calculation
Note * and ** indicate significant at 5% and 1% levels, respectively

Whatever way we look at it, there seems to be some association between the extents of mobility achieved in the three spheres. Are there macrovariables that support mobility? Two factors were examined in this context—(a) State's effort at expanding education and (b) Economic performance of the state. The former is measured as Per capita Plan Expenditure on Education over the previous 5 years (PCPEE). Economic performance is measured as average Per Capita Net State Domestic Product over previous 5 years (PCNSDP) and 5 years of average growth in PCNSDP (PCNSDPGR).

It is observed that expenditure on education boosts education and occupation mobility, especially educational mobility, to a great extent, whereas the link between this and income mobility is positive but insignificant (Table 5.12). Economic performance of the state also seems to act as an impetus for social upliftment. A significant positive correlation between PCNSDP and both educational and occupational mobility is observed, indicating that states which perform well in economic frontier have higher educational and occupational mobility. The positive association between economic performance and educational and occupational mobility is also supported by their positive association with growth in PCNSDP. Income mobility is observed to have a significant association with PCNSDP but not with its growth.

We have also attempted to link mobility with livelihood outcome of the people. Higher mobility is supposed to lead to progress and lowering of poverty. On the other, the regions which

Table 5.12 Association of mobility indices with possible correlates 2021

Dimension	Per capita plan expenditure on education	Per Capita NSDP	Growth in PCNSDP	Poverty HCR
Educational mobility	0.375** (0.01)	0.585** (0.01)	0.305* (0.06)	–0.325** (0.01)
Occupational mobility	0.320* (0.06)	0.524** (0.01)	0.512** (0.01)	–0.332** (0.01)
Income mobility	0.138 (0.15)	0.308* (0.08)	0.245 (0.50)	–0.485* (0.05)

Source Author's calculation
Note * and ** indicate significant at 5% and 1% levels, respectively. Figures in parenthesis are significance levels. Poverty HCR are derived from household consumption expenditure data of PLFS

suffer from lack of basic entitlement may create impediment towards further development and such cycle reduces intergenerational mobility. To examine this interrelationship we use Poverty Head Count Ratio (HCR) as a (inverse) measure of livelihood situation. It is observed that there exists a significant negative association between incidence of poverty (HCR) and educational, occupational and income mobility. It suggests that states with lower mobility on all three fronts are also the states with higher incidence of poverty. It is quite likely that there exists bi-directional causality between mobility and poverty as lack of mobility reinforces poverty and poverty binds people within parental levels, not allowing them to move up the ladder.

5.8 Process of Stickiness: Some Working Hypothesis

Though we find both micro and macro level linkage between three dimensions of mobility, we also observe that success in educational mobility is not being translated into occupational and income mobility up to similar extent; which makes us think what is the likely process of transformation of one type of mobility into another and why occupational and income stickiness persists even when educational mobility is a reality? The answer perhaps lies in working of the education market and the labour market. The education market in India is still mostly in the public domain with state playing a key role in augmenting supply of education. Over the last three decades, there has been a massive upscaling of educational infrastructure in terms of reach, coverage, physical assets and personnel. This has been complemented by increased demand for education, as people, especially from hitherto excluded groups, see education as the key to improvement in the livelihood options for their children. This has resulted in significant educational mobility as brought out in this paper. However, translating this into occupational and income mobility faces roadblock in the labour market which is almost entirely in private domain, largely unorganised and lacks minimum standards. As educational achievements have increased across the board, relative position of individuals as well as social groups in the highly competitive excess supply labour market has remained more or less unchanged. This is evident in the form of improved educational level among workers in the same occupation over time.[2] Simply put, even with better educational standard than one's parents, one is still in an occupation similar to his parents as everyone else has also improved their educational level. Microstudies undertaken in several sectors and regions indicate that such a hypothesis is in fact quite possible.[3] The link becomes further weaker when we move from occupation to income and the possible explanation may be in the wage conditions over the last three decades. A preliminary exploration of Rural Labour Enquiry (RLE) and Occupational Wage Survey (OWS) Reports (GOI, various years) suggest that since 1993–94 the condition of wage workers, especially in the blue-collar jobs in rural areas, have deteriorated. Recent studies also show that proportion of workers who are Chronically Unemployed or have Irregular employment has increased while that in stable employment has decreased (Majumder 2023). Along with that, there has been a decrease in numbers of days worked in a year and a rise in non-working due to non-availability of work. There has been a rise in self-employment of children—a key indicator pointing to inadequacy of adult income to sustain the household. Though daily earnings have increased marginally in real terms, it has declined for women and STs. Indebtedness has increased and more alarmingly the share of money lenders in providing loans has increased substantially. Also, loans are increasingly being required for non-productive purposes—another sign of distress among vast majority of wage earners. It is not surprising therefore that we find low-wage income mobility in our study.

A broader regional comparison shows that earning mobility in India is higher than China, Japan, Bangladesh and Malaysia, but lower than most of the countries in Asia–Pacific region for which such estimates are available

(Narayan et al. 2018; Huang and Saxena 2021). It is evident from regional international experience that the dominant channel of transmission of intergenerational persistence of earning level is education. However, education plays a dual and divergent role in shaping persistence and mobility. High-earning parents are able to invest more in their children's education, resulting in relatively superior earning capability for these children when they grow up. This channel is strengthened when education is mostly privately financed, e.g. in Japan, Malaysia and Bangladesh or centrally politically determined as in China. In New Zealand and Taiwan too relatively lower mobility, as compared to other developed countries, has been attributed to better education levels of children of high-income parents. The opposite role that education plays is through expansion of the human capital base and broadening the capability domain, thereby increasing earning opportunity. This has been the case in Australia, Singapore, the Philippines and more remarkably in Indonesia (Hertz and Jayasundera 2007). It has been observed in these country studies that public funded education, especially primary and tertiary education, leads to greater income mobility by disassociating the process of capability formation from parental income and affordability levels. India's standing somewhere in the middle of the regional range stems from the fact that education has been mostly funded by the State leading to substantial educational mobility across generations (Majumder 2013). But contrary to the Philippines, this has not resulted in earning mobility of comparable magnitude. Key to this seeming paradox may lie in the abysmally poor quality of public education in India—India ranked second last among 73 countries that participated in the Programme for International Student Assessment 2013, conducted annually to evaluate education systems worldwide by the OECD Secretariat and none of the Indian institutes of higher learning feature in the top 100 institutions list published by annual Times Higher Education magazine's 2023 World Reputation Rankings (The 2023).

It would imply that the current generation is no better endowed than their parents in terms of skill and capability even though the average years of formal education are higher. This makes their occupational pattern and earnings similar to their parents resulting in lower mobility. Another channel of transmission, as evident from regional experience, has been through occupation—industrialisation and occupational shifts facilitating income mobility in South Korea, Singapore, Philippines and occupational stagnancy thwarting income mobility in China. In India too, occupational continuity across generations has been observed (Ray and Majumder 2011). This may also be an outcome of the unfinished industrialisation process in India where it appears that the middle phase of development process, where industries become the dominant sector, has been skipped and we have moved directly from primary sector-led growth to tertiary sector-led growth. This has perhaps impacted the occupational transformation process as well. These explanations behind the absence of any significant occupational and income mobility in India and exclusion of marginalised sections of the society even in face of educational mobility need further analysis, preferably with field studies, and can be an important research agenda for future researchers.

5.9 Summary and Policy Interlinkages

We thus find evidence of interlinkage between the three dimensions of educational, occupational and income mobility both at the micro (or household) level and at the macro (regional) level, and it would not be wrong to comment that mobility from one dimension do transmit to the other two, though the strength of such transmission is weak for the income level. In addition, mobility is substantially lower for the excluded social groups, especially the STs. There are, however, signs of convergence in terms of the gap between the advanced class and the excluded class narrowing down over time and stickiness in 2021 being lower for the excluded groups compared to the advanced class. This is in sharp contrast to earlier findings (Majumder 2013) where the gap was observed to have increased between 1993 and 2004. Thus the convergence, if any, has been witnessed in the current century. However, there is no scope for euphoria as the inequality in basic indicators of education, occupation and income are still very high and increasing over time, indicating that the economic growth-led story in India is not as rosy as thought to be. On the contrary, it has been exclusionary and has aggravated socio-economic inequality. This calls in for integrated policy where educational upliftment, occupational transformation and income improvements are targeted in a holistic manner so that complementarities between them can be tapped for overall development of the lagging sections of the society, especially the tribals. At the same time, we find evidence of the role played by causal factors like expansion of educational infrastructure and state's economic performance in accentuating upward mobility. This would imply another set of complementary policies—where the state develops the skeleton of facilities for human capital formation and market forces capitalise on that to expand occupational and income opportunities. However, the weak transmission from educational mobility to occupational and income mobility suggests that the labour market is still ridden with various barriers and discriminatory processes. This perpetuates occupational segregation in India, and even within occupations wage discrimination is substantially high (Mukherjee and Majumder 2011 and others). Occupational and income mobility therefore would be hard to achieve up to desirable level if the labour market is left entirely to the market. Some kind of affirmative policy for the private sector, enabling a National Minimum Wage legislation (as discussed in Ghosh 1997; Belser and Rani 2010) and monitoring its compliance are some of the policies that may help income mobility.

Relationship between the three dimensions and the proximate determinants is intricate and complex and needs comprehensive policies. In this regard, integrated policies to target all three dimensions of education, occupation and income can be helpful. One can mention some of the field experiences in this context. Several visits were made in three districts of West Bengal during this study to understand the process of mobility and stagnancy of backward social classes and the role of state interventions. We observed substantial educational progress in the region with lesser number of illiterates and more school passouts among the present generation compare to the parental generation. However, the relative position of the backward classes is still behind that of the upper castes. Various government schemes are operational for supporting the members of SC and ST communities in livelihood promotion. To understand the role played by the State, we visited households that have benefitted from such schemes and examined whether state support has played any role in intergenerational mobility among them. State support in terms of financial grants for livelihood promotion among the backward classes seems to have a great role to play in this regard. In most of the cases where communities/households have obtained such support, the occupational stability has helped the present generation to achieve higher educational standards compared to their parents. Thus the vicious cycle of (low) parental occupation–parental education leading to low current education and occupation could be broken and intergenerational mobility was achieved. However the household members were reluctant to diversify from traditional family occupations and this created occupational stagnancy and vulnerabilities for them. Only in some cases, state interventions were able to initiate occupational shifts through providing information, technical knowhow, financial capital and marketing support (for example, promoting sericulture among tribals in Purulia district, animal husbandry in Burdwan district and handloom in Hooghly district). These successful cases were the ones where the local officials took the pains to educate and support the tribals in search for a better source of livelihood. Such cases need to be upscaled while preserving flexibility at the ground level to bring a comprehensive improvement in the lives of the people, especially those from the backward classes so that intergenerational upward mobility becomes a reality and not a dream or cosmetic change for them.

The policy options that one can suggest against this backdrop should necessarily be multipronged. While the current thrust to expand public education system (through Right to Education legislation and infrastructural provisioning) should be continued, focus should now shift to enhancing quality of education and market-ready skill formation. Employability should be the guiding factor to assess success of educational programmes rather than the sheer volume of students that enrol or pass out with degrees. This may bring about greater occupational mobility in the labour market and improve the income mobility situation as well. A policy regime that crosses between Singapore's economic growth and competitive skill formation, Indonesia's primary education expansion, Australia's thrust on easy, but quality higher education, non-farm sector growth of Philippines and South Korea's industrialisation drive would be ideal to facilitate higher income mobility in India. Policymakers should immediately look into these issues and take steps so that economic growth translates into a more visible and inclusive improvement of the lives of the working mass. In absence of perceptible improvements, social disparity may spiral out of control, sporadic discontent may well give way to mass uprising and Indian spring may transform into a summer of discontent, ripping apart the fabric of Indian society, polity and economy.

Authors are thankful to G. Omkarnath and Maitreyi Bordia Das for comments on an earlier version of the paper presented at the Global Conference on 'Prosperity, Equality and Sustainability: Perspectives and Policies for a Better World', organised by the Institute for Human Development, New Delhi, in 2016. We also acknowledge comments from an anonymous referee which added substantial value to the paper. The pre-print version of this paper is available at the pre-print repository at MPRA (https://mpra.ub.uni-muenchen.de/71182/1/MPRA_paper_71182.pdf). MPRA retains the right to use the pre-print version.

Notes

1. This period, starting loosely from 1989–90, has seen Structural Adjustment Program and large-scale economic reforms, substantial withdrawal of State from the business domain, unprecedented macroeconomic growth, implementation of reservations for Other Backward Castes in education and jobs, phenomenal increase in tertiary sector jobs, along with large scale displacement of mainly tribals for development projects and building up of stiff resistance to such displacement and disempowerment in the recent past, thereby providing a structural break in the social and economic trajectory of India
2. Average years of education of all workers increased from 3.8 years in 1993 to 5.7 years in 2011—an increase by 1.9 years or 50 per cent. At a disaggregated level, this increase ranges from 1.7 years for clerical workers to 6.3 years for technical and professional workers.
3. Field studies were undertaken in Bardhaman, Bankura and Purulia districts of West Bengal, an eastern state of India during 2012–13 to understand the process of mobility and stagnancy of backward social classes and the role of state interventions in promoting improvement in education, occupation and income. Most of the surveyed households were beneficiaries of one or other government schemes for socially disadvantaged classes. This study was partly funded by the University Grants Commission of India.

References

Akerlof George (1976) The economics of caste and of rat race and other woeful tales, *Quarterly Journal of Economics*, XC 4 November

Alderman H, Behrman JR, Kohler H-P, Maluccio JA, Watkins SC (2001) Attrition in longitudinal household survey data: some tests for three developing-country samples. Demogr Res 5(4):79–124

Altham PME, Ferrie JP (2007) Comparing contingency tables tools for analyzing data from two groups cross-classified by two characteristics. Hist Methods J Quant Interdisc History 40(1):3–16

Altham PME (1970) The measurement of association of rows and columns for an r × s contingency table. J Roy Stat Soc B 32:63–73

Ambedkar BR (1936) Annihilation of Caste, in *Bheem Patrica*, Jallunder

Ambedkar BR (1987) The Hindu social order - its essential features, Vasant Moon (ed) *Dr Babasaheb Ambedkar Writing and Speeches*, Vol 3, Education Department, Bombay

Atkinson AB (1998) Social exclusion, poverty and unemployment, in Hills, J. (ed.) Exclusion, Employment and Opportunity, Centre for Analysis of Social Exclusion (CASE), London School of Economics and Political Science, London, pp 1–20. www.sticerd.lse.ac.uk/dps/case/cp/Paper4.pdf, accessed on 16–12–2014

Azam Mehtabul, Bhatt Vipul (2014) Like Father, Like Son? Intergenerational Education Mobility in India. SSRN: http://ssrn.com/abstract=2050742 or https://doi.org/10.2139/ssrn.2050742

Barooah VK, Dubey A, Iyer S (2007) The effectiveness of jobs reservation: caste, religion, and economic status in India. http://mpra.ub.uni-muenchen.de/19421/ MPRA Paper No. 19421, accessed on 12–10–2014

Becker GS, Tomes N (1979) An equilibrium theory of the distribution of income and intergenerational mobility. J Polit Econ 87(6):1153–1189

Behrman Jere R, Alejandro Gaviria, Miguel Székely (2001) Intergenerational mobility in Latin America. J Lacea Econ Lacea - Latin Am Carib Econ Assoc

Belser P, Rani U (2010) Extending the coverage of minimum wages in India: Simulations from household data. Conditions of Work and Employment Series No. 26, ILO. Geneva. http://wiego.org/sites/wiego.org/files/publications/files/Amara_Minimum_Wages.pdf, accessed on 12–11–16

Bhowmik S (1992) Caste and class in India. Econ Polit Weekly 27(24/25, June 13–20):1246–1248

Borooah VK, Iyer S (2005) Vidya, Veda, and Varna: the influence of religion and caste on education in rural India. J Dev Stud 41(8):1369–1404

Bourdieu J, Ferrie J, Kesztenbaum L (2006) Vive la différence? Intergenerational Occupational Mobility in France and the U.S. in the 19th and 20th Centuries. http://mauricio.econ.ubc.ca/pdfs/ferrie.pdf

Checchi Daniele Flabbi Luca (2007) Intergenerational mobility and schooling decisions in Germany and Italy: the impact of secondary school tracks, *IZA Discussion Paper No. 2876*. SSRN: https://ssrn.com/abstract=999373, accessed on 12/12/2014

Deshpande and Palshikar (2008) Occupational mobility: how much does caste matter? Economic and Political Weekly 43(34, August, 23):61–70

Driver Edwin D (1962) Caste and occupational structure in Central India. Social Forces 41(1, October):26–31

Gang Ira, Zimmermann Klaus (2000) Is Child like Parent? Educational attainment and ethnic origin. J Human Res 35(3):550–569. https://EconPapers.repec.org/RePEc:uwp:jhriss:v:35:y:2000:i:3:p:550-569 accessed on 12/12/2015

Ghose AK (1997) Should there be a national minimum wage in India? Indian J Lab Econ 40(4):697–713

GoI (1993) Rural labour enquiry: report on general characteristics of rural labour households (51stRound of N.S.S.) 1993–94, Government of India, Ministry of Labour, Labour Bureau, Shimla/Chandigarh

GoI (1993a) Rural labour enquiry: report on indebtedness among rural labour households (51stRound of N.S.S.) 1993–94, Government of India, Ministry of Labour, Labour Bureau, Shimla/Chandigarh

GoI (1999) Rural labour enquiry: report on general characteristics of rural labour households (55thRound of N.S.S.) 1999–2000, Government of India, Ministry of Labour, Labour Bureau, Shimla/Chandigarh

GoI (1999a) Rural labour enquiry: report on indebtedness among rural labour households (55thRound of N.S.S.) 1999–2000, Government of India, Ministry of Labour, Labour Bureau, Shimla/Chandigarh

GoI (2011) Rural labour enquiry: report on general characteristics of rural labour households (61stRound of N.S.S.) 2004–05, Government of India, Ministry of Labour, Labour Bureau, Shimla/Chandigarh

GoI (2011a) Rural labour enquiry: report on indebtedness among rural labour households (61stRound of N.S.S.) 2004–05, Government of India, Ministry of Labour, Labour Bureau, Shimla/Chandigarh

GoI (2015) Wage Rates in Rural India 2013–14, Government of India, Ministry of Labour& Employment, Labour Bureau, Shimla/Chandigarh

Hertz Tom, Tamara Jayasundera (2007) School construction and intergenerational mobility in Indonesia. Working Papers 2007–18, American University, Department of Economics. https://doi.org/10.17606/qt3z-3c16

Hnatkovska V, Paul Saurav, Lahiri A (2013) Breaking the caste barrier: intergenerational mobility in India. J Human Res Spring 48(2):435–473

Huang Zhenqian, Sweta Saxena (2021) Will I be rich if I come from an impoverished family? Improving intergenerational mobility in Asia and the Pacific. MPFD Policy Brief, No. 122. Bangkok: ESCAP

Kumar S, Heath A, Heath O (2002a) Determinants of social mobility in India. Econ Pol Weekly 37(29)

Kumar S, Heath A, Heath O (2002b) Changing patterns of social mobility: some trends over time. Econ Pol Weekly 37(40)

Lal Deepak (1988) Hindu equilibrium. Oxford Clarendon Press

Madheswaran S, Attewell P (2007) Caste discrimination in the Indian urban labor market: evidence from the national sample survey. Econ Pol Wkly 42(41):4146–4153

Madheswaran S (2009) Labour market discrimination in India: issues in theory, methodological development and empirical evidence" keynote paper presented at 51st annual conference of Indian society of labour economics, Punjab, India 11–13 December

Majumder Rajarshi (2013) Intergenerational mobility: a study of social classes in India, Springer Briefs in Economics, Springer, London

Majumder Rajarshi (2010) Intergenerational mobility in occupational and occupational attainment: a comparative study of social classes in India. Margin 4(4)

Majumder R (2023) The employment challenge in India: hundred years from ten days that shook the world. Indian J Labour Econ 66(1):37–59

Mazumder B (2001) Earnings mobility in the US: a new look at intergenerational inequality. Working Paper Series WP-01–18, Federal Reserve Bank of Chicago. http://www.chicagofed.org/digital_assets/publications/working_papers/2001/Wp2001-18.pdf, accessed on 16–12–2014

Minicozzi Alexandra L (2003) Estimation of sons' intergenerational earnings mobility in the presence of censoring. J Appl Econom John Wiley & Sons, Ltd. 18(3):291–314. http://qed.econ.queensu.ca:80/jae/2003-v18.3/ accessed on 16–12–2015

Motiram S, Singh A (2012) How close does the apple fall to the tree? Some Estimates on Intergenerational Occupational Mobility for India," Economic and Political Weekly, XLVII (40), October 06

Munshi K, Rosenzweig M (2006) Traditional institutions meet the modern world: caste, gender and schooling choice in a globalizing economy. Am Econ Rev 96(4):1225–1252

Mukherjee Dipa, Rajarshi Majumder (2011) Occupational pattern, wage rates & earning disparities in India: a decomposition analysis. Indian Econ Rev 46(1):131–152

Narayan Ambar, Van Der Weide Roy, Cojocaru Alexandru, Lakner Christoph, Redaelli Silvia, Mahler Daniel Gerszon, Nichanametla Ramasubbaiah Rakesh Gupta, Thewissen Stefan Hubert (2018) Fair progress : economic mobility across generations around the World, World Bank. http://documents.worldbank.org/curated/en/663711528200431822/Fair-progress-economic-mobility-across-generations-around-the-World

NSSO (2004) Employment and unemployment survey. Round 64th, GOI

NSSO (1997) Unit level records on sixth quinquennial survey on employment and unemployment in India 1993–1994 (Data on CD-ROM), NSS Fiftieth Round—July 1993–June

NSSO (2009) Unit level records on eighth quinquennial survey on employment and unemployment in India 2009 (Data on CD-ROM), NSS 66th Round—July 2009–June 2010, National Sample Survey Organisation, New Delhi: Government of India

Ray Jhilam, Majumder R (2010) Occupational and occupational mobility across generations in India: social and regional dimensions. Indian J Lab Econ 53(4)

Ray Jhilam, Majumder R (2013) Snakes & ladders : intergenerational income mobility in India. Indian J Human Develop 7(2):257–274

Ray Jhilam, Majumder R (2014) Structural change or social Fluidity?- Examining intergeneration educational mobility in India. J Educ Plan Admin 28(3):249–264, 2014ISSN 0971–3859

Ray Jhilam, Majumder R (2015) Occupation and income mobility in India : evidences from recent NSSO surveys: Sarvekshana, 99th Issue ISSN -2249–197X

Scoville James GL (1991) Status influences in third world labour markets, caste, gender and custom. New York

Solon GR (1999) Intergenerational mobility in the labour market. In: Ashenfelter OC, Card D (eds) Handbook of labour economics, vol 3A. North Holland, Amsterdam, pp 1761–1800

Solon GR (1992) Intergenerational income mobility in the United States. Am Econ Rev 82(3):393–408

Takahiro Ito (2007) Caste discrimination and transaction costs in the labor market: evidence from rural north India. Discussion Paper Series No. 200, Institute of Economic Research, Hitotsubashi University. https://hermes-ir.lib.hit-u.ac.jp/hermes/ir/re/16275/0100802201.pdf, accessed on 16–12–2015

The (2023) Times higher education world reputation rankings 2023. https://www.timeshighereducation.com/world-university-rankings/2023/reputation-ranking accessed on 25–02–2023

Thomas D, Frankenberg E, Smith JP (2001) Lost but not forgotten: attrition and follow-up in the Indonesia family life survey. J Human Res 36(3):556–592

Thorat S, Newman KS (2009) Blocked by caste: economic discrimination and social exclusion in modern India. Oxford University Press, New Delhi

Intergenerational Transfer of Education Among Social Groups in India

6

Palashpriya Halder and Ishita Mukhopadhyay

Abstract

In this paper, it is examined whether inequality in human capital in today's generations reflects unequal opportunities which individuals inherit from their parents due to their belonging to a particular social group or caste category. To do so, we specifically examine the role of parental education on child's educational attainment. We have taken India Human Development Survey data (2011–2012) in this paper and tried to find out the effect of father's/mother's education on the educational attainment level of son/daughter particularly for different social groups, General Caste (GC), Other Backward Castes (OBC), Scheduled Castes (SC), and Scheduled Tribes (ST) in India. Intergenerational transfer can be understood in three ways: Persistence (same attainment of education between generations/no improvement), Downward Mobility (if education attainment falls from one generation to another), and Upward Mobility (if education attainment increases from one generation to the other). We focus particularly on upward mobility and persistence because only through these mechanisms human capability formation takes place. We further analyse this upward mobility to identify which pathway (Father-Son/Father-Daughter/Mother-Son/Mother-Daughter) is playing a more significant role in this transfer mechanism, whether father's education or mother's education play the same or different role, whether son's level of educational attainment is the same or different for that of daughter to capture any gender differential in the transfer mechanism.

Keywords

Intergeneration · Education · Transfer · Mobility

JEL Classification

O12 · I21 · C31

6.1 Introduction

One natural way to assess inequality of access to opportunities to be availed against investment in human capital is to examine how educational outcomes vary across different social groups that affect the level of education of a child within the household. An additional lens to study the

P. Halder
Department of Economics, Vijaygarh Jyotish Ray College, Kolkata, India

I. Mukhopadhyay (✉)
Department of Economics, University of Calcutta, Kolkata, India
e-mail: mukhopadhyayishita@gmail.com

A. Alakshendra et al. (eds.), *Development, Inclusion and Sustainability*,
Sustainable Development Goals Series, https://doi.org/10.1007/978-981-97-6863-9_6

impact of parents' education on child's education is to ask whether inequality in human capital in today's generations reflects very unequal opportunities that individuals inherit from their parents due to their belonging to a particular social group or caste category. It is particularly important for India which stands out as a society deeply stratified by caste which has historically been associated with poor outcomes. Therefore, in this paper we focus on one dimension of inequality of opportunities—acquisition of human capital due to a different social category or caste of an individual like General Caste (GC), Other Backward Castes (OBC), Scheduled Castes (SC), and Scheduled Tribes (ST). There have been discussions of intergenerational transfer of education and social exclusion by Munshi and Rosenzweig (2006), Ray Jhilam and Majumder, Rajarshi (2010), Jalan and Murgai (2008), and others. This paper focuses on the association of social deprivation with intergenerational capability enhancement in terms of education.

6.2 Objective

The objective is to look at inequalities in educational outcomes across groups of individuals and the perpetuation of these inequalities across generations. We pay special attention to the roles of caste or social group, asking whether caste hierarchy has influenced intergenerational transfer of education across gender lines.

6.3 Data

We use data from the India Human Development Survey, 2011–12 (IHDS), a national representative survey of households jointly organised by the National Council of Applied Economic Research (NCAER) and the University of Maryland. The IHDS covers 42,554 households in 1503 villages and 971 urban neighbourhoods located throughout India. There are two distinct advantages of using the IHDS data for an intergenerational education mobility study like IHDS asks additional questions that allows to identify the parents' education for almost the entire youth population. There are notations like "ID of each individual", "ID of each Household" along with "The Head of the Household" which are used to match Father-Son, Father-Daughter, Mother-Son, and Mother-Daughter pair to study the Intergenerational effect. Therefore, this dataset is appropriate to capture the effect of over-generation transmission for each of the above-mentioned pairs.

6.4 Descriptive Statistics

The descriptive statistics table is computed based on IHDS Round II data for the four broad social groups: General Caste (GC), Other Backward Castes (OBC), Scheduled Castes (SC), and Scheduled Tribe (ST) (Table 6.1).

The main variable of interest is to focus on the **educational attainment** of a son or daughter which is measured by the number of completed years of schooling as an effect of father's or mother's level of educational attainment. The summary statistics derived from the IHDS data Round II (2011–2012) is given through graphical representation of different social groups for each of the Father-Son, Father-Daughter, Mother-Son, and Mother-Daughter pairs.

There can be three ways of intergenerational transfer: **Persistence (P)/Upward Mobility (UM)/Downward Mobility (DM)**

1. Diagonal elements shows the **Persistence (P)** in the same level of education over generations.

Table 6.1 Descriptive statistics

Total observations: 2,04,569			
Category	Observation	Mean	Standard deviation
GC	57,115	0.28	0.44
OBC	86,482	0.42	0.49
SC	43,138	0.21	0.40
ST	17,380	0.09	0.27

Source IHDS round II. Results based on author's calculation

2. The elements to the right-hand side of the diagonal elements show the **Upward Mobility (UM)** of education over generations.
3. The elements to the left-hand side of the diagonal elements shows the **Downward Mobility (DM)** of education over generations.

The following tables describe the level of education of the child (son/daughter) corresponding to each level of education of the parent (father/mother). Each cell in the table represents the **percentage** of son/daughter attaining the level of education (given in the column) corresponding to father/mother's level of education (given in row) for each Social Category: General Caste (GC), Other Backward Castes (OBC), Scheduled Castes (SC), and Scheduled Tribe (ST), respectively (Table 6.2).

The percentage values of the above table and the corresponding graph show that for each level of father's education there is a trend of upward mobility or persistence in the level of education of the son over generations, for category General Caste. Therefore, it signifies that the process leads to capability enhancement over generations (Table 6.3).

The percentage values of the above table and the corresponding graph shows that for each level of father's education there is a trend of upward mobility or persistence in the level of education of the daughter over generations, for category General Caste. Therefore, it signifies that the process leads to capability enhancement over generations (Table 6.4).

The percentage values of the above table and the corresponding graph show that for each level

Table 6.2 Level of son's education corresponding to father's level of education (given in %): Progression of Son's education measured in percentage (%) for different levels of father's education in General Caste households

GC	Son illiterate	Son elementary	Son secondary	Son HS & above	Row total
Father illiterate	**11.82**	29.15	28.63	30.39	100
Father elementary	2.16	**22.22**	30.52	45.1	100
Father secondary	0.45	5.91	**21.65**	71.99	100
Father HS & above	0.56	2.38	8.9	**88.16**	100

Source IHDS round II. Results based on author's calculation

Table 6.3 Level of daughter's education corresponding to father's level of education (given in %): progression of daughter's education measured in percentage (%) for different levels of father's education in general caste household

GC	Daughter illiterate	Daughter elementary	Daughter secondary	Daughter HS & above	Row total
Father illiterate	**20.96**	32.78	15.3	30.96	100
Father elementary	10.48	**20.49**	16.71	52.33	100
Father secondary	3.8	4.48	**14.23**	77.49	100
Father HS & above	0	2.63	4.36	**93.01**	100

Source IHDS round II. Results based on author's calculation

Table 6.4 Level of son's education corresponding to mother's level of education (given in %): Progression of Son's education measured in percentage (%) for different levels of mother's education in General Caste households

GC	Son illiterate	Son elementary	Son secondary	Son HS & above	Row total
Mother illiterate	**6.34**	26.52	26.45	40.68	100
Mother elementary	1.01	**11.31**	26.34	61.34	100
Mother secondary	0.44	2.13	**14.75**	82.69	100
Mother HS & above	0	0.47	4.13	**95.4**	100

Source IHDS round II. Results based on author's calculation

of mother's education there is a trend of upward mobility or persistence in the level of education of the son over generations, for category General Caste. Therefore, it signifies that the process leads to capability enhancement over generations (Table 6.5).

The percentage values of the above table and the corresponding graph show that for each level of mother's education there is a trend of upward mobility or persistence in the level of education of the daughter over generations, for category General Caste. Therefore, it signifies that the process leads to capability enhancement over generations (Table 6.6).

The percentage values of the above table and the corresponding graph show that for each level of father's education there is a trend of upward mobility or persistence in the level of education of the son over generations, for category Other Backward Castes. Therefore, it signifies that the process leads to capability enhancement over generations (Table 6.7).

The percentage values of the above table and the corresponding graph show that for each level of father's education there is a trend of upward mobility or persistence in the level of education of the daughter over generations, for category Other Backward Castes. Therefore, it signifies

Table 6.5 Level of Daughter's education corresponding to father's level of education (given in %): Progression of Daughter's education measured in percentage (%) for different levels of mother's education in General Caste households

GC	Daughter illiterate	Daughter elementary	Daughter secondary	Daughter HS & above	Row total
Mother illiterate	**15.96**	27.43	14.95	41.66	100
Mother elementary	1.41	**8.74**	16.71	73.15	100
Mother secondary	1.77	1.21	**8.78**	88.24	100
Mother HS & above	0.68	1.07	3.21	**95.03**	100

Source IHDS round II. Results based on author's calculation

Table 6.6 Level of son's education corresponding to father's level of education (given in %): Progression of Son's education measured in percentage (%) for different levels of father's education in other backward caste households

OBC	Son illiterate	Son elementary	Son secondary	Son HS & above	Row total
Father illiterate	**15.39**	40.2	22.49	21.92	100
Father elementary	2.5	**26.67**	32.14	38.69	100
Father secondary	0.74	8.83	**24.12**	66.31	100
Father HS & above	0.57	4.83	13.78	**80.82**	100

Source IHDS round II. Results based on author's calculation

Table 6.7 Level of daughter's education corresponding to father's level of education (given in %): progression of daughter's education measured in percentage (%) for different levels of father's education in other backward caste households

OBC	Daughter illiterate	Daughter elementary	Daughter secondary	Daughter HS & above	Row total
Father illiterate	**34.64**	32.72	15.03	17.61	100
Father elementary	7.25	**24.23**	22.48	46.03	100
Father secondary	2.62	16.43	**10.68**	70.27	100
Father HS & above	2.35	4.86	10.02	**82.76**	100

Source IHDS round II. Results based on author's calculation

that the process leads to capability enhancement over generations (Table 6.8).

The percentage values of the above table and the corresponding graph show that for each level of mother's education there is a trend of upward mobility or persistence in the level of education of the son over generations, for category Other Backward Castes. Therefore, it signifies that the process leads to capability enhancement over generations (Table 6.9).

The percentage values of the above table and the corresponding graph show that for each level of mother's education there is a trend of upward mobility or persistence in the level of education of the daughter over generations, for category Other Backward Castes. Therefore, it signifies that the process leads to capability enhancement over generations (Table 6.10).

The percentage values of the above table and the corresponding graph show that for each level of father's education there is a trend of upward mobility or persistence in the level of education of the son over generations, for category Scheduled Caste. Therefore, it signifies that the process leads to capability enhancement over generations (Table 6.11).

The percentage values of the above table and the corresponding graph show that for each level of father's education there is a trend of upward mobility or persistence in the level of education of the daughter over generations, for category Scheduled Caste. Therefore, it signifies that the

Table 6.8 Level of son's education corresponding to mother's level of education (given in %): progression of Son's education measured in percentage (%) for different levels of mother's education in Other Backward Caste households

OBC	Son illiterate	Son elementary	Son secondary	Son HS & above	Row total
Mother illiterate	**9.45**	33.84	25.79	30.93	100
Mother elementary	0.8	**15.76**	28.47	54.98	100
Mother secondary	0.51	3	**16.99**	79.49	100
Mother HS & above	0	2.68	8.18	**89.14**	100

Source IHDS round II. Results based on author's calculation

Table 6.9 Level of daughter's education corresponding to mother's level of education (given in %): progression of daughter's education measured in percentage (%) for different levels of mother's education in other backward caste households

OBC	Daughter illiterate	Daughter elementary	Daughter secondary	Daughter HS & above	Row total
Mother illiterate	**23.31**	32	17.49	27.2	100
Mother elementary	1.74	**16.25**	21.39	60.62	100
Mother secondary	2.61	5.44	**5.98**	85.96	100
Mother HS & above	0.65	1.74	3.69	**93.93**	100

Source IHDS round II. Results based on author's calculation

Table 6.10 Level of son's education corresponding to father's level of education (given in %): progression of Son's education measured in percentage (%) for different levels of father's education in scheduled caste households

SC	Son illiterate	Son elementary	Son secondary	Son HS & above	Row total
Father illiterate	**14.57**	41	23.03	21.4	100
Father elementary	2.33	**30.82**	27.99	38.85	100
Father secondary	0.34	12.36	**27.08**	59.22	100
Father HS & above	0.3	4.24	13.2	**82.27**	100

Source IHDS round II. Results based on author's calculation

Table 6.11 Level of daughter's education corresponding to father's level of education (given in %): progression of daughter's education measured in percentage (%) for different levels of father's education in scheduled caste households

SC	Daughter illiterate	Daughter elementary	Daughter secondary	Daughter HS & above	Row total
Father illiterate	**24.33**	24.4	21.91	29.37	100
Father elementary	8.46	**22.49**	21.9	47.15	100
Father secondary	0.93	9.61	**15.22**	74.23	100
Father HS & above	0	2.14	10.04	**87.82**	100

Source IHDS round II. Results based on author's calculation

Table 6.12 Level of son's education corresponding to mother's level of education (given in %): progression of Son's education measured in percentage (%) for different levels of mother's education in scheduled caste households

SC	Son illiterate	Son elementary	Son secondary	Son HS & above
Mother illiterate	**10.41**	36.1	24.34	29.16
Mother elementary	1.47	**19.63**	26.04	52.86
Mother secondary	0	3	**19.97**	77.03
Mother HS & above	0	0	5.45	**94.55**

Source IHDS round II. Results based on author's calculation

process leads to capability enhancement over generations (Table 6.12).

The percentage values of the above table and the corresponding graph show that for each level of mother's education there is a trend of upward mobility or persistence in the level of education of the son over generations, for category Scheduled Caste. Therefore, it signifies that the process leads to capability enhancement over generations (Table 6.13).

The percentage values of the above table and the corresponding graph show that for each level of mother's education there is a trend of upward mobility or persistence in the level of education of the daughter over generations, for category Scheduled Caste. Therefore, it signifies that the process leads to capability enhancement over generations (Table 6.14).

The percentage values of the above table and the corresponding graph show that for each level of father's education there is a trend of upward mobility or persistence in the level of education of the son over generations, for category Scheduled Tribe. Therefore, it signifies that the process leads to capability enhancement over generations (Table 6.15).

The percentage values of the above table and the corresponding graph show that for each level of father's education there is a trend of upward mobility or persistence in the level of education

Table 6.13 Level of mother's education corresponding to daughter's level of education (given in %): progression of daughter's education measured in percentage (%) for different levels of mother's education in SC households

SC	Daughter illiterate	Daughter elementary	Daughter secondary	Daughter HS & above	Row total
Mother illiterate	**20.39**	25.63	20.24	33.74	100
Mother elementary	2.98	**13.65**	20.37	62.99	100
Mother secondary	0	3.93	**14.4**	81.68	100
Mother HS & above	0	0	0	**100**	100

Source IHDS round II. Results based on author's calculation

Table 6.14 Level of son's education corresponding to father's level of education (given in %): progression of son's education measured in percentage (%) for different levels of father's education in scheduled tribe households

ST	Son illiterate	Son elementary	Son secondary	Son HS & above	Row total
Father illiterate	**20.13**	37.82	22.81	19.25	100
Father elementary	6.35	**34.74**	29.15	29.76	100
Father secondary	3.75	9.82	**22.32**	64.11	100
Father HS & above	0	10	11.25	**78.75**	100

Source IHDS round II. Results based on author's calculation

Table 6.15 Level of daughter's education corresponding to father's level of education (given in %): progression of daughter's education measured in percentage (%) for different levels of father's education in scheduled tribe households

ST	Daughter illiterate	Daughter elementary	Daughter secondary	Daughter HS & above	Row total
Father illiterate	**33.72**	34.76	12.01	19.52	100
Father elementary	9.78	**29.33**	26.22	34.67	100
Father secondary	2.26	6.21	**24.01**	67.51	100
Father HS & above	2.78	3.7	11.57	**81.94**	100

Source IHDS round II. Results based on author's calculation

Table 6.16 Level of son's education corresponding to mother's level of education (given in %): progression of son's education measured in percentage (%) for different levels of mother's education in scheduled tribe households

ST	Son illiterate	Son elementary	Son secondary	Son HS & above	Row total
Mother illiterate	**15.37**	35.85	24.4	24.38	100
Mother elementary	1.87	**23.49**	29.18	45.46	100
Mother secondary	3.19	12.77	**9.22**	74.82	100
Mother HS & above	0	5.47	12.5	**82.03**	100

Source IHDS round II. Results based on author's calculation

of the daughter over generations, for category Scheduled Tribe. Therefore, it signifies that the process leads to capability enhancement over generations (Table 6.16).

The percentage values of the above table and the corresponding graph show that for each level of mother's education there is a trend of upward mobility or persistence in the level of education of the son over generations, for category Scheduled Tribe. Therefore, it signifies that the process leads to capability enhancement over generations (Table 6.17).

The percentage values of the above table and the corresponding graph show that for each level of mother's education there is a trend of upward mobility or persistence in the level of education of the daughter over generations, for category Scheduled Tribe. Therefore, it signifies that the process leads to capability enhancement over generations.

6.5 Testing of Hypothesis and Results

In order to check the validity of the outcomes obtained from each of the above tables, the following propositions are tested by using **multinomial logistic regression.**

Table 6.17 Level of daughter's education corresponding to mother's level of education (given in %): progression of daughter's education measured in percentage (%) for different levels of mother's education in scheduled tribe households

ST	Daughter illiterate	Daughter elementary	Daughter secondary	Daughter HS & above	Row total
Mother illiterate	**26.81**	28.83	15.52	28.83	100
Mother elementary	4.91	**26.19**	26.19	42.72	100
Mother secondary	0	0	**26.09**	73.91	100
Mother HS & above	0	0	12.5	**87.5**	100

Source IHDS round II. Results based on author's calculation

Table 6.18 Predicted Probability values obtained from multinomial logistic regression: each cell showing son's level of education (given in column) corresponding to each level of education of the father (given in row) respectively

GC	Son illiterate	Son elementary	Son secondary	Son HS & above	Upward mobility
Father illiterate	**0.02 (0.981)**	0.18 (0.189)	0.33 (0.188)	0.47 (0.188)	0.98
Father elementary	0.00 (0.982)	**0.16*** (0)**	0.30*** (0)	0.54*** (0)	0.84***
Father secondary	0.00 (0.982)	0.05*** (0)	**0.21*** (0)**	0.73*** (0)	0.73***
Father HS & above	0.00 (0.982)	0.03*** (0)	0.12*** (0)	**0.85*** (0)**	

Source IHDS round II. Results based on author's calculation
*, **, *** refer to 10%, 5%, 1% levels of significance respectively

Proposition 1: Father's level of education has a significant effect on son's educational attainment at each level of education for General Caste households:

$$Y_{ij} = \alpha + \beta_{1ij}X_{1ij} + \beta_{2ij}X_{2ij} + \beta_{4ij}X_{4ij} + \beta_{5ij}X_{5ij} + \beta_{6ij}X_{6ij} + \beta_{7ij}X_{7ij}\varepsilon_1$$

where Y_{ij} represents Son's level of education, X_1 represents Father's level of education, X_2 represents Mother's level of education, X_4 represents Income, X_5 represents Number of siblings, X_6 represents Birth order, and X_7 represents Religion.

Here i stands for the ith individual and j stands for Illiterate, Elementary, Secondary, and HS & Above.

Null Hypothesis: $H_{o:}\ \beta_{1ij} = 0$.

Alternative Hypothesis: H_1: $\beta_{1ij} \neq 0$ (Table 6.18).

Observations:

Therefore, we accept the null hypothesis: H_o, that is, the above-mentioned proposition is valid, in the following cases:

1. When father is elementary educated, the predicted probability is significant at son's elementary level of education showing persistence over generations, and the predicted probability of upward mobility is significant at the higher levels of education, whereas the downward mobility is insignificant. Even if father is receiving minimum education, this is capability enhancing for the son in this case.
2. When father is receiving secondary level of education, the predicted probability is significant at son's secondary level of education showing intergenerational persistence, and the predicted probability of upward mobility is significant at the higher levels of education,

Table 6.19 Predicted probability values obtained from multinomial logistic regression: each cell showing daughter's level of education (given in column) corresponding to each level of education of the father (given in row) respectively

GC	Daughter illiterate	Daughter elementary	Daughter secondary	Daughter HS & above	Upward mobility
Father illiterate	**0.11*** (0)**	0.13*** (0)	0.18*** (0)	0.57*** (0)	0.89***
Father elementary	0.06*** (0)	**0.11*** (0)**	0.17*** (0)	0.66*** (0)	0.83***
Father secondary	0.02*** (0)	0.03*** (0)	**0.14*** (0)**	0.81*** (0)	0.81***
Father HS & above	0.00 (0.998)	0.03*** (0)	0.06*** (0)	**0.91*** (0)**	

Source IHDS round II. Results based on author's calculation
*, **, *** refers to 10%, 5%, 1% levels of significance respectively

along with significant downward mobility at the elementary level of education. So here also we find the intergenerational capability enhancing effect, but this is also not true sometimes.

3. When father is attaining the highest level of education, there is a significant predicted probability that his son will achieve the same level of education, that is, there is persistence in the educational level over generations., but also there is significant downward mobility at the lower levels of education. This is capability enhancing, but, however, this is not true in all cases.

Proposition 2: Father's level of education has a significant effect on daughter's educational attainment at each level of education for General Caste households:

$$Y_{ij} = \alpha + \beta_{1ij}X_{1ij} + \beta_{2ij}X_{2ij} + \beta_{4ij}X_{4ij} + \beta_{5ij}X_{5ij} + \beta_{6ij}X_{6ij} + \beta_{7ij}X_{7ij} + \varepsilon_2$$

where Y_{ij} represents Daughter's level of education, X_1 represents Father's level of education, X_2 represents Mother's level of education, X_4 represents Income, X_5 represents Number of siblings, X_6 represents Birth order, and X_7 represents Religion.

Here i stands for the ith individual and j stands for Illiterate, Elementary, Secondary, and HS & Above.

Null Hypothesis: H_o: $\beta_{1ij}=0$.

Alternative Hypothesis: H_1: $\beta_{1ij}\neq 0$ (Table 6.19).

Observations:

Therefore, we accept the null hypothesis: H_o, that is, the above-mentioned proposition is valid, in the following cases:

1. When father is illiterate, the predicted probability of persistence is significant at this level of education, but intergenerational transfer is also significant for higher levels of education which shows significant upward mobility over generations. This leads to capability enhancement. The result shows difference with transfer from father to son.
2. When father is educated up to the elementary level, the predicted probability of persistence is significant at this level of education, but intergenerational transfer is also significant for higher levels of education which shows significant upward mobility over generations and downward mobility is also significant at a lower level. So here also we find the intergenerational capability enhancing effect, but this is also not true always. This result is same as that of father to son.
3. When father is educated up to the secondary level, the predicted probability of persistence is significant, and is also significant at the highest level of education showing upward mobility. However, downward mobility is also significant at lower levels of education.

Table 6.20 Predicted probability values obtained from multinomial logistic regression: each cell showing son's level of education (given in column) corresponding to each level of education of the mother (given in row) respectively

GC	Son illiterate	Son elementary	Son secondary	Son HS & above	Upward mobility
Mother illiterate	**0.02*** (0)**	0.16*** (0)	0.26*** (0)	0.56*** (0)	0.98***
Mother elementary	0.01*** (0)	**0.09*** (0)**	0.27*** (0)	0.63*** (0)	0.90***
Mother secondary	0.01*** (0)	0.03*** (0)	**0.20*** (0)**	0.76*** (0)	0.76***
Mother HS & above	0.00 (0.999)	0.01*** (0.003))	0.08*** (0)	**0.91*** (0)**	

Source IHDS round II. Results based on author's calculation
*, **, *** refers to 10%, 5%, 1% levels of significance respectively

Here also we find the intergenerational capability enhancing effect, but this is not true in case of lower level of education. This result is same as that of father to son.

4. When father has attained the highest level of education, the predicted probability of persistence is significant. This is capability enhancing. This result is same as that of father to son.

Proposition 3: Mother's level of education has a significant effect on son's educational attainment at each level of education for General Caste households:

where $Y_{ij=}\ \alpha+\beta_{1ij}X_{1ij}+\beta_{2ij}\ X_{2ij}+\beta_{4ij}X_{4ij}+\beta_{5ij}X_{5ij}+\beta_{6ij}X_{6ij}+\beta_{7ij}X_{7ij}+\varepsilon_3$.

where Y_{ij} represents Son's level of education, X_1 represents Father's level of education, X_2 represents Mother's level of education, X_4 represents Income, X_5 represents Number of siblings, X_6 represents Birth order, and X_7 represents Religion.

Here i stands for the ith individual and j stands for Illiterate, Elementary, Secondary, and HS & Above.

Null Hypothesis: H_o: $\beta_{2ij}=0$.

Alternative Hypothesis: H_1: $\beta_{2ij}\neq 0$ (Table 6.20).

Observations:

Therefore, we accept the null hypothesis: H_o, that is, the above-mentioned proposition is valid, in the following cases:

1. When mother is illiterate, the predicted probability of persistence is significant at this level of education, but intergenerational transfer is also significant for higher levels of education which shows significant upward mobility over generations. This leads to capability enhancement.
2. When mother is educated up to the elementary level, the predicted probability of persistence is significant at this level of education, but intergenerational transfer is also significant for higher levels of education which shows significant upward mobility over generations and downward mobility is also significant at a lower level. So here also we find the intergenerational capability enhancing effect, but this is also not true always. This result is same as that of father to son.
3. When mother is secondary educated up to the secondary level, the predicted probability of persistence is significant, and is also significant at the highest level of education showing upward mobility. However, downward mobility is also significant at lower levels of education. Here also we find the intergenerational capability enhancing effect, but this is also not true in some cases. This result is same as that of father to son.
4. When mother has attained the highest level of education, the predicted probability of persistence is significant. This is capability enhancing.

Proposition 4: Mother's level of education has a significant effect on daughter's educational attainment at each level of education for General Caste households:

$$Y_{ij} = \alpha + \beta_{1ij}X_{1ij} + \beta_{2ij}X_{2ij} + \beta_{4ij}X_{4ij} + \beta_{5ij}X_{5ij} + \beta_{6ij}X_{6ij} + \beta_{7ij}X_{7ij} + \varepsilon_4$$

where Y_{ij} represents Daughter's level of education, X_1 represents Father's level of education, X_2 represents Mother's level of education, X_4 represents Income, X_5 represents Number of siblings, X_6 represents Birth order, and X_7 represents Religion.

Here i stands for the ith individual and j stands for Illiterate, Elementary, Secondary, and HS & Above.

Null Hypothesis: H_o: $\beta_{2ij}=0$.

Alternative Hypothesis: H_1: $\beta_{2ij}\neq 0$ (Table 6.21).

Observations:

Therefore, we accept the null hypothesis: H_o, that is, the above-mentioned proposition is valid, in the following cases:

1. When mother is illiterate, the predicted probability of intergenerational transfer at higher levels of education indicates upward mobility. This leads to capability enhancement over generations.
2. When mother is educated up to the elementary level, the predicted probability of persistence is significant at this level of education, but intergenerational transfer is also significant for higher levels of education which shows significant upward mobility over generations and downward mobility is also significant at a lower level. So here also we find the intergenerational capability enhancing effect, but this is also not true always. This result is same as that of father to son.
3. When mother is educated up to the elementary level, the predicted probability of persistence is significant, and is also significant at the highest level of education showing upward mobility. However, downward mobility is also significant at lower levels of education. Here also we find the intergenerational capability enhancing effect, but this is also not true in some cases. This result is same as that of father to son.
4. When mother has attained the highest level of education, the predicted probability of persistence is significant. This is capability enhancing.

Proposition 5: Father's level of education has a significant effect on son's educational attainment at each level of education for Other Backward Caste households:

Table 6.21 Predicted probability values obtained from multinomial logistic regression: each cell showing daughter's level of education (given in column) corresponding to each level of education of the mother (given in row) respectively

GC	Daughter illiterate	Daughter elementary	Daughter secondary	Daughter HS & above	Upward mobility
Mother illiterate	**0.00 (0.993)**	0.18*** (0)	0.14*** (0)	0.67*** (0)	1.00***
Mother elementary	0.00 (0.993)	**0.08*** (0)**	0.15*** (0)	0.77*** (0)	0.92***
Mother secondary	0.00 (0.993)	0.01*** (0.001)	**0.11*** (0)**	0.88*** (0)	0.88***
Mother hs & above	0.00 (0.993)	0.00*** (0.048)	0.07*** (0)	**0.92*** (0)**	

Source IHDS round II. Results based on author's calculation

*, **, *** refers to 10%, 5%, 1% levels of significance respectively

Table 6.22 Predicted probability values obtained from multinomial logistic regression: each cell showing son's level of education (given in column) corresponding to each level of education of the father (given in row) respectively

OBC	Son illiterate	Son elementary	Son secondary	Son HS & above	Upward mobility
Father illiterate	**0.06 (0.945)**	0.35 (0.257)	0.28 (0.257)	0.31 (0.257)	0.94
Father elementary	0.01 (0.947)	**0.25*** (0)**	0.33*** (0)	0.41*** (0)	0.74***
Father secondary	0.00 (0.948)	0.10*** (0)	**0.25*** (0)**	0.65*** (0)	0.65***
Father HS & above	0.00 (0.948)	0.07*** (0)	0.17*** (0)	**0.75** (0)**	

Source IHDS round II. Results based on author's calculation
*, **, *** refers to 10%, 5%, 1% levels of significance respectively

$Y_{ij=}\ \alpha+\beta_{1ij}X_{1ij}+\beta_{2ij}\ X_{2ij}+\beta_{4ij}X_{4ij}+\beta_{5ij}X_{5ij}+\beta_{6ij}X_{6ij}+\beta_{7ij}X_{7ij}+\varepsilon_5.$

where Y_{ij} represents Son's level of education, X_1 represents Father's level of education, X_2 represents Mother's level of education, X_4 represents Income, X_5 represents Number of siblings, X_6 represents Birth order, and X_7 represents Religion.

Here i stands for the ith individual and j stands for Illiterate, Elementary, Secondary, and HS & Above.

Null Hypothesis: H_o: $\beta_{1ij}=0$.

Alternative Hypothesis: H_1: $\beta_{1ij}\neq 0$ (Table 6.22).

Observations:

Therefore, we accept the null hypothesis: H_o, that is, the above-mentioned proposition is valid, in the following cases:

1. When father is educated up to the elementary level, the predicted probability of persistence is significant at this level of education, but intergenerational transfer is also significant for higher levels of education which shows significant upward mobility over generations and downward mobility is also significant at a lower level. So here also we find the intergenerational capability enhancing effect, but this is also not true always.
2. When father is educated up to the secondary level, the predicted probability of persistence is significant, and is also significant at the highest level of education showing upward mobility. However, downward mobility is also significant at lower levels of education. Here also we find the intergenerational capability enhancing effect, but this is also not true in some cases.
3. When father has attained the highest level of education, the predicted probability of persistence is significant though there is also a significant downward mobility but the probability of downward mobility is much less compared to the probability of upward mobility. Therefore this is capability enhancing but not always.

Proposition 6: Father's level of education has a significant effect on daughter's educational attainment at each level of education for Other Backward Caste households:

$$Y_{ij} = \alpha + \beta_{1ij}X_{1ij} + \beta_{2ij}X_{2ij} + \beta_{4ij}X_{4ij} + \beta_{5ij}X_{5ij} + \beta_{6ij}X_{6ij} + \beta_{7ij}X_{7ij} + \varepsilon_6$$

where Y_{ij} represents Daughter's level of education, X_1 represents Father's level of education, X_2 represents Mother's level of education, X_4 represents Income, X_5 represents Number of siblings, X_6 represents Birth order, and X_7 represents Religion.

Table 6.23 Predicted probability values obtained from multinomial logistic regression: each cell showing daughter's level of education (given in column) corresponding to each level of education of the father (given in row) respectively

OBC	Daughter illiterate	Daughter elementary	Daughter secondary	Daughter HS & above	UM
Father illiterate	**0.19***** **(0)**	0.30*** (0)	0.18*** (0)	0.32*** (0)	0.81***
Father elementary	0.06*** (0)	**0.24***** **(0)**	0.22*** (0)	0.49*** (0)	0.71***
Father secondary	0.02*** (0)	0.17*** (0)	**0.12***** **(0)**	0.69*** (0)	0.69***
Father HS & above	0.03*** (0)	0.07*** (0)	0.15*** (0)	**0.76***** **(0)**	

Source IHDS round II. Results based on author's calculation
*, **, *** refers to 10%, 5%, 1% levels of significance respectively

Here i stands for the ith individual and j stands for Illiterate, Elementary, Secondary, and HS & Above.

Null Hypothesis: $H_{o:}$ $\beta_{1ij}=0$.

Alternative Hypothesis: H_1**:** $\beta_{1ij}\neq 0$ (Table 6.23).

Observations:

Therefore, we accept the null hypothesis: H_o, that is, the above-mentioned proposition is valid, in the following cases:

1. When father is illiterate, the predicted probability of persistence is significant at this level of education, but intergenerational transfer is also significant for higher levels of education which shows significant upward mobility over generations. This leads to capability enhancement.
2. When father is educated up to the elementary level, the predicted probability of persistence is significant at this level of education, but intergenerational transfer is also significant for higher levels of education which shows significant upward mobility over generations and downward mobility is also significant at a lower level. So here also we find the intergenerational capability enhancing effect, but this is also not true always.
3. When father is educated up to the secondary level, the predicted probability of persistence is significant, and is also significant at the highest level of education showing upward mobility. However, downward mobility is also significant at lower levels of education. Here also we find the intergenerational capability enhancing effect, but this is also not true in some cases.
4. When father has attained the highest level of education, the predicted probability of persistence is significant. This is capability enhancing.

Proposition 7: Mother's level of education has a significant effect on son's educational attainment at each level of education for Other Backward Caste households:

$$Y_{ij} = \alpha + \beta_{1ij}X_{1ij} + \beta_{2ij}X_{2ij} + \beta_{4ij}X_{4ij} + \beta_{5ij}X_{5ij} + \beta_{6ij}X_{6ij} + \beta_{7ij}X_{7ij} + \varepsilon_7$$

where Y_{ij} represents Son's level of education, X_1 represents Father's level of education, X_2 represents Mother's level of education, X_4 represents Income, X_5 represents Number of siblings, X_6 represents Birth order, and X_7 represents Religion.

Here i stands for the ith individual and j stands for Illiterate, Elementary, Secondary, and HS & Above.

Null Hypothesis: H_o**:** $\beta_{2ij}=0$.

Alternative Hypothesis: H_1**:** $\beta_{2ij}\neq 0$ (Table 6.24).

Table 6.24 Predicted probability values obtained from multinomial logistic regression: each cell showing son's level of education (given in column) corresponding to each level of education of the mother (given in row) respectively

OBC	Son illiterate	Son elementary	Son secondary	Son HS & above	UM
Mother illiterate	**0.041***** **(0)**	0.29*** (0)	0.29*** (0)	0.38*** (0)	0.96***
Mother elementary	0.00*** (0)	**0.17***** **(0)**	0.32*** (0)	0.51*** (0)	0.83***
Mother secondary	0.00*** (0.002)	0.05*** (0)	**0.24***** **(0)**	0.71*** (0)	0.71***
Mother HS & above	0.00 (0.999)	0.09*** (0)	0.14*** (0)	**0.77***** **(0)**	

Source IHDS round II. Results based on author's calculation
*, **, *** refers to 10%, 5%, 1% levels of significance respectively

Observations:

Therefore, we accept the null hypothesis: H_o, that is, the above-mentioned proposition is valid, in the following cases:

1. When the mother is illiterate, the predicted probability shows that there is significant upward mobility at elementary, secondary, and higher levels of education. So, this leads to complete capability enhancement.
2. When mother is educated up to the elementary level, the predicted probability of persistence is significant at this level of education, but intergenerational transfer is also significant for higher levels of education which shows significant upward mobility over generations. So here also we find the intergenerational capability enhancing effect.
3. When mother is educated up to the secondary level, the predicted probability of persistence is significant, and is also significant at the highest level of education showing upward mobility. However, downward mobility is also true and significant at lower levels of education. Here also we find the intergenerational capability enhancing effect, but this is not true in case of lower level of education.
4. When mother is highly educated, the predicted probability shows there is significant persistence at that level and significant downward mobility at the secondary level. So, this leads to capability enhancing effect except in a few cases.

Proposition 8: Mother's level of education has a significant effect on daughter's educational attainment at each level of education for Other Backward Caste households:

$$Y_{ij} = \alpha + \beta_{1ij}X_{1ij} + \beta_{2ij}X_{2ij} + \beta_{4ij}X_{4ij} + \beta_{5ij}X_{5ij} + \beta_{6ij}X_{6ij} + \beta_{7ij}X_{7ij} + \varepsilon_8$$

where Y_{ij} represents Daughter's level of education, X_1 represents Father's level of education, X_2 represents Mother's level of education, X_4 represents Income, X_5 represents Number of siblings, X_6 represents Birth order, and X_7 represents Religion.

Here i stands for the ith individual and j stands for Illiterate, Elementary, Secondary, and HS & Above.

Null Hypothesis: $H_{o:}$ $\beta_{2ij} = 0$.

Alternative Hypothesis: H_1: $\beta_{2ij} \neq 0$ (Table 6.25).

Observations:

Therefore, we accept the null hypothesis: H_o, that is, the above-mentioned proposition is valid, in the following cases:

1. When the mother is illiterate, the predicted probability shows that there is significant upward mobility at elementary, secondary, and higher levels of education. So, this leads to complete capability enhancement.
2. When mother is educated up to the elementary level, the predicted probability of persistence is significant at this level of education, but intergenerational transfer is also

Table 6.25 Predicted probability values obtained from multinomial logistic regression: each cell showing daughter's level of education (given in column) corresponding to each level of education of the mother (given in row) respectively

OBC	Daughter illiterate	Daughter elementary	Daughter secondary	Daughter HS & above	UM
Mother illiterate	**0.16*** (0)**	0.30*** (0)	0.20*** (0)	0.35*** (0)	0.84***
Mother elementary	0.01*** (0)	**0.15*** (0)**	0.21*** (0)	0.63*** (0)	0.83***
Mother secondary	0.04*** (0)	0.10*** (0)	**0.07*** (0)**	0.79*** (0)	0.79**
Mother HS & above	0.02* (0.084)	0.05*** (0.004)	0.07*** (0)	**0.86*** (0)**	

Source IHDS round II. Results based on author's calculation
*, **, *** refers to 10%, 5%, 1% levels of significance respectively

significant for higher levels of education which shows significant upward mobility over generations. So here also we find the intergenerational capability enhancing effect.

3. When mother is educated up to the secondary, the predicted probability of persistence is significant, and is also significant at the highest level of education showing upward mobility. However, downward mobility is also true and significant at lower levels of education. Here also we find the intergenerational capability enhancing effect, but this is not true in case of lower level of education.
4. When mother is highly educated, the predicted probability shows there is significant persistence at that level and significant downward mobility at secondary level. So, this leads to capability enhancing effect except in a few cases.

Proposition 9: Father's level of education has a significant effect on son's educational attainment at each level of education for Scheduled Caste households:

$$Y_{ij} = \alpha + \beta_{1ij}X_{1ij} + \beta_{2ij}X_{2ij} + \beta_{4ij}X_{4ij} + \beta_{5ij}X_{5ij} + \beta_{6ij}X_{6ij} + \beta_{7ij}X_{7ij} + \varepsilon_9$$

where Y_{ij} represents Son's level of education, X_1 represents Father's level of education, X_2 represents Mother's level of education, X_4 represents Income, X_5 represents Number of siblings, X_6 represents Birth order, and X_7 represents Religion.

Here i stands for the ith individual and j stands for Illiterate, Elementary, Secondary, and HS & Above.

Null Hypothesis: H_0: $\beta_{1ij}=0$.

Alternative Hypothesis: H_1: $\beta_{1ij}\neq 0$ (Table 6.26).

Observations:

Here, we reject the null hypothesis H_0 for all the levels of education, that is, the above-mentioned proposition is not valid at any level of education. This result shows that the above pathway does not lead to capability enhancement.

Proposition 10: Father's level of education has a significant effect on daughter's educational attainment at each level of education for Scheduled Caste households:

$$Y_{ij} = \alpha + \beta_{1ij}X_{1ij} + \beta_{2ij}X_{2ij} + \beta_{4ij}X_{4ij} + \beta_{5ij}X_{5ij} + \beta_{6ij}X_{6ij} + \beta_{7ij}X_{7ij} + \varepsilon_{10}$$

where Y_{ij} represents Daughter's level of education, X_1 represents Father's level of education, X_2 represents Mother's level of education, X_4 represents Income, X_5 represents Number of siblings, X_6 represents Birth order, and X_7 represents Religion.

Table 6.26 Predicted probability values obtained from multinomial logistic regression: each cell showing son's level of education (given in column) corresponding to each level of education of the father (given in row) respectively

SC	Son illiterate	Son elementary	Son secondary	Son HS & above	UM
Father illiterate	**0.04** **(0.992)**	0.36 (0.955)	0.28 (0.925)	0.31 (0.925)	0.96
Father elementary	0.00 (0.992)	**0.26** **(0.96)**	0.31 (0.886)	0.43 (0.886)	0.74
Father secondary	0.00 (0.992)	0.12 (0.966)	**0.30** **(0.76)**	0.57 (0.76)	0.57
Father HS & above	0.00 (0.992)	0.06 (0.969)	0.16 (0.525)	**0.78** **(0.525)**	

Source IHDS round II. Results based on author's calculation
*, **, *** refers to 10%, 5%, 1% levels of significance respectively

Table 6.27 Predicted probability values obtained from multinomial logistic regression: each cell showing daughter's level of education (given in column) corresponding to each level of education of the father (given in row) respectively

SC	Daughter illiterate	Daughter elementary	Daughter secondary	Daughter HS & above	Upward mobility
Father illiterate	**0.04** **(0.994)**	0.19 (0.986)	0.21 (0.987)	0.56 (0.963)	0.96
Father elementary	0.02 (0.994)	**0.18** **(0.986)**	0.16 (0.988)	0.65 (0.954)	0.80
Father secondary	0.00 (0.994)	0.08 (0.987)	**0.09** **(0.999)**	0.82 (0.913)	0.82
Father HS & above	0.00 (0.999)	0.03 (0.988)	0.07 (0.999)	**0.90** **(0.865)**	

Source IHDS round II. Results based on author's calculation
*, **, *** refers to 10%, 5%, 1% levels of significance respectively

Here i stands for the ith individual and j stands for Illiterate, Elementary, Secondary, and HS & Above.

Null Hypothesis: H_0: $\beta_{1ij} = 0$.

Alternative Hypothesis: H_1: $\beta_{1ij} \neq 0$ (Table 6.27).

Observations:

Here, we reject the null hypothesis H_0 for all the levels of education, that is, the above-mentioned proposition is not valid at any level of education. This result shows that the above pathway does not lead to capability enhancement.

Proposition 11: Mother's level of education has a significant effect on son's educational attainment at each level of education for Scheduled Caste households:

$$Y_{ij} = \alpha + \beta_{1ij}X_{1ij} + \beta_{2ij}X_{2ij} + \beta_{4ij}X_{4ij} + \beta_{5ij}X_{5ij} + \beta_{6ij}X_{6ij} + \beta_{7ij}X_{7ij} + \varepsilon_{11}$$

where Y_{ij} represents Son's level of education, X_1 represents Father's level of education, X_2 represents Mother's level of education, X_4 represents Income, X_5 represents Number of siblings, X_6 represents Birth order, and X_7 represents Religion.

Here i stands for the ith individual and j stands for Illiterate, Elementary, Secondary, and HS & Above.

Null Hypothesis: H_0: $\beta_{2ij} = 0$.

Alternative Hypothesis: H_1: $\beta_{2ij} \neq 0$ (Table 6.28).

Table 6.28 Predicted probability values obtained from multinomial logistic regression: each cell showing son's level of education (given in column) corresponding to each level of education of the mother (given in row) respectively

SC	Son illiterate	Son elementary	Son secondary	Son HS & above	UM
Mother illiterate	**0.05 (0.968)**	0.34 (0.412)	0.27 (0.412)	0.34 (0.412)	0.95
Mother elementary	0.01 (0.969)	**0.21*** (0.008)**	0.27*** (0.007)	0.50*** (0.007)	0.77***
Mother secondary	0.00 (1)	0.06*** (0)	**0.31*** (0)**	0.63*** (0)	0.63***
Mother HS & above	0.00 (1)	0.00 (1)	0.09 (0.985)	**0.91 (0.897)**	

Source IHDS round II. Results based on author's calculation
*, **, *** refers to 10%, 5%, 1% levels of significance respectively

Observations:
Therefore, we accept the null hypothesis: H_o, that is, the above-mentioned proposition is valid, in the following cases:

1. When mother is elementary educated, the predicted probability is significant at that level of education showing persistence, significant upward mobility at higher level of education, and downward mobility is insignificant. Here, it is capability enhancing.
2. When mother is secondary educated, the predicted probability of persistence at the same level is significant, significant upward mobility for higher level of education, and significant downward mobility at son's elementary level of education. The result shows the pathway leads to capability enhancement but not for all cases.

Proposition 12: Mother's level of education has a significant effect on daughter's educational attainment at each level of education for Scheduled Caste households:

$$Y_{ij}= \alpha+\beta_{1ij}X_{1ij}+\beta_{2ij}\ X_{2ij}+\beta_{4ij}X_{4ij}+\beta_{5ij}X_{5ij}+\beta_{6ij}X_{6ij}+\beta_{7ij}X_{7ij}+\varepsilon_{12}.$$

where Y_{ij} represents Daughter's level of education, X_1 represents Father's level of education, X_2 represents Mother's level of education, X_4 represents Income, X_5 represents Number of siblings, X_6 represents Birth order, and X_7 represents Religion.

Here i stands for the ith individual and j stands for Illiterate, Elementary, Secondary, and HS & Above.

Null Hypothesis: $H_{o:}$ $\beta_{2ij}=0$.

Alternative Hypothesis: H_1: $\beta_{2ij}\neq0$ (Table 6.29).

Observations:
Here, we reject the null hypothesis H_o for all the levels of education, that is, the above-mentioned proposition is not valid at any level of education. This result shows that the above pathway does not lead to capability enhancement.

Proposition 13: Father's level of education has a significant effect on son's educational attainment at each level of education for Scheduled Tribe households:

$$Y_{ij} = \alpha + \beta_{1ij}X_{1ij} + \beta_{2ij}X_{2ij} + \beta_{4ij}X_{4ij} + \beta_{5ij}X_{5ij} + \beta_{6ij}X_{6ij} + \beta_{7ij}X_{7ij} + \varepsilon_{13}$$

where Y_{ij} represents Son's level of education, X_1 represents Father's level of education, X_2 represents Mother's level of education, X_4 represents Income, X_5 represents Number of siblings, X_6 represents Birth order, and X_7 represents Religion.

Here i stands for the ith individual and j stands for Illiterate, Elementary, Secondary, and HS & Above.

Table 6.29 Predicted probability values obtained from multinomial logistic regression: each cell showing daughter's level of education (given in column) corresponding to each level of education of the mother (given in row) respectively

SC	Daughter illiterate	Daughter elementary	Daughter secondary	Daughter HS & above	UM
Mother illiterate	**0.02 (0.994)**	0.27 (0.907)	0.20 (0.975)	0.51 (0.907)	0.98
Mother elementary	0.00 (0.994)	**0.11 (0.907)**	0.21 (0.975)	0.68 (0.907)	0.89
Mother secondary	0.00 (0.999)	0.04 (0.887)	**0.18 (0.976)**	0.79 (0.887)	0.79
Mother HS & above	0.00 (1)	0.00 (1)	0.00 (1)	**1*** (0)**	

Source IHDS round II. Results based on author's calculation
*, **, *** refers to 10%, 5%, 1% levels of significance respectively

Table 6.30 Predicted probability values obtained from multinomial logistic regression: each cell showing son's level of education (given in column) corresponding to each level of education of the father (given in row) respectively

ST	Son illiterate	Son elementary	Son secondary	Son HS & above	UM
Father illiterate	**0.08 (0.974)**	0.41 (0.696)	0.27 (0.696)	0.25 (0.696)	0.92
Father elementary	0.03 (0.975)	**0.36 (0.347)**	0.30 (0.348)	0.31 (0.348)	0.61
Father secondary	0.03 (0.976)	0.12 (0.317)	**0.28 (0.314)**	0.57 (0.313)	0.57
Father HS & above	0.00 (0.999)	0.15*** (0)	0.17*** (0)	**0.67*** (0)**	

Source IHDS round II. Results based on author's calculation
*, **, *** refers to 10%, 5%, 1% levels of significance respectively

Null Hypothesis: H_o: $\beta_{1ij}=0$.

Alternative Hypothesis: H_1: $\beta_{1ij}\neq 0$ (Table 6.30).

Observation:

Therefore, we accept the null hypothesis: H_o, that is, the above-mentioned proposition is valid, only in the following case:

1. When father is highly educated, the predicted probability is significant at son's highest level of education showing persistence over generations; however, the downward mobility is also significant in lower levels of education. Therefore, when the father is receiving high education, the pathway becomes capability enhancing for the son in this case.

Proposition 14: Father's level of education has a significant effect on daughter's educational attainment at each level of education for Scheduled Tribe households:

$$Y_{ij} = \alpha + \beta_{1ij}X_{1ij} + \beta_{2ij}X_{2ij} + \beta_{4ij}X_{4ij} + \beta_{5ij}X_{5ij} + \beta_{6ij}X_{6ij} + \beta_{7ij}X_{7ij} + \varepsilon_{14}$$

where Y_{ij} represents Daughter's level of education, X_1 represents Father's level of education, X_2 represents Mother's level of education, X_4 represents Income, X_5 represents Number of siblings, X_6 represents Birth order, and X_7 represents Religion.

Here i stands for the ith individual and j stands for Illiterate, Elementary, Secondary, and HS & Above.

Table 6.31 Predicted probability values obtained from multinomial logistic regression: each cell showing daughter's level of education (given in column) corresponding to each level of education of the father (given in row) respectively

ST	Daughter illiterate	Daughter elementary	Daughter secondary	Daughter HS & above	UM
Father illiterate	**0.07 (0.991)**	0.10 (0.988)	0.30 (0.923)	0.53 (0.923)	0.93
Father elementary	0.02 (0.991)	**0.07 (0.989)**	0.43 (0.866)	0.48 (0.866)	0.91
Father secondary	0.00 (0.992)	0.02 (0.989)	**0.31 (0.55)**	0.67 (0.549)	0.67
Father HS & above	0.01 (0.992)	0.02 (0.989)	0.14 (0.561)	**0.83 (0.559)**	

Source IHDS round II. Results based on author's calculation
*, **, *** refers to 10%, 5%, 1% levels of significance respectively

Null Hypothesis: H_o: $\beta_{1ij}=0$.

Alternative Hypothesis: H_1: $\beta_{1ij}\neq 0$ (Table 6.31).

Observations:

Here, we reject the null hypothesis H_o for all the levels of education, that is, the above-mentioned proposition is not valid at any level of education. This result shows that the above pathway does not lead to capability enhancement.

Proposition 15: Mother's level of education has a significant effect on son's educational attainment at each level of education for Scheduled Tribe households:

$$Y_{ij} = \alpha + \beta_{1ij}X_{1ij} + \beta_{2ij}X_{2ij} + \beta_{4ij}X_{4ij} + \beta_{5ij}X_{5ij} + \beta_{6ij}X_{6ij} + \beta_{7ij}X_{7ij} + \varepsilon_{15}$$

where Y_{ij} represents Son's level of education, X_1 represents Father's level of education, X_2 represents Mother's level of education, X_4 represents Income, X_5 represents Number of siblings, X_6 represents Birth order, and X_7 represents Religion.

Here i stands for the ith individual and j stands for Illiterate, Elementary, Secondary, and HS & Above.

Null Hypothesis: H_o: $\beta_{2ij}=0$.

Alternative Hypothesis: H_1: $\beta_{2ij}\neq 0$ (Table 6.32).

Observation:

Therefore, we accept the null hypothesis: H_o, that is, the above-mentioned proposition is valid, only in the following case:

1. When mother is elementary educated, the predicted probability of persistence is significant at this level of education, but intergenerational transfer is also significant for higher levels of education which shows significant upward mobility over generations. So here also we find the intergenerational capability enhancing effect.
2. When mother is highly educated, the predicted probability shows there is significant persistence at that level and significant downward mobility at secondary level. So, this leads to capability enhancing effect except in a few cases.
3. When mother is illiterate or secondary educated, there is no significant transfer taking place. Therefore these pathways are not capability enhancing.

Proposition 16: Mother's level of education has a significant effect on daughter's educational attainment at each level of education for Scheduled Tribe households:

$$Y_{ij} = \alpha + \beta_{1ij}X_{1ij} + \beta_{2ij}X_{2ij} + \beta_{4ij}X_{4ij} + \beta_{5ij}X_{5ij} + \beta_{6ij}X_{6ij} + \beta_{7ij}X_{7ij} + \varepsilon_{16}$$

Table 6.32 Predicted probability values obtained from multinomial logistic regression: each cell showing son's level of education (given in column) corresponding to each level of education of the mother (given in row) respectively

ST	Son illiterate	Son elementary	Son secondary	Son HS & above	UM
Mother illiterate	**0.03** **(0.986)**	0.37 (0.563)	0.29 (0.563)	0.31 (0.563)	0.97
Mother elementary	0.00 (0.987)	**0.30**** **(0.017)**	0.33** (0.017)	0.37** (0.016)	0.70**
Mother secondary	0.02 (0.987)	0.28 (0.42)	**0.18** **(0.421)**	0.52 (0.418)	0.52
Mother HS & above	0.00 (0.999)	0.15** (0.015)	0.33*** (0)	**0.52***** **(0)**	

Source IHDS round II. Results based on author's calculation
*, **, *** refers to 10%, 5%, 1% levels of significance respectively

Table 6.33 Predicted probability values obtained from multinomial logistic regression: each cell showing daughter's level of education (given in column) corresponding to each level of education of the mother (given in row) respectively

ST	Daughter illiterate	Daughter elementary	Daughter secondary	Daughter HS & above	UM
Mother illiterate	**0.14** **(0.976)**	0.21 (0.967)	0.24 (0.902)	0.42 (0.902)	0.86
Mother elementary	0.05 (0.978)	**0.21** **(0.967)**	0.26 (0.88)	0.48 (0.88)	0.74
Mother secondary	0.00 (0.999)	0.00 (0.999)	**0.41***** **(0)**	0.59*** (0)	0.59***
Mother HS & above	0.00 (1)	0.00 (0.999)	0.20** (0.013)	**0.80***** **(0)**	

Source IHDS round II. Results based on author's calculation
*, **, *** refers to 10%, 5%, 1% levels of significance respectively

where Y_{ij} represents Daughter's level of education, X_1 represents Father's level of education, X_2 represents Mother's level of education, X_4 represents Income, X_5 represents Number of siblings, X_6 represents Birth order, and X_7 represents Religion.

Here i stands for the ith individual and j stands for Illiterate, Elementary, Secondary, and HS & Above.

Null Hypothesis: H_0: $\beta_{2ij} = 0$.

Alternative Hypothesis: H_1: $\beta_{2ij} \neq 0$ (Table 6.33).

Observations:

Therefore, we accept the null hypothesis: H_0, that is, the above-mentioned proposition is valid, in the following cases:

1. When mother is educated up to the secondary level, the predicted probability of persistence at the same level is significant, significant upward mobility for higher level of education, and in significant downward mobility at daughter's elementary level of education. The result shows the pathway leads to capability enhancement but not for all cases.
2. When mother is highly educated, the predicted probability of persistence is significant and also significant downward mobility at the secondary level. The result shows the pathway leads to capability enhancement.

Therefore, the above results have identified the pathways which turn out to be significant in

Table 6.34 The following table represents the pathways which turn out to be **significant** in the **transfer** mechanism at each level of education: Illiteracy (Illit), Elementary (Elem), Secondary (Secon), HS, and Above (HS&Abv) for each religious community: Hindu, Muslim, and Other Religions (OR)

Father/Mother level of education—Illiteracy				
	GC	OBC	SC	ST
Father-son	–	–	–	–
Father-daughter	Illit, Elem, Secon, HS&Abv	Illit, Elem, Secon, HS&Abv	–	–
Mother-son	Illit, Elem, Secon, HS&Abv	Illit, Elem, Secon, HS&Abv	–	–
Mother-daughter	Elem, Secon, HS&Abv	Illit, Elem, Secon, HS&Abv	–	–
Father/mother level of education—elementary				
	GC	OBC	SC	ST
Father-son	Elem, Secon, HS&Abv	Elem, Secon, HS&Abv	–	–
Father-daughter	Illit, Elem, Secon, HS&Abv	Illit, Elem, Secon, HS&Abv	–	–
Mother-son	Illit, Elem, Secon, HS&Abv	Illit, Elem, Secon, HS&Abv	Elem, Secon, HS&Abv	Elem, Secon, HS&Abv
Mother-daughter	Elem, Secon, HS&Abv	Illit, Elem, Secon, HS&Abv	–	–
Father/mother level of education—Secondary				
	GC	OBC	SC	ST
Father-son	Elem, Secon, HS&Abv	Elem, Secon, HS&Abv	–	–
Father-daughter	Illit, Elem, Secon, HS&Abv	Illit, Elem, Secon, HS&Abv	–	–
Mother-son	Illit, Elem, Secon, HS&Abv	Illit, Elem, Secon, HS&Abv	Elem, Secon, HS&Abv	–
Mother-daughter	Elem, Secon, HS&Abv	Illit, Elem, Secon, HS&Abv	–	Secon, HS&Abv
Father/mother level of education—HS & Above				
	GC	OBC	SC	ST
Father-son	Elem, Secon, HS&Abv	Elem, Secon, HS&Abv	–	Elem, Secon, HS&Abv
Father-daughter	Elem, Secon, HS&Abv	Illit, Elem, Secon, HS&Abv	–	–
Mother-son	Elem, Secon, HS&Abv	Elem, Secon, HS&Abv	–	Elem, Secon, HS&Abv
Mother-daughter	Elem, Secon, HS&Abv	Illit, Elem, Secon, HS&Abv	HS&Abv	Secon, HS&Abv

Source IHDS round II. Results based on author's calculation

the transfer of education over generations. All the three social groups GC, OBC, SC, and ST have their respective pathways which turn out to be significant in the transfer mechanism for different levels of education. These results are summarised in the following tables: Table 8.19 showing the significant pathways, Table 8.20 showing the predicted probability values of the mobility/persistence, and Table 8.21 the tabular representation of pathways indicating significant upward mobility. Table 8.22, Table 8.23, Table 8.24, and Table 8.25 show the summary tree of Upward Mobility/Persistence for General Caste, Other Backward Caste, Scheduled Caste, and Scheduled Tribe, respectively (Tables 6.34, 6.35 and 6.36).

Table 6.35 Table showing predicted probability values of **Upward Mobility/Persistence** of son's/daughter's educational attainment for each level of father/mother education (obtained from Tables 6.18–6.33)

Father/Mother level of education—illiteracy—upward mobility				
	GC	OBC	SC	ST
Father-son	0.98	0.94	0.96	0.92
Father-daughter	0.89***	0.81***	0.96	0.93
Mother-son	0.98***	0.96***	0.95	0.97
Mother-daughter	1.00***	0.84***	0.98	0.86
Father/mother level of education—elementary—upward mobility				
	GC	OBC	SC	ST
Father-son	0.84***	0.74***	0.74	0.61
Father-daughter	0.83***	0.71***	0.80	0.91
Mother-son	0.90***	0.83***	0.77***	0.70**
Mother-daughter	0.92***	0.83***	0.89	0.74
Father/mother level of education—secondary—upward mobility				
	GC	OBC	SC	ST
Father-son	0.73***	0.65***	0.57	0.57
Father-daughter	0.81***	0.69***	0.82	0.67
Mother-son	0.76***	0.71***	0.63***	0.52
Mother-daughter	0.88***	0.79**	0.79	0.59***
Father/mother level of education—HS & above—persistence				
	GC	OBC	SC	ST
Father-son	0.85***	0.75**	0.78	0.67***
Father-daughter	0.91***	0.76***	0.90	0.83
Mother-son	0.91***	0.77***	0.91	0.52***
Mother-daughter	0.92***	0.86***	1***	0.80***

Source IHDS round II. Results based on author's calculation

Table 6.36 Table showing precise result of predicted probability of **Upward Mobility/Persistence** of son's/daughter's educational attainment for each level of father/mother education (obtained from Tables 6.18–6.33) where "√" represents significant upward mobility/persistence and "X" represents insignificant

Father/mother level of education—illiteracy—upward mobility				
	GC	OBC	SC	ST
Father-son	X	X	X	X
Father-daughter	√	√	X	X
Mother-son	√	√	X	X
Mother-daughter	√	√	X	X
Father/mother level of education—elementary—upward mobility				
	GC	OBC	SC	ST
Father-son	√	√	X	X
Father-daughter	√	√	X	X
Mother-son	√	√	X	√
Mother-daughter	√	√	X	X
Father/mother level of education—secondary—upward mobility				
	GC	OBC	SC	ST
Father-son	√	√	X	X
Father-daughter	√	√	X	X
Mother-son	√	√	√	X
Mother-daughter	√	√	X	√
Father/mother level of education—HS & above—persistence				
	GC	OBC	SC	ST
Father-son	√	√	X	√
Father-daughter	√	√	X	X
Mother-son	√	√	X	√
Mother-daughter	√	√	√	√

Source IHDS round II. Results based on author's calculation

Therefore, the results of significant transfer pathways resulting in upward mobility are summarised by the following summary tree for each religion community: GC, OBC, SC, and ST (Figs. 17, 18, 19 and 20).

6.6 Conclusion

The above results indicate that for privileged castes there is intergenerational transfer of education for all the pathways, whereas for the under-privileged castes there is lack of mobility in education over generations. However, for the under-privileged castes in particular, it is the mother's education that gets transferred more towards the daughter than the son. This implies that if the mother is educated, capability enhancement over generation is taking place and daughter's educational attainment is higher than that of the son. Mother's education leads to more awareness and the child gets more motivation and support if the mother is educated, as the mother acts as the role model for her children.

The results also highlight the fact that historical discrimination and social exclusion have a long-run effect, thus it is the privileged castes who are benefitted and capability enhancement is mainly taking place over generations for the higher social caste individuals and the excluded castes are lagging behind in educational development over generations. Long-run persisting intersectional inequality in educational attainment caused by caste and gender discrimination finds some evidence backing in the empirical estimates presented in the paper.

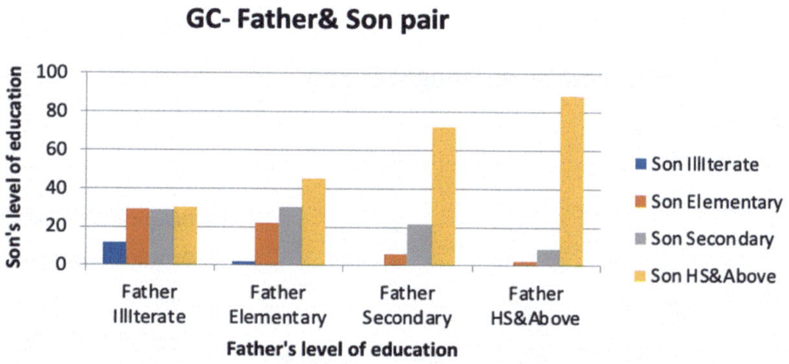

Fig. 1 Graphical Representation of Level of son's education corresponding to father's level of education (given in %): Progression of Son's education measured in percentage (%) for different levels of father's education in General Caste households. *Source* IHDS round II. Results based on author's calculation

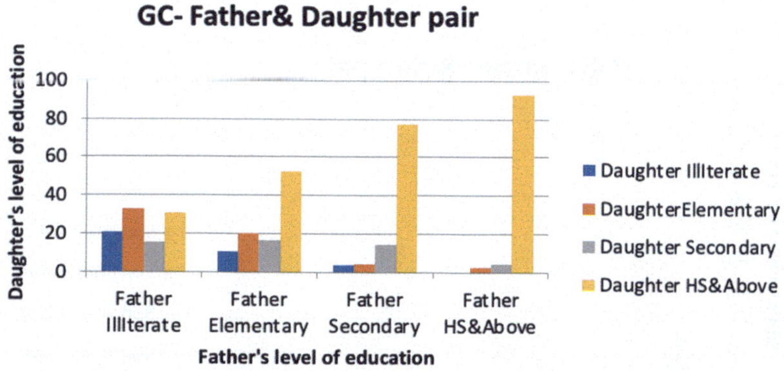

Fig. 2 Graphical Representation of Level of daughter's education corresponding to father's level of education (given in %): Progression of Daughter's education measured in percentage (%) for different levels of father's education in General Caste household. *Source* IHDS round II. Results based on author's calculation

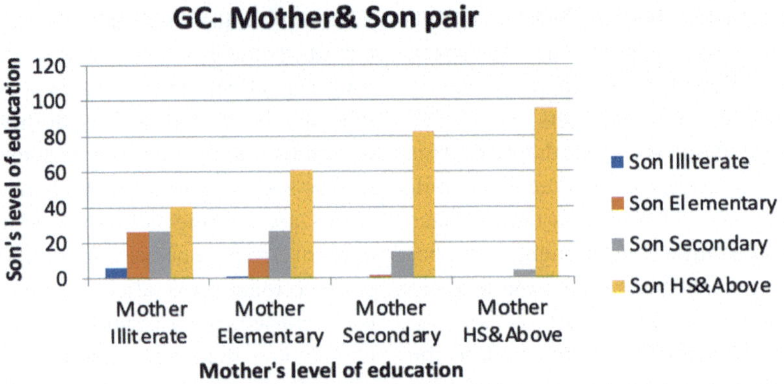

Fig. 3 Graphical Representation of Level of son's education corresponding to mother's level of education (given in %): Progression of Son's education measured in percentage (%) for different levels of mother's education in General Caste households. *Source* IHDS round II. Results based on author's calculation

Government of India (2006) Social, economic and

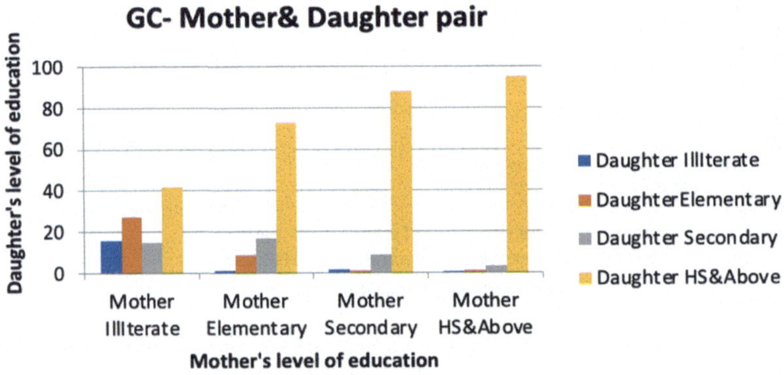

Fig. 4 Graphical Representation of Level of son's education corresponding to father's level of education (given in %): Progression of Daughter's education measured in percentage (%) for different levels of mother's education in General Caste households. *Source* IHDS round II. Results based on author's calculation

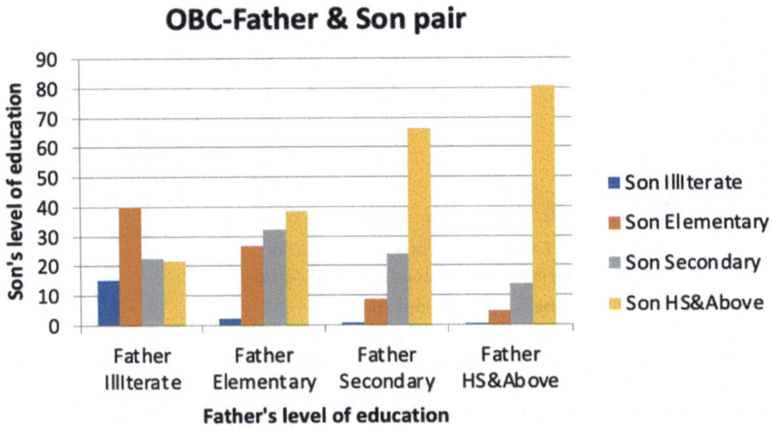

Fig. 5 Graphical Representation of Level of son's education corresponding to father's level of education (given in %): Progression of Son's education measured in percentage (%) for different levels of father's education in Other Backward Caste households. *Source* IHDS round II. Results based on author's calculation

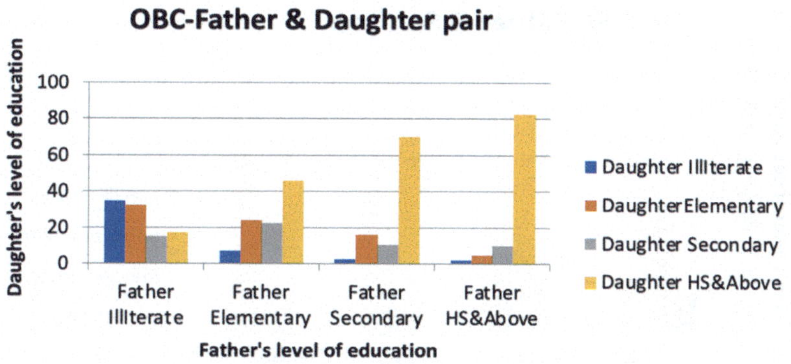

Fig. 6 Graphical representation of Level of daughter's education corresponding to father's level of education (given in %): Progression of Daughter's education measured in percentage (%) for different levels of father's education in Other Backward Caste households. *Source* IHDS round II. Results based on author's calculation

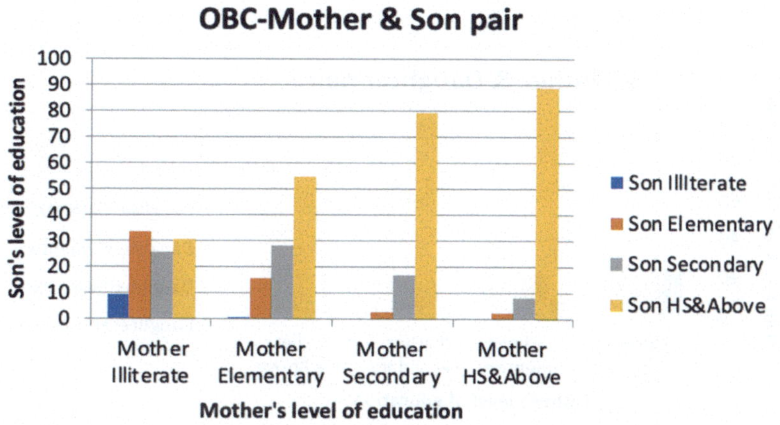

Fig. 7 Graphical representation of level of son's education corresponding to mother's level of education (given in %): progression of Son's education measured in percentage (%) for different levels of mother's education in Other Backward Caste households. *Source* IHDS round II. Results based on author's calculation

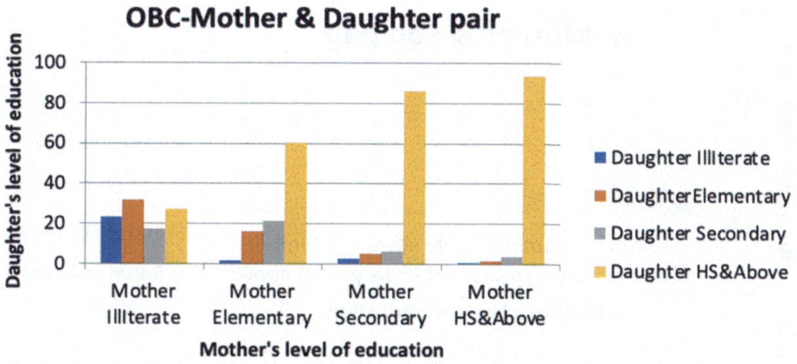

Fig. 8 Graphical representation of level of daughter's education corresponding to mother's level of education (given in %): Progression of Daughter's education measured in percentage (%) for different levels of mother's education in other backward caste households. *Source* IHDS round II. Results based on author's calculation

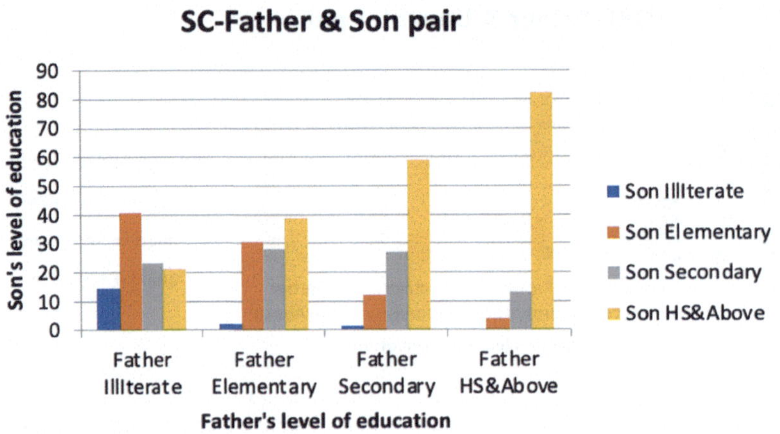

Fig. 9 Graphical representation of level of son's education corresponding to father's level of education (given in %): Progression of Son's education measured in percentage (%) for different levels of father's education in scheduled caste households. *Source* IHDS round II. Results based on author's calculation

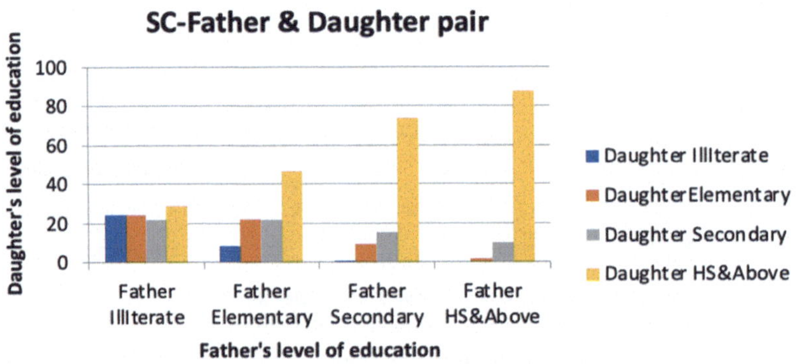

Fig. 10 Graphical Representation of Level of daughter's education corresponding to father's level of education (given in %): progression of daughter's education measured in percentage (%) for different levels of father's education in scheduled caste households. *Source* IHDS round II. Results based on author's calculation

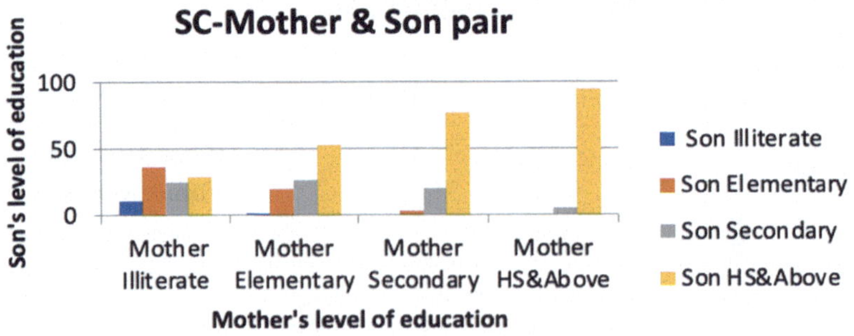

Fig. 11 Graphical representation of level of son's education corresponding to mother's level of education (given in %): progression of Son's education measured in percentage (%) for different levels of mother's education in SC household. *Source* IHDS round II. Results based on author's calculation

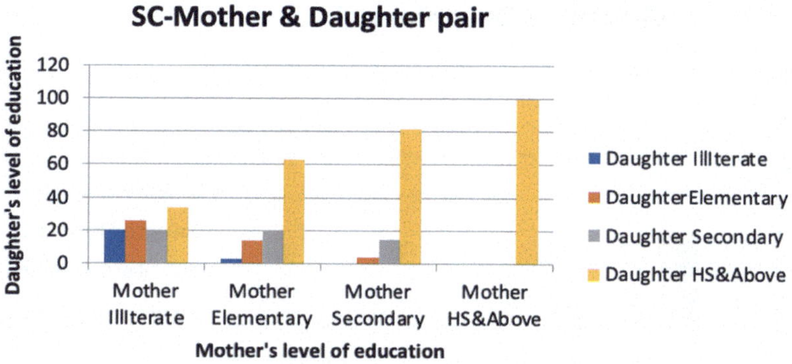

Fig. 12 Graphical representation of level of mother's education corresponding to daughter's level of education (given in %): progression of daughter's education measured in percentage (%) for different levels of mother's education in SC households. *Source* IHDS round II. Results based on author's calculation

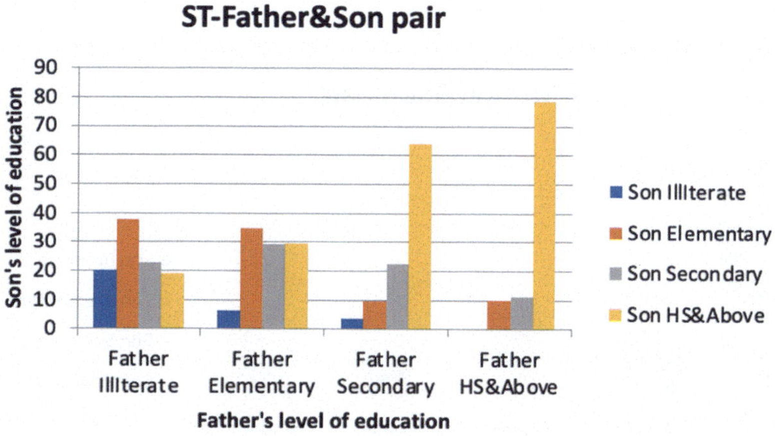

Fig. 13 Graphical representation of level of son's education corresponding to father's level of education (given in %): Progression of Son's education measured in percentage (%) for different levels of father's education in scheduled tribe households. *Source* IHDS round II. Results based on author's calculation

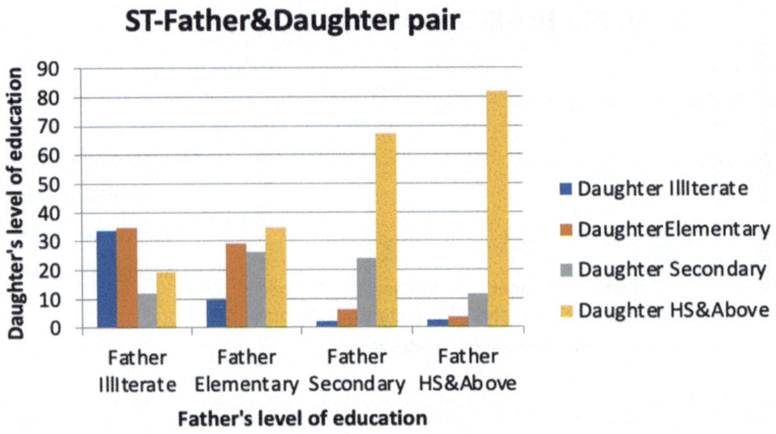

Fig. 14 Graphical representation of level of daughter's education corresponding to father's level of education (given in %): progression of daughter's education measured in percentage (%) for different levels of father's educations in scheduled Tribe households. *Source* IHDS round II. Results based on author's calculation

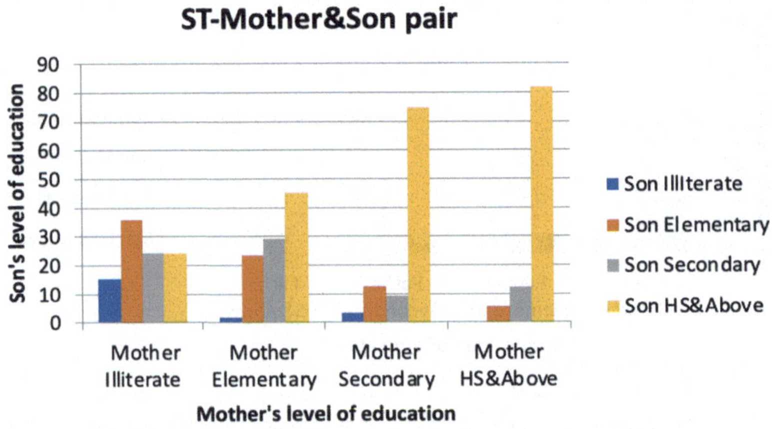

Fig. 15 Graphical representation of level of son's education corresponding to mother's level of education (given in %): progression of Son's education measured in percentage (%) for different levels of mother's education in Scheduled Tribe households. *Source* IHDS round II. Results based on author's calculation

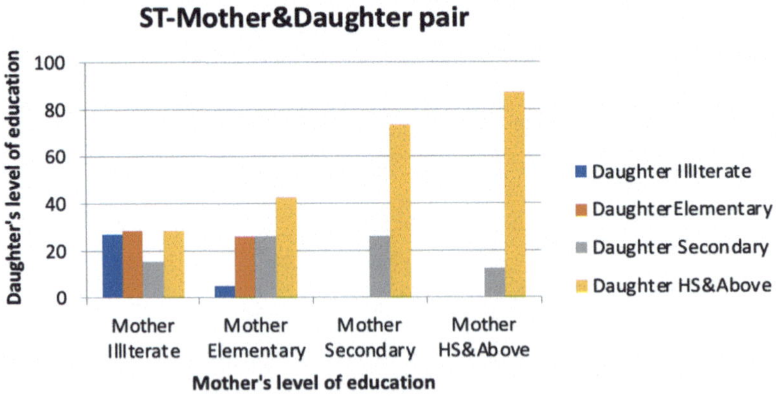

Fig. 16 Graphical representation of level of daughter's education corresponding to mother's level of education (given in %): progression of daughter's education measured in percentage (%) for different levels of mother's education in scheduled tribe households. *Source* IHDS round II. Results based on author's calculation

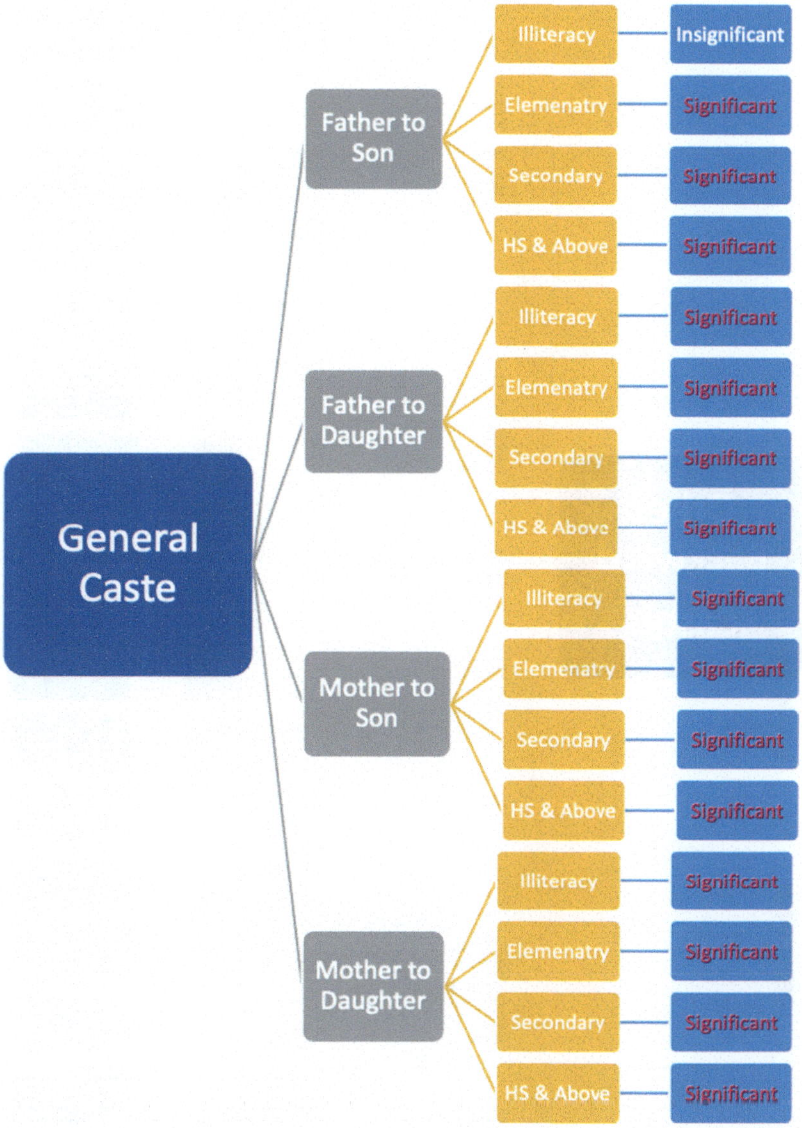

Fig. 17 Summary tree showing the **significant upward mobility** of the child's (Son/Daughter) educational level corresponding to each level of **parental (father/mother) education** (Illiteracy, Elementary, Secondary, and HS & Above) for GC households. *Source* IHDS round II. Results based on author's calculation

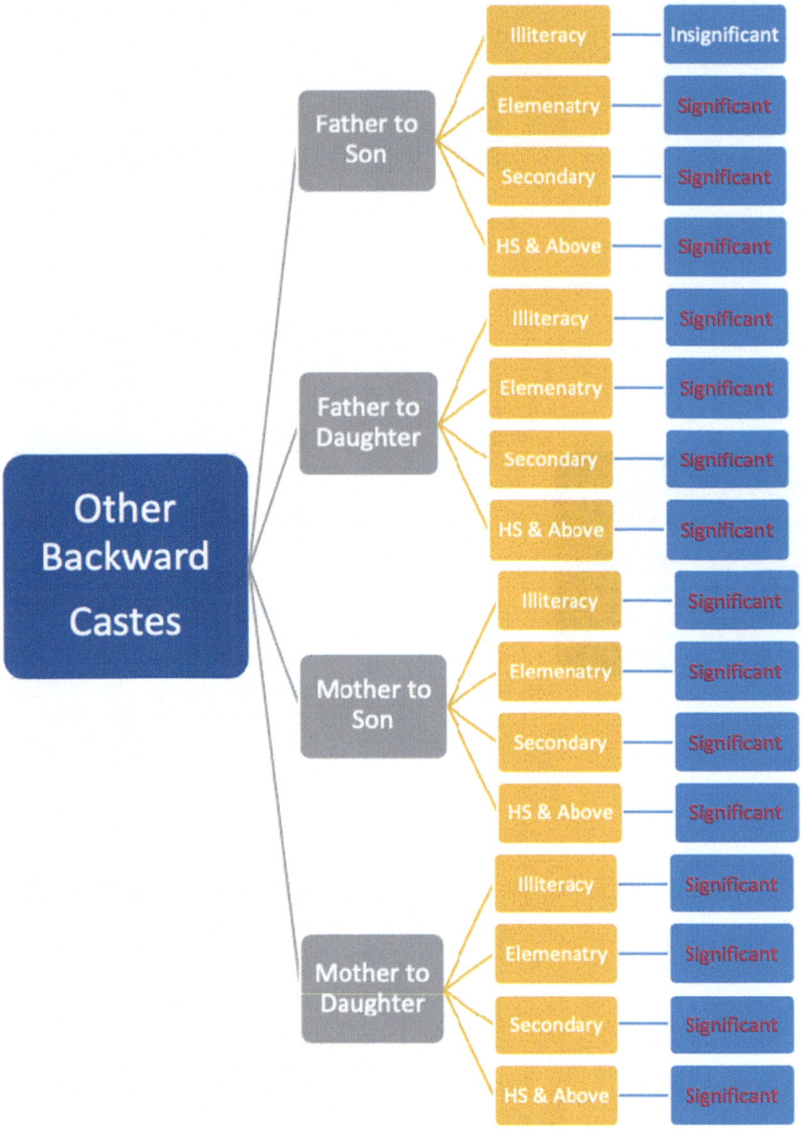

Fig. 18 Summary tree showing the **significant upward mobility** of the child's (Son/Daughter) educational level corresponding to each level of **parental (father/mother) education** (Illiteracy, Elementary, Secondary, and HS & Above) for OBC households. *Source* IHDS round II. Results based on author's calculation

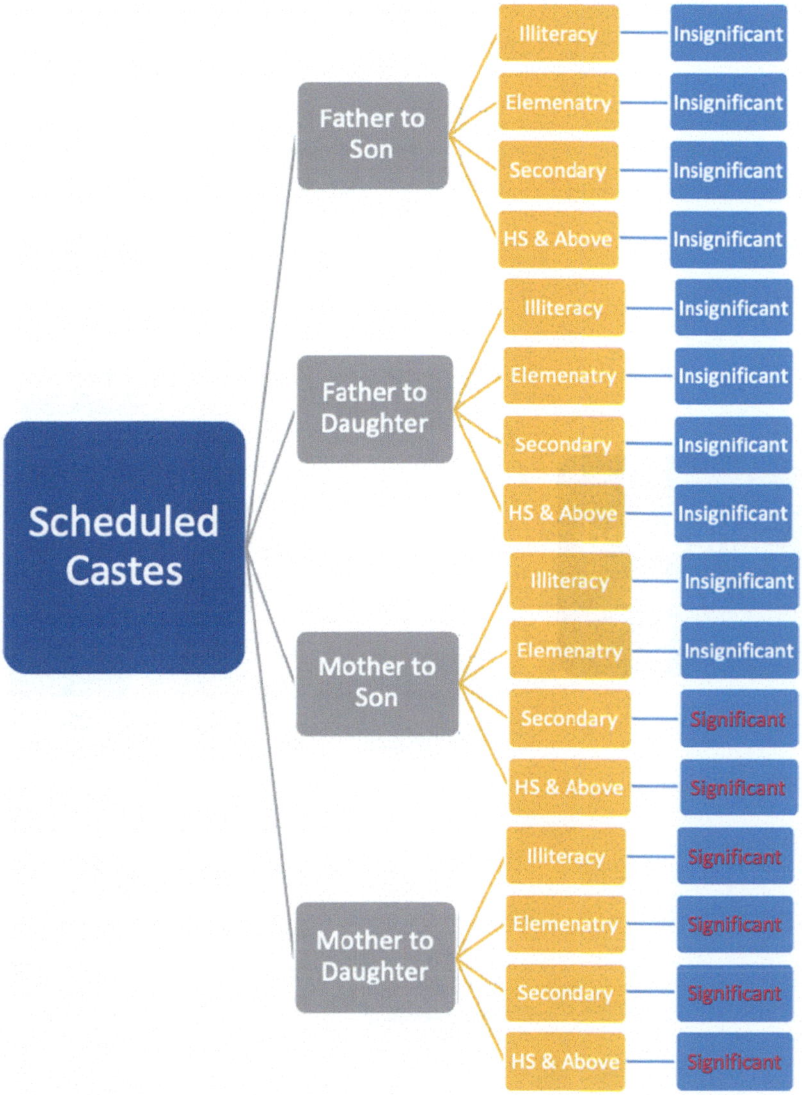

Fig. 19 Summary tree showing the **significant upward mobility** of the child's (Son/Daughter) educational level corresponding to each level of **parental (father/mother) education** (Illiteracy, Elementary, Secondary, and HS & Above) for SC households. *Source* IHDS round II. Results based on author's calculation

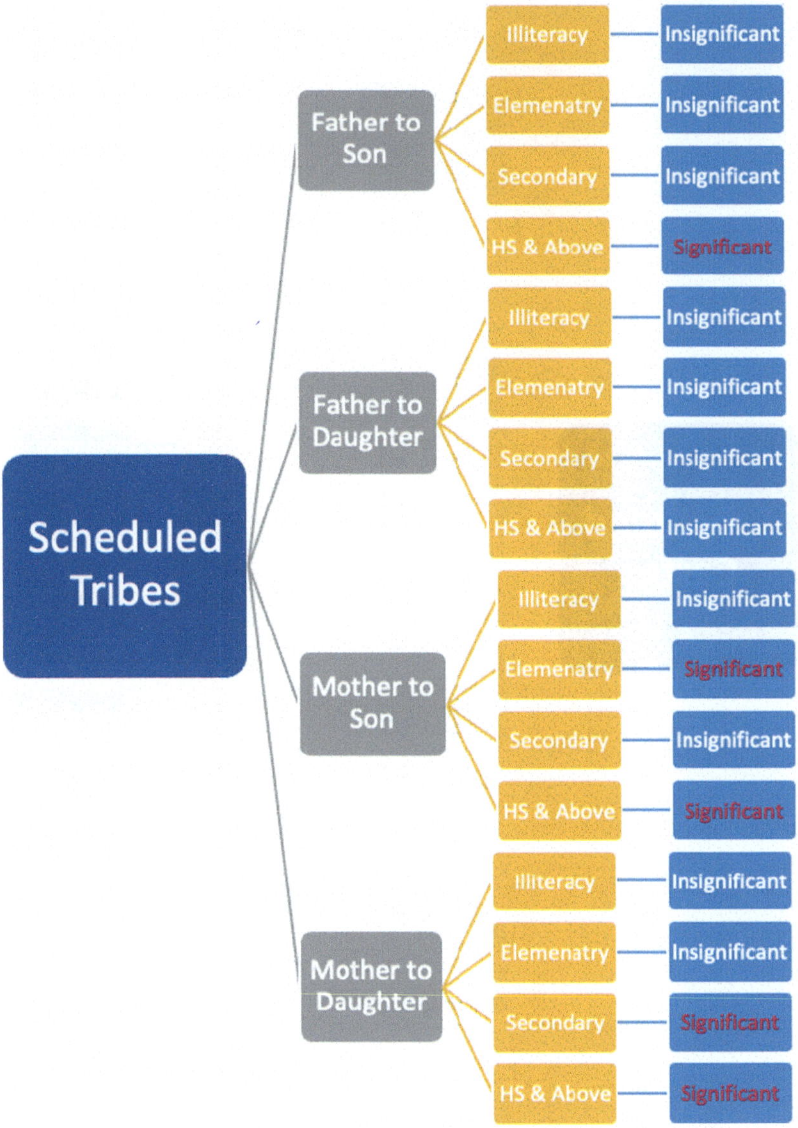

Fig. 20 Summary tree showing the **significant upward mobility** of the child's (Son/Daughter) educational level corresponding to each level of **parental (father/mother) education** (Illiteracy, Elementary, Secondary, and HS & Above) for ST households. *Source* IHDS round II. Results based on author's calculation

References

Azam M, Bhatt V (2012) Like father like son-intergenerational educational mobility in India. IZA DP No.6549

Becker GS, Tomes N (1979) An equilibrium theory of the distribution of income and intergenerational mobility. J Pol Econ 87(6):1153–1189

Becker GS, Tomes N (1986) Human capital and the rise and fall of families. J Labor Econ 4(3):S1-S39

Behrman J, Gaviria A, Szkely M (2001) Intergenerational mobility in Latin America. Economia 2(1):1–44

Besley T, Burgess R, Esteve-Volart B (2007) The policy origins of poverty and growth in India. In Tim Besley and Louise Cord (Eds), Delivering on the promise of pro-poor growth. London: Palgrave Macmillan and the World Bank

Black S, Devereux P (2011) Recent developments in intergenerational mobiliity. Handbook Labor Econ 4(B):Ch-16, 1487-1541

Chaudhuri Shubham, Ravallion Martin (2006) Partially awakened giants: uneven growth in China and India. Dancing with giants: China, India and the global economy. Ed. L. Alan Winters and Shahid Yusuf. Washington DC: World Bank

Daude Christian (2011) Ascendance by descendants? On Intergenerational Education Mobility in Latin America. *OECD Development Centre Working Papers*, Paper, No. 297

Fessler P, Mooslechner P, Schurz M (2012) Intergenerational transmission of educational attainment in Austria. Empirica 39:65–86

education status of the Muslim community of India. New Delhi

Hertz, Tom, Jayasunderay, Tamara, Pirainoz, Patrizio, Selcuk, Sibel, Smithyy, Nicole, and Verashchagina, Alina. 2007. "The Inheritance of Educational Inequality: Interna- tional Comparisons and Fifty-Year Trends." *The B.E. Journal of Economic Analysis and Policy (Advances)*, 7 (2), 1–46

Hnatkovskay V, Lahiri A and Paul SB (2012) Breaking the caste barrier: intergenerational mobility in India. Unpublished manuscript, University of British Columbia

Jalan J, Murgai R (2008) Intergenerational mobility in education in India, Paper Presented at the Indian Statistical Institute, Delhi

Kingdon G (2007) The progress of school education in India. Oxf Rev Econ Policy 23(2):168195

Maitra P, Sharma A (2009) Parents and children: education across generations in India, unpublished manuscript, department of economics, Monash University, 35

Munshi K, Rosenzweig M (2009) Why is mobility in India so low? Social Insurance, Inequality, and Growth. Unpublished Manuscript

Munshi K, Rosenzweig M (2006) Traditional institutions meet the modern world: caste, gender, and schooling choice in a globalizing economy. Am Econ Rev 96(4):1225–1252

Ray Jhilam, Majumder Rajarshi (2010) Educational and occupational mobility across generations in India: social and regional dimensions. Unpublished Manuscript, 36

Public Goods Provision and Upward Intergenerational Occupational Mobility: Empirical Evidence from China

7

Shengfeng Lu, Ziming Li and Abhinav Alakshendra

Abstract

Public goods provision influences a vast majority of the population in China. It directly affects the quality of life in general and has a strong effect on future socio-economic outcomes, directly affecting the prosperity of the next generation of Chinese citizens. However, the different rural and urban public welfare institutions challenge the public policy goal of inclusiveness. Levels of access to public goods vary a great deal for residents of urban sub-districts and rural villages and there are prominent differences in public goods provision even among urban sub-districts and rural villages across China. This paper investigates the impact of public goods provision on social mobility, particularly on upward intergenerational occupational mobility between 1989 and 2009 in China. This paper utilizes China Health and Nutrition Survey (CHNS) questionnaire data to calculate the upward intergenerational occupational mobility by categorizing occupation hierarchies in (1) from lower to middle or to advanced class and from the middle to the advanced class occupation, and (2) from a non-public to a public occupation. Regression analysis using 60,836 representative household samples covering 9 provinces demonstrates low intergenerational occupational mobility within the study period. The upward intergenerational occupational mobility is reduced annually and in underdeveloped regions, it is lower than in the prosperous regions. Further, our empirical analysis shows that despite the differences in public goods provision (education, hospital/health services, sanitation, tap water, transit services, etc.) between rural and urban regions and the differences across four regions, public goods can facilitate upward intergenerational occupational mobility. However, the difference among sub-districts or among villages matters. Residents with public goods available in their sub-districts or villages experience between 2 and 30% higher upward

We sincerely acknowledges the financial support provided by the National Natural Science Foundation of China (No. 72073103, No. 72473108, No. 72073145), the Major Project of Key Research Base for Humanities and Social Sciences of the Ministry of Education (No. 22JJD790060) and Wuhan University-Duke Kunshan University joint research funding (XXWHUDKUZZJJ202302) for this research.

S. Lu (✉) · Z. Li
Department of Public Finance and Taxation, School of Economic and Management, Wuhan University, Wuhan, China
e-mail: shengflu@whu.edu.cn

Z. Li
e-mail: liziming_idcc@whu.edu.cn

A. Alakshendra
Department of Urban and Regional Planning, University of Florida, Gainesville, USA
e-mail: alakshendra@dcp.ufl.edu

A. Alakshendra et al. (eds.), *Development, Inclusion and Sustainability*,
Sustainable Development Goals Series, https://doi.org/10.1007/978-981-97-6863-9_7

intergenerational occupational mobility compared to those with limited access to public goods and the difference is starker in rural areas. The findings imply the significance of an equitable distribution mechanism of public goods and its role in achieving a more inclusive society in developing countries in general and China in particular. The local governments should focus more on enlarging the coverage of public-goods beneficiaries at the micro (i.e., household/neighborhood) level than on smoothing the regional gap or rural–urban gap in terms of the total amount of public goods provision at the macro level.

7.1 Introduction

Social mobility, the changes in social status between different generations within the same family, reflects the equality of opportunities (OECD 2010). The positive impact of equality of opportunity incentivizes laborers to work hard and encourages households to continuously invest in human capital accumulation. This micro impact, converging above gradually, is supposed to sustain economic growth (Heckman 2012). Therefore, for developing countries, especially those confronting a middle-income trap, the question of how to increase social mobility draws the attention of policymakers. It is well accepted that increasing the provision of public goods and increased public expenditure can facilitate upward social mobility. However, the evidence from China suggests the lack of a significant impact on upward mobility of increased public expenditure. Chinese expenditure on public goods has increased steadily from 1989 to 2009[1] but at the same time period China has witnessed, class solidification and social stratification which is usually associated with low intergenerational social mobility (Bian 2002; Dollar 2013). It seems that increasing the provision of public goods alone cannot benefit the majority in the society. This study aims to investigate the institutions-induced heterogeneity in China and discusses the differences between the intended and actual effect of public policies.

The institutional factors are an important explanation of the above-mentioned contradiction. Dollar (2013) observes that the *'hukou'* registration system and locally funded education system hinders the general population from enjoying an equal level of public goods and thereby contributes to low social mobility. Further, he categorizes the heterogeneity of public goods provision in three dimensions.

The first heterogeneity arises when neighborhoods and villages experience a different level of provisions within the same city or province respectively. This is defined as heterogeneity at the community level.

The hierarchies of Chinese cities consist of districts (*qu*), sub-districts (*jiedao*), and neighborhoods (*xiaoqu*). The quantity of public goods such as public schools, health care services, roads, and bus stations are determined by the total population of districts. The physical locations of these services such as hospitals or schools are determined based on the population density of sub-districts and as per local land-use restrictions. Therefore, unequal provision of public goods becomes a certainty.

Meanwhile, the heterogeneity in the provision of public goods across villages is obvious (Tsai 2007). Two unique characteristics of rural local governing can account for the variations in the provision of the public good across villages. One is the differences in local governance (Zhang et al. 2004), or more precisely, informal institutions of accountability (Tsai 2007). The other is the impact of political connections between a village and its upper-level local government (Qu 2012).

In China, usually, the local government's tax base is not big enough to cover planned expenditures. The "project system" (*XiangmuZhi*) grants competitive opportunities for villages

[1] According to 2014 *Financial Year Book of China* and the report from National Bureau of Statistics of China, during 1990 to 2012, the percentage of public expenditure in terms of public education, health, and infrastructure in total fiscal investment increased from 20 to 66.87% in general; the percentage of expenditure on education in GDP increased from 2.71 to 4.28% in particular.

Table 7.1 Heterogeneous provision of public goods: RRBAN/rural areas and different income groups

Public goods	Urban/rural areas			High/low income groups		
	Mean		Difference	Mean		Difference
	Urban	Rural		High	Low	
Primary	0.6810	0.7363	14.87***	0.6863	0.7381	14.67***
Middle	0.2928	0.2944	0.41	0.3190	0.2811	−10.56***
Senior-high	0.1629	0.1220	−14.51***	0.1414	0.1282	−4.92***
Hospital	3.0228	2.8237	−17.93***	2.7325	2.9367	18.76***
Medicare	0.5500	0.3086	−55.47***	0.5544	0.3128	−54.79***
Clean-toilet	0.5942	0.2464	−99.36***	0.5261	0.2560	−78.27***
Sanitation	0.9035	0.8168	−30.38***	0.9106	0.8082	−37.21***
Water-supply	0.8940	0.5697	−92.02***	0.78950	0.59790	−54.46***
Highroad	0.7425	0.5634	−47.36***	0.7405	0.5554	−50.51***
Bus-station	0.7080	0.5347	−45.72***	0.6572	0.5487	−29.15***

Notes The data is from the "China Health and Nutrition Survey" (CHNS) in 1989, 1991, 1993, 1997, 2000, 2004, 2006 and 2009. See (http://www.cpc.unc.edu/projects/china). The first two columns are the means of their responses. The third column is the differences between urban and rural residences' average responses, the significance of which are estimated under a t-test. Significance levels are denoted by * $p<0.1$, ** $p<0.05$, *** $p<0.01$

in seeking vertical intergovernmental transfers from the upper-level government. Villages can apply for two types of projects: the first is called "*XiaokangCun*" (granted to 'model villages', usually wealthy villages with good economic performance and administrative management based on pre-defined standards), and the second is called "*PinkunCun*" (an under-developed village). Villages are classified and the financial support or administrative priorities from the upper governments in providing public goods of all kinds.

The second heterogeneity is the difference between urban and rural areas, which are shown in Table 7.1. Public goods include primary schools, middle schools, high schools, hospitals, medical insurance, clean toilets, sanitation, water supply, paved roads, and bus stations. Respondents were asked whether they have public goods provision in their neighborhoods or villages. Excluding primary and middle school, urban residents had more public goods than rural residents.

The gap between urban and rural areas is exacerbated by many institutional barriers. The policy-induced segregation of rural–urban sectors associated with the *hukou* system, cadre system, work-unit system, and one-child policy accounts for the difference between public-goods provision between rural and urban areas (Yang and Cai 2003; Lu and Chen 2006). The "urban-bias policies" give fiscal priority to urban citizens (Yang 1999). Public goods in cities are funded and organized by municipal governments and supported by provincial governments. In contrast, village governments hold primary responsibility for the provision of public goods, such as roads within villages, sanitation, primary school facilities, irrigation works, trash disposal services, drainage systems, and community recreational facilities, since fiscal decentralization began in 1994 (Tsai 2007).Regarding separated health care systems between urban and rural areas, the gap in health care utilization and income between rural and urban sectors has grown since the 1990s (Y. Liu, Hsiao, and Eggleston 1999). This finding also correlates with intergenerational health inequality between urban and rural areas through 2009 (Eriksson, Pan, and Qin 2014), despite the introduction of New Cooperative Health Insurance in 2003.

The third heterogeneity is regional differences among provinces in different stages of economic

Table 7.2 Heterogeneous provision of public goods among regions

Public goods	Northeast China		Eastern China		Central China		Western China	
	Mean	Difference	Mean	Difference	Mean	Difference	Mean	Difference
Primary	0.664	14.61***	0.598	38.17***	0.754	−14.75***	0.798	−29.59***
Middle	0.329	−8.72***	0.158	41.56***	0.286	3.77***	0.382	−33.89***
Senior-high	0.151	−6.02***	0.070	25.52***	0.129	2.23**	0.172	−20.22***
Hospital	2.285	50.42***	3.175	−29.59***	2.622	39.64***	3.303	−55.73***
Medicare	0.434	−11.06***	0.567	−49.84***	0.324	20.92***	0.271	32.97***
Clean-toilet	0.301	10.24***	0.388	−14.30***	0.345	−1.20	0.326	6.15***
Sanitation	0.916	−24.68***	0.908	−27.34***	0.837	2.43**	0.756	41.87***
Water-supply	0.592	16.72***	0.667	−2.71***	0.611	21.90***	0.749	−34.28***
Highroad	0.619	−1.74*	0.629	−5.18***	0.619	−2.74***	0.589	8.86***
Bus-station	0.803	−54.07***	0.522	17.99***	0.478	46.61***	0.646	−23.77***

Notes The data is from the "China Health and Nutrition Survey" (CHNS) in 1989, 1991, 1993, 1997, 2000, 2004, 2006 and 2009. See (http://www.cpc.unc.edu/projects/china). According to the classification method adopted by the National Bureau of Statistics, the nine provinces in the survey are classified into four regions. Liaoning and Heilongjiang represent Northeast China. Jiangsu and Shandong represent Eastern China. Henan, Hubei and Hunan represent Central China. Guangxi and Huizhou represent Western China. The second, fourth, sixth and eighth columns are the difference between the region (North, Eastern, Central, and Western China, respectively) and the average of other regions. Significance levels are denoted by * $p<0.1$, ** $p<0.05$, *** $p<0.01$

development (in Table 7.2). The regional disparities in economic performance and local fiscal capacity originate from differentiating central policies for western, eastern, middle, and northeast China (Fan et al. 2011; Kanbur and Zhang 1999; Yang 1999; Yang and Cai 2003).[2] Correspondingly, Table 7.2 suggests that in the developed regions (Northeast and Eastern China), when compared to the developing regions (Middle and West China), the provision of public goods such as middle schools, high schools, medical care, and insurance, sanitation and public facilities are more comprehensively enjoyed by residents.

Considering this institutions-induced heterogeneity of public-goods provision at the local level, this study also intends to find the role of public goods at the local level in increasing intergenerational mobility. Most existing studies focused on one type of public goods and their impact on employment or income, such as public schools (Causa et al. 2009; Mayer and Lopoo 2008; Solon 2004), health care facilities and services (Frost, Forste, and Haas 2005; Chou et al. 2010), or public transportation (Sanchez 2007; Tomer 2012; Sanchez and Schweitzer 2008; Sanchez, Shen, and Peng 2004). However, the all-round human development that guarantees the individuals' comprehensive welfare cannot be separated from each other (Sen 2000).

Using intergenerational occupational mobility as an index over intergenerational income mobility has certain advantages. First, it would

[2]The preferential policies of the opening-up in the 1980s gave priority in terms of economic growth to the eastern regions. Years of rapid growth promised the sufficient public income from the local economy. However, the backwards western regions has also drawn the attention of the central government since 2000. From 2000 to 2012, the total inter-governmental fiscal transfer for western regions was 8.5 trillion yuan, and the national budget investment for them was 1 trillion yuan. Both account for 40% of the respective total amount. See http://www.npc.gov.cn/npc/xinwen/jdgz/2013-10/22/content_1810727.htm. The increment of public goods in Northwest is larger than in the other regions, but there is still a gap in the total amount. As for Northeast China, reviving policies towards the old industrial zones have been carried out since 2003. To catch up with the other regions, middle regions have gained the attention of the central government since 2006 and received fiscal transfers of 1.35 trillion yuan during the 11th "five-year plan."

more accurately represent the socioeconomic status than income could. Respondents are usually reluctant to report their true income (Moore and Welniak 2000). In contrast, asking about occupation is less sensitive. Second, apart from the monetary benefit, occupation also brings out often invisible social resources associated with positions (Bian et al. 2005; Zhou and Suhomlinova 2001). For example, regarding the civil servant positions in China, although some are low paid, they are highly sought-after jobs in Chinese society. Also, statistically, using occupation rather than income excludes endogenous effects from regional income discrepancies and the influence of possible monetary volatility.

The rest of the paper is organized as follows: Sect. 7.2 describes the data and evaluates the intergenerational occupational mobility from 1989–2009 based on the household data from the China Health and Nutrition Survey (CHNS); Sect. 7.3 provides the empirical model and results; Sect. 7.4 includes three robustness checks, in which population mobility and genetic factors are excluded. Also, Multinomial Logit Models (Mlogit) and Partial Proportional Odds Models (PPOM) are applied for comparison; Sect. 7.5 concludes.

7.2 Data

The data used in this paper comes from the "China Health and Nutrition Survey" (CHNS) in 1989, 1991, 1993, 1997, 2000, 2004, 2006, and 2009. CHNS includes nine representative provinces from different regions of China. This paper combines three subsets of the CHNS dataset ("Household Survey", "Adult Survey" and "Community Survey") to form paired samples of parents and children. Specifically, four types of observations are eliminated or reclassified: (1) the observations of respondents with post-retirement employment are eliminated, (2) respondents still in school and performing unpaid housework are dropped from the sample, (3) respondents still in school along with respondents who lost their jobs due to disability are also dropped from the sample, and 4) respondents older than 65 and younger than 30 are not considered in the estimation.[3]

7.2.1 Categorizations of Occupation and Intergenerational Mobility

In CHNS, occupations are categorized into 14 types. Bian et al. (2005) conclude that because of China's great social transition after 1978, the linear hierarchy of occupation-based class structure was changing dramatically. Therefore, the evaluation of occupation ranks based on a 14-class schema by assuming that the occupation rank is stable between 1989 and 2009 turns to be problematic.

Given the sensitivity of occupational mobility to how occupations are defined, previous studies created different metrics of occupational or prestige ranks and measured parent and child occupation continuously according to different standards (Black and Devereux 2011). To avoid this problem, this paper classifies these 14 types of occupations into two categories and employs a unique method to rank them.

As Table 7.3 demonstrates, in the first method, all occupations are categorized into three sub-types in terms of social reputation, socioeconomic power, and the security of expected income: advanced, middle, and lower employment. Advanced employment, consists of high paying jobs, come with social, economic, and political privileges, whereas, middle-level employment includes some of the middle level white and pink-collar jobs, and employment requiring physical labor falls under the low-level employment category.

[3] Individuals' occupational choice at the early stage of a career cannot precisely define their social status; thus, using the data in such situations in estimations can lead to biased results (Black and Devereux, 2011; Böhlmark and Lindquist, 2006). Mandatory age for retirement in Western countries is 65. To maintain comparability, this paper also adopts such a standard.

Table 7.3 Description of occupation

Variables	Descriptions
Advanced	No.01 senior technicians (doctor, professor, lawyer, architect, engineer, etc.), No.03 executive officer and manager (factory director, public official, administration cadre, village cadre, etc.), No. 08 military officer and police officer
Middle	No.02 technicians (midwife, nurse, teacher, editor, photographer, etc.), No.04 office worker (secretary and office clerk), No.06 skilled workers (section foreman and craftsmen), No.07 non-skilled workers (housekeeper, cook, doorman, barber, salesman, laundryman, child-care worker, etc.), No.12 athlete, actor and performer as manual work
Lower	No.05 peasant, fisherman, and hunter, No.13 unemployment, No.14 others

Note The description of occupations are categorized based on the economic and social hierarchy

Table 7.4 General situation of intergenerational occupational mobility

Parent's employment	Child's employment		
	Lower employment (%)	Middle employment (%)	Advanced employment (%)
Lower employment	73.50	25.12	1.38
Middle employment	28.56	68.00	3.44
Advanced employment	23.70	65.68	10.62
	Non-public employment (%)	Public employment (%)	
Non-public employment	87.75	12.25	
Public employment	36.00	64.00	

Note The sample of informal employment comes from the lower employment (low paying jobs requiring physical labor)

Given the influence of cadre-dominated social hierarchy which still exists in China (Bian, Breiger, Davis, and Galaskiewicz 2005), the second method re-categorizes higher and middle employment into public employment and non-public employment. Public employment includes positions in government, state-owned institutions, and state-owned enterprises. The rest represents non-public employment.

We use a 3 × 3 matrix to calculate the probabilities of occupational mobility between parents and children for the first category of employment. In this matrix, the diagonal elements are the probabilities of unchanged occupational status, which is denoted by "Mobility = 0"; the off-diagonal elements indicate intergenerational occupational mobility. The elements in the upper triangle of the matrix are the probabilities for upward mobility, which is denoted by Mobility > 0; similarly, the elements in the lower triangle are the probabilities of downward mobility, which is denoted by Mobility < 0. Based on the above calculations, the probabilities of intergenerational occupational mobility are shown in Table 7.5.

In Table 7.4, 25.12% of children whose parents were employed in lower employment, moved to middle employment, and 1.38% moved to advanced employment. As for workers who have parents employed at the middle level, 68% of the workers remained in the same occupational class as their parents. 3.44% of them moved to an advanced level. In contrast, 28.56% experienced downward mobility. On the other hand, only 12.25% of children coming from families employed in the non-public sector now

Table 7.5 Intergenerational occupational mobility based on employment

Mobility	Advanced/middle/lower employment (%)	Public/ nonpublic employment
Mobility = –2	1.85	–
Mobility = –1	11.52	7.77%
Mobility = 0	67.37	82.62%
Mobility = 1	18.30	9.61%
Mobility = 2	0.96	–

Note The results denote probalitlities of occupational mobility

Table 7.6 Intergenerational occupational mobility in urban and rural areas

Mobility (%)	Advanced/middle/ lower employment		Public/nonpublic employment	
	Urban	Rural	Urban	Rural
=0	58.55	70.42	76.52	84.76
>0	21.49	18.49	12.56	8.58

Note The calculations takes into account 3 time periods between – 1989–1993, 1997–2004 and 2006–2009

have public employment. Overall, these results imply fairly low upward intergenerational occupational mobility and quite strong class consolidation in China.

Table 7.5 demonstrates intergenerational occupational mobility based on employment, across time. In Table 7.5, intergenerational occupational mobility is indexed by the occupational distance between parents and children. In our calculation, the data of the ones whose occupations rank is in a higher category than their spouses are selected. For all occupations, the values of occupational distance range between –2 and 2. A value of –2 means that a child's occupation is lower than his parents' by 2 classes; it indicates downward intergenerational occupational mobility. A value of –1 means that when children are in the middle class, their parents are in the advanced class. A value of 0 indicates that children and parents have the same type of occupation. In sum, values of 2 and 1 indicate upward intergenerational occupational mobility; –2 and –1, downward intergenerational occupational mobility.

For public/non-public employment, the occupational distance varies between –1 and 1. Where, –1 denotes downward mobility as in the case of a child with non-public employment, while their parents have public employment. When the value equals 1, the situation is the opposite, denoting upward mobility. When it is 0, there is no intergenerational change. Table 7.6 reports the intergenerational occupational mobility in three periods (1989 ~ 1993, 1997 ~ 2004, 2006 ~ 2009). The situations of all occupations and the two categorizations of occupations are reported. Similarly, Table 7.6 demonstrates rural–urban differences in intergenerational occupational mobility.

These tables also imply four aspects of intergenerational occupational mobility in China. In Table 7.5, Intergenerational occupational solidification (i.e., when mobility = 0) is observed in all occupations (67.37% children have the same types of occupations as their parents) and is higher for public employment (82.62%).

Table 7.6 implies that intergenerational occupational solidification in rural areas is higher than in urban areas; and the upward mobility in rural areas is lower than in urban areas.

7.2.2 Public Goods

Public education is categorized as primary schools (*Primary*), Middle schools (*Middle*), and senior high schools (*Senior-high*).[4] The respondents are asked whether they have enrolled in one of these levels of school in their rural villages or urban neighborhood. Also,

[4]Although senior high school education is not compulsory in China's education system, we still consider it because access to senior high school education determines whether an individual has the opportunity to receive college education, thereby exerting a huge influence on his occupational choices in the future. Because most of the neighborhoods and villages do not have a college or university, we ignore college education in this paper.

the distance from the home to the nearest senior high school (*Distance-high*) is considered a proxy for the provision of public education.[5]

Health care services are measured in terms of the quantity of health care service organizations (*Hospital*) provided in neighborhoods and villages and the distance between homes and the nearest hospital (*Distance-hospital*).

Public medical insurance is measured indirectly. Although respondents aged 18 or older were all asked whether they are covered by medical insurance. In the questionnaires, medical insurance consists of five types: commercial medical insurance, free medical service for public servants, basic medical insurance for urban workers, basic medical insurance for non-worker urban residents, and new cooperative medical insurance for rural residents. Only the first one is paid privately and a minor proportion of respondents fall in this category. Thus, for simplification purposes, medical insurance (*Medicare*) is considered as a public good in this study.

The facility of public health is represented by the availability of clean toilets (*Clean-toilet*) and sanitation (*Sanitation*). Both variables are dummy variables. Also, the clean toilet variable does not represent the toilet system in individual homes. This variable is a proxy for the extensive network of public sewage systems which facilitates clean toilets.

Public facilities consist of three types in neighborhoods or villages: paved roads (*Highroad*), public transportation (*Bus-station*), and tap water (*Water-supply*).[6] Specifically, public transportation is measured as the existence of bus stations near the communities. All of them are dummy variables (i.e., if there is a paved road, public bus station, or tap water system, the datum is 1; if there is none, the datum is 0).

Detailed descriptions are shown in Table 7.7.

7.2.3 Control Variables

The control variables refer to the following three aspects: household characteristics, parents' characteristics, and children's characteristics. Specifically, household characteristics include two variables, institutional urban–rural discrepancies, and the household's economic situation. The former is represented by a dummy variable of residency status (*urban identity = 1, rural identity = 0*) called *Urban* and the latter by income per capita (*Hhincpc*), which is calculated in 2009 prices. Also, the fixed effect of provinces (*t1*) and years (*wave*) are controlled.

Some studies have verified the impact of that parents' education on their children's (Pronzato 2012) and parental life circle on intergenerational mobility (Ermisch and Francesconi 2002). Thus, controlling for the influence of parents, their age (*Age_famo*), education (*Education_famo*), and occupational class are taken into account.

According to Restuccia and Urrutia (2004), innate ability, early education, and college education are the keys to intergenerational human capital transmission. Thus, in this paper, children's age (*Age*), education (*Education*), and gender (*Gender*) fall into the categories of children's characteristics.

7.3 Empirics

7.3.1 Model

To estimate how the provision of public goods affects the upward intergenerational occupational mobility of individuals, we utilize the following model:

$$Mobility_{it} = \alpha_0 + \sum_{n=1}^{l} \beta_n Public_{nit} + \gamma_m' X' + \sigma_\eta + \varphi_t + \varepsilon_{it} \quad ,$$

[5]The distance to the nearest primary school and the distance to the nearest middle school are not suitable proxies. For one thing, primary and middle schools are planned in most neighborhoods and villages. For another, an abundance of respondents do not provide information on this.

[6]Most of households in China use electricity; thus electricity supply is not taken into analysis.

Table 7.7 Description of variables and summary statistics

Variables	Description	Mean	Std. dev	Obs
Occupation				
Advanced/middle/lower employments	3 for higher, 2 for middle, 1for lower	1.4794	0.6190	60,534
Public employment	2 for public employment, 1for not	1.2721	0.4450	60,836
Provision of public goods				
primary	Dummy- primary school in neighborhood/ village	0.7215	0.4483	73,871
Middle	Dummy- junior high school in urban neighbor-hood /rural village	0.2939	0.4556	73,967
Senior-high	Dummy- senior high school in urban neighborhood/rural village	0.1329	0.3395	74,127
Distance-high	Distance to nearest senior high school (in ten kilometers)	0.6812	1.2061	69,800
Hospital	Number of medical institutions nearby	2.8785	1.4015	79,611
Distance-hospital	Distance between the most often used medical institution and urban neighborhood/rural village	0.8636	3.1483	79,449
Medicare	Dummy- whether interviewee has health insurance	0.3824	0.4860	55,712
Clean-toilet	Dummy- whether flush toilet is available at home	0.3418	0.4743	82,398
Sanitation	Dummy- whether sanitary condition is good	0.8406	0.3660	82,053
Water-supply	Dummy- whether drinking water is available at home	0.6586	0.4742	82,450
Highroad	Dummy- whether high quality roads are in neighborhood/village	0.6121	0.4873	81,577
Bus-station	Dummy- whether bus stop is available in neighborhood/village	0.5820	0.4932	83,404
Control variables				
Gender	Dummy- whether interviewee is a man	0.5142	0.4998	81,739
Education	Years of schooling completed (in years)	7.4647	4.1689	60,072
Age	Years of age	43.76	9.225	52,238
Hhincpc	Annual household income per capita (ten thousands yuan)	0.5500	0.8315	82,485
Urban	Dummy- whether the investigation areas is urban	0.2736	0.4458	85,163

where the subscript ***I*** and ***t*** represent the *i-th* individuals and the *t-th* period, respectively. α_0 is the constant term. σ and φ capture the province-fixed effects and year-fixed effects, respectively. *Public* represents the public goods received by the individual. X' is a set of control variables.

Mobility is the individuals' upward intergenerational occupational mobility. It represents the difference between the ranks of individuals' occupation and their parents' occupations. *Mobility* is a dummy variable with two values: when there is upward intergenerational occupational mobility, it equals 1. Otherwise, it equals 0. Thus, downward mobility and unchanged status in occupation both fall into *"Mobility=0"*. Because the dependent variable *Mobility* is binary, we adopt logit and probit models to estimate the parameters.

7.3.2 Results

The empirical results in Table 7.8 prove the influence of public goods on individuals' intergenerational occupational upward mobility.

We employ logit and probit regressions on the model. In Models (1) and (3) of Table 7.8, only household characteristics, parents' characteristics, and children's characteristics are controlled. In Models (2) and (4), the dependent variables are also considered. Public employment is used as the dependent variable in models 5 through 8.

In general, nearly all the models show the positive significant impact of public goods provision on intergenerational mobility, but with some exceptions. Hospitals are not significant, but when it comes to distance, the nearer it is to the hospital, the higher the impact. It also works regarding middle schools. In contrast, primary schools demonstrate a negative impact. We assume it maybe because of the reform of small primary schools under remerging in the recent two decades.

The results of Models (1), (3), (5), and (7) report significant impacts of control variables when the fixed effects of province and year are controlled. First, the impacts of households are significant in terms of families' economic situation and their residency in urban areas. A lower income per capita of households implies a greater difference between children and their parents in terms of the type of occupation. Also, children from urban households have a greater chance to have different types of occupations than their counterparts from rural households. Specifically, the scales of the impacts varied when different classifications of occupations were compared. The smallest impact of income per capita of households is observed when all occupations are classified as public/non-public ones. The classification of public/non-public occupations has the largest impact on urban residency and classification of advanced/middle/lower occupations has the smallest impact.

Second, the education of parents is not a significant influential factor for changes in children's occupation, but the education of children is important for their social mobility. This encouraging finding implies that the aid for schooling from third parties could be effective in enhancing children's mobility regardless of parents' education. However, the age of parents influences children's occupational mobility positively, except with the classification of advanced/middle/lower occupations. It provides another research question: whether late marriage and late childbirth are beneficial for the social development of children since their parents are more ready for supporting them, or, the early type happens because their parents are in low social status.

Third, in terms of the individual characteristics of children, only children's education is significant, while their genders and ages are not. In Fan, Yi, and Zhang (2013), however, gender disparity in intergenerational mobility was obvious when the children were born before the 1980s. Our study indicates an improvement in the aspect of gender equality in recent decades. Models (2), (4), (6), and (8) estimate the influence of public goods. Compared with (1), (3), (5), and (7) respectively, the impact of the control variables is verified after adding the independent variables. In general, public goods concerning public education, health care services (except hospitals in neighborhoods and villages), medical insurance, facilities of public health, and some public facilities (excluding paved roads) significantly influence the intergenerational occupation mobility for both classifications of occupations.[7]

[7] Several special findings relevant to Chinese education are also obvious. First, the influence of primary school in neighborhoods or villages is negative, which is in contrast to the influence of other levels of public education. This finding means that the more residents have primary schools nearby, the less upward intergenerational mobility there is. One possible reason is that the average quality trades off total number of primary schools under the constraint of constant total public budget. When the number of schools decrease, the quality of education increase; therefore, children will attain greater benefits from primary education for their future career. This trend is obvious in rural areas. Unlike urban primary schools,

Table 7.8 Provision of public goods and intergenerational occupational mobility

Variable	Dep. var. Advanced/middle/lower employment				Dep. var.Public/non-public employment			
	(1) Logit	(2) Logit	(3) Probit	(4) Probit	(5) Logit	(6) Logit	(7) Probit	(8) Probit
Primary		−0.288*** (−3.98)		−0.169*** (−4.07)		−0.700*** (−3.88)		−0.393*** (−3.81)
Middle		0.273*** (3.68)		0.160*** (3.76)		0.715*** (3.47)		0.405*** (3.39)
Senior-high		0.204** (2.09)		0.123** (2.20)		−0.0997 (−0.38)		−0.0478 (−0.31)
Distance-high		−0.0461** (−2.36)		−0.0242** (−2.31)		−0.202** (−2.12)		−0.114** (−2.06)
Hospital		0.0169 (0.78)		0.0126 (1.01)		0.0947 (1.41)		0.0549 (1.41)
Distance-hosp		−0.0348*** (−3.69)		−0.0192*** (−3.96)		−0.0520** (−2.11)		−0.0246** (−2.35)
Medicare		1.110*** (13.68)		0.644*** (13.86)		0.497*** (2.73)		0.299*** (2.84)
Clean-toilet		0.381*** (4.93)		0.225*** (5.04)		0.051 (0.29)		0.0353 (0.35)
Sanitation		0.292*** (3.88)		0.180*** (4.20)		0.320* (1.66)		0.179 (1.62)
Water-supply		0.1380** (2.14)		0.0886** (2.38)		0.0833 (0.50)		0.0354 (0.37)
Highroad		0.0907 (1.47)		0.050 (1.41)		0.0394 (0.24)		0.0339 (0.37)
Bus-station		0.1330** (2.05)		0.0655* (1.77)		0.302* (1.94)		0.164* (1.85)
Gender	0.0307 (0.64)	0.0298 (0.53)	0.0339 (1.24)	0.0261 (0.81)	0.0799 (0.66)	0.0171 (0.13)	0.0554 (0.79)	0.0215 (0.28)
Education	0.211*** (24.08)	0.195*** (18.56)	0.123*** (25.16)	0.112*** (18.98)	0.273*** (11.51)	0.249*** (9.44)	0.155*** (11.71)	0.143*** (9.59)
Age	−0.0018 (−0.52)	−0.0052 (−1.21)	0.0015 (0.79)	−0.0014 (−0.58)	−0.0395*** (−3.18)	−0.0422*** (−3.08)	−0.0221*** (−3.14)	−0.0239*** (−3.05)
Education_famo	0.0087 (1.14)	0.0045 (0.49)	0.0052 (1.19)	0.0022 (0.41)	0.0246 (1.29)	0.0189 (0.87)	0.0128 (1.13)	0.0094 (0.74)
Age_famo	0.0091*** (2.93)	0.0020 (−0.53)	0.0040** (2.37)	−0.0001 (−0.05)	0.0274** (2.51)	0.0253** (2.10)	0.0153** (2.47)	0.0142** (2.07)
Hhincpc	0.576*** (11.4)	0.577*** (9.82)	0.269*** (11.39)	0.297*** (9.66)	0.313*** (4.89)	0.456*** (4.90)	0.190*** (5.63)	0.264*** (5.36)
Urban	0.448*** (7.81)	0.194*** (2.69)	0.263*** (7.99)	0.109*** (2.62)	0.609*** (3.51)	0.410** (2.05)	0.368*** (3.64)	0.248** (2.17)
_cons	−3.524*** (−16.81)	−4.062*** (−13.76)	−2.049*** (−17.21)	−2.334*** (−13.96)	−3.843*** (−6.33)	−3.770*** (−4.94)	−2.218*** (−6.31)	−2.200*** (−5.01)
Parents' occupational class	Yes	Yes	Yes	Yes	Yes	Yes	Yes	Yes
Province FE	Yes	Yes	Yes	Yes	Yes	Yes	Yes	Yes
Year FE	Yes	Yes	Yes	Yes	Yes	Yes	Yes	Yes
Pseudo R2	0.2101	0.2493	0.2075	0.2477	0.1842	0.2103	0.1842	0.2103

(continued)

Table 7.8 (continued)

Variable	Dep. var. Advanced/middle/lower employment				Dep. var.Public/non-public employment			
	(1) Logit	(2) Logit	(3) Probit	(4) Probit	(5) Logit	(6) Logit	(7) Probit	(8) Probit
LR/Wald chi2()	2999.03***	2710.07***	2961.93***	2692.51***	398.62***	384.22***	398.65***	384.19***
Log likelihood	−5636.37	−4079.28	−5654.92	−4088.05	−882.71	−721.33	−882.69	−721.34
Observations	13,654	10,085	13,654	10,085	1881	1599	1881	1599

Note Significance levels are denoted by * p<0.1, ** p<0.05, *** p<0.01

Fourth, comparing models (1) and (5); (2) and (6); (3) and (7); and (4) and (8), we find that the coefficients are larger in the public/non-public occupation model, which means public goods play a significant role in the attainment of a public job.

7.4 Identification and Robustness Checks

7.4.1 Population Mobility

The underlying assumption in the above analysis is that residents do not move to other places to seek out better public goods. It omits situations where people's preferences on residences are associated with population mobility across the communities. For example, some public goods, such as schools, are not equally distributed in cities. The prices of real estate are related to the quality of public goods nearby. To have access to good schools for children before high school, some well-off parents choose to buy certain school district housing. After children go to college, parents can move out or sell the school district housing. Households who can afford the high-price houses in good school districts may also be able to afford good education or provide other financial support for their children's successful career. Thus, not only public goods but also households' finances contribute to the accumulation of children's human capital and, therefore, influence their employment in the future.

China has seen unprecedented rural-to-urban migration in the last 20 years. Although rural migrants live in cities, they are still not completely included in the urban social welfare system. Children of migrants do not necessarily get to attend nearby schools which may be part of their school district. Instead, they go to special schools for migrants in cities. This situation potentially introduces the problem of endogeneity. To detect and eliminate endogeneity, a small sample including only rural residents is selected in our robustness check (see Table 7.9).

In rural areas, the unique pattern of public funding poses another problem. Public goods such as health insurance, sanitation, tap water systems, and roads are jointly financed by residents, the village committee, and local governments (township or prefectural government). The influence of households should not be neglected in estimating the impact of public goods. Thus, the second robustness check concerns only schools, medical institutions, and public transportation services. Models (2) and (4)in Table 7.9 address such concerns.

which are supported by municipal finance, rural primary schools are financially supported by prefectural governments. Given the delegation of financial responsibility and centralized fiscal rights in Chinese fiscal decentralization (Qian and Weingast, 1996), prefectural governments have limited budgets for compulsory education. For the sake of improving the quality of primary schools under financial strain, joining several primary schools scattered in neighboring villages is one solution. Such a trend began in the 1990s, and official reform was initiated in 2001, which falls within the period of our sample. Because the rural sample and urban sample are not separated here, such an assertion should be argued when only a rural sample is used (The results are illustrated in Table 7.9). In addition, in regard to the situation of senior high schools, the result is less significant than those of other two types of public education. In fact, it is insignificant when the occupations are classified into public and non-public.

Table 7.9 Robustness check: population mobility

Variable	Dep. var. Advanced/middle/lower employment		Dep. var.Public/non-public employment	
	(1)	(2)	(3)	(4)
Primary	–0.385*** (–4.37)	–0.267*** (–3.91)	–0.805*** (–3.94)	–0.618*** (-3.52)
Middle	0.171* (1.94)	0.337*** (4.81)	0.772*** (3.28)	0.647*** (3.26)
Senior-high	0.155 (1.32)	0.198** (2.13)	0.0426 (0.15)	0.114 (0.45)
Distance-high	–0.0252 (–1.27)	–0.0705*** (–3.53)	–0.143 (–1.33)	–0.198** (–2.08)
Hospital	0.0424 (1.55)	0.0013 (0.06)	0.0844 (1.11)	0.0876 (1.32)
Distance-hos	–0.0377*** (–3.36)	–0.0305*** (–3.49)	–0.0545** (–2.19)	–0.0537** (–1.99)
Medicare	0.854*** (8.23)		0.413** (1.99)	
Clean-toilet	0.455*** (4.87)		0.118 (0.60)	
Sanitation	0.214** (2.53)		0.422** (1.97)	
Water-supply	0.0304 (0.42)		0.0948 (0.53)	
Highroad	0.0374 (0.52)		0.256 (1.35)	
Bus-station	0.235*** (3.15)	0.185*** (3.08)	0.495*** (2.82)	0.315** (2.10)
_cons	–4.090*** (–11.90)	–2.983*** (–11.12)	–4.631*** (–5.53)	–3.667*** (–5.11)
Control variables	Yes	Yes	Yes	Yes
Parents' occupational class	Yes	Yes	Yes	Yes
Province FE	Yes	Yes	Yes	Yes
Year FE	Yes	Yes	Yes	Yes
Pseudo R2	0.2371	0.2136	0.2141	0.2051
LR/Wald chi2()	1896.85***	2465.53***	327.98***	377.24***
Log likelihood	-3052.08	–4537.86	–601.96	–730.82
Observations	7748	10,642	1384	1605

Note. Significance levels are denoted by * $p<0.1$, ** $p<0.05$, *** $p<0.01$

The impact of most public goods shown in Table 7.9 corresponds to the results in Table 7.8, except for some differences in the sample of rural residents. In Models (1) and (3), the impacts of middle and high school provisions, health care service provisions, water supply provisions, and roads are insignificant in the sample of only rural residents. The coefficient of clean toilets is insignificant in Model (3). These results demonstrate that rural residents did not benefit from the provision of public goods in terms of upward intergenerational mobility. Apart from that, the influence of primary schools is still negative, and the absolute values of the coefficients are larger than in Table 7.8, which verifies the assertion about primary education above. Also, the fixed effect of year and provinces are controlled in Table 7.8. It's extremely hard to control public goods provisions for mobile populations. To counter this problem, we take into account if public goods in the former residences of parents are different from those in the places where their children were born or grew up. There are intergenerational differences in time and location. However, thanks to the small percentage of interprovincial mobility in the survey data (2.09% of total households), this problem is trivial. Moreover, due to the dual social welfare system between urban and rural areas, although rural migrant workers account for a large percentage of provincial migration, rural migrant workers' young children rarely live in cities with parents or enjoy the same health services and education as urban residents (Hu, Li, and Liu 2011; Antman 2012; Lee 2011; Ye and Lu 2011; Y. Lu 2012). Thus, this proxy we adopt is still effective.

7.4.2 Alternative Methods and Assignment

In the initial design, *Mobility* has three values,[8] but the latter two situations are merged into one in the logit and probit model in Table 7.8. Merging information about downward mobility and unchanged cases may lead to unreliable results. Therefore, we employ Mlogit in the robustness test and categorize mobility into three categories. In the Mlogit model, dependent variables are not treated in a sequence according to the value. Thus, another alternative method is introduced, and Mlogit models serve as the control group. An Ologit model treats dependent variables as a sequence. As an alternative method, PPOM could solve both problems (Peterson and Harrell 1990; Lall et al. 2002; Williams 2006).

Table 7.10 reports the estimation of Mlogit and PPOM. Applying the Mlogit method, Models (1) and (3) demonstrate that upward intergenerational mobility is more visible when people have access to high schools, medical institutions, public health services, health insurance, and public facilities such as public health and transportation. Besides, a shorter distance from home to high school and medical institutions means more upward intergenerational mobility. In contrast, downward intergenerational mobility more easily occurs when people do not have such public goods nearby. In contrast, Models (2) and (4) employ the sequencing PPOM estimation method. The results are the same as with Models (1) and (3).

7.4.3 Genetic Influence

Genetic factors such as IQ and physical qualities are important for individuals' career development. Neglecting such factors would exaggerate the influence of public goods. According to Plug (2004), Björklund, Lindahl, and Plug (2006), and H. Liu and Zeng (2009), using paired data between adoptive parents and children eliminates disturbances from genetic factors. Therefore, we select the paired data between parents-in-law and children-in-law as well as between adoptive parents and children from the CHNS database in the third robustness check. Table 7.11 suggests that excluding the effect of high school, sanitation, and public transportation services, the results correspond to the original estimation.

In summary, through three robustness checks, the original conclusion from Table 7.8 is verified. Most public goods are conducive to upward intergenerational mobility.

7.5 Evaluation and Discussion

To understand the influence of heterogeneous provision on intergenerational occupational mobility, we use three methods: nearest matching method, half-matching method, and kernel matching method to address this question.

[8] *Mobility* >0: upward mobile; *Mobility* <0: downward mobility, *Mobility* $=0$: no mobility.

Table 7.10 Robustness check: Mlogit and PPOM

Variable	Dep. var. Advanced/middle/lower employment				Dep. var.Public/non-public employment			
	(1) Mlogit		(2) PPOM		(3) Mlogit		(4) PPOM	
	Mobility=–1	*Mobility=1*	*Mobility=–1*	*Mobility=0*	*Mobility=–1*	*Mobility=1*	*Mobility=–1*	*Mobility=0*
Primary	–0.005 (–0.04)	–0.287*** (–3.96)	–0.000 (–0.23)	0.318*** (–4.34)	0.169 (1.23)	–0.160 (–1.61)	–0.160** (–2.00)	–0.160** (–2.00)
Middle	0.342*** (2.67)	0.276*** (3.72)	–0.402*** (–3.72)	0.292*** (4.06)	0.403** (2.54)	0.426*** (4.25)	–0.432*** (–3.14)	0.447*** (4.74)
Senior-high	–0.018 (–0.11)	0.203** (2.09)	0.151* (1.82)	0.151* (1.82)	–0.059 (0.30)	0.067 (0.51)	0.064 (0.59)	0.064 (0.59)
Distance-high	0.064 (1.09)	–0.046** (–2.36)	–0.048*** (–2.62)	–0.048*** (–2.62)	0.021 (0.19)	0.030 (1.09)	0.027 (0.96)	0.027 (0.96)
Hospital	–0.077** (–2.05)	0.016 (0.75)	0.033* (1.74)	0.033* (1.74)	0.033 (0.73)	0.052* (1.76)	–0.049 (–1.16)	0.063** (2.16)
Distance-hosp	–0.008 (–0.43)	–0.035*** (–3.70)	0.011 (0.65)	–0.036*** (–3.81)	0.029 (0.96)	–0.051*** (–3.16)	–0.047*** (–3.46)	–0.047*** (–3.46)
Medicare	–1.025*** (–8.29)	1.095*** (13.49)	1.147*** (16.58)	1.147*** (16.58)	–2.018*** (14.42)	2.168*** (20.79)	2.115*** (25.54)	2.115*** (25.54)
Clean-toilet	–0.583*** (–5.01)	0.375*** (4.84)	0.486*** (7.47)	0.486*** (7.47)	–0.264* (–1.83)	0.410*** (3.96)	0.363*** (4.36)	0.363*** (4.36)
Sanitation	–0.335* (–1.92)	2.292*** (3.89)	0.276*** (4.03)	0.276*** (4.03)	–0.286 (–1.30)	0.289*** (2.65)	0.313*** (3.23)	0.313*** (3.23)
Water-supply	–0.242* (–1.87)	0.139** (2.15)	0.148** (2.56)	0.148** (2.56)	–0.257 (–1.54)	0.156* (1.70)	0.193** (2.41)	0.193** (2.41)
Highroad	–0.198* (–1.74)	0.089 (1.45)	0.121** (2.22)	0.121** (2.22)	–0.242* (–1.70)	0.111 (1.29)	0.150** (2.06)	0.150** (2.06)
Bus-station	–0.106 (–0.94)	0.132** (2.04)	0.128** (2.26)	0.128** (2.26)	0.159 (1.15)	0.142 (1.56)	0.058 (0.76)	0.058 (0.76)
_cons	–17.674 (–0.03)	–4.006*** (–13.56)	1.932*** (4.54)	–3.451*** (–12.77)	–15.848 (–0.02)	–8.943*** (–18.81)	15.986 (0.02)	–7.663*** (–17.84)
Control variables	Yes	Yes	Yes	Yes	Yes	Yes	Yes	Yes
Parents' occupational class	Yes	Yes	Yes	Yes	Yes	Yes	Yes	Yes
Province FE	Yes	Yes	Yes	Yes	Yes	Yes	Yes	Yes
Year FE	Yes	Yes	Yes	Yes	Yes	Yes	Yes	Yes
Pseudo R2	0.4272	0.4201	0.3726	0.3711				
LR/Wald chi2()	8159.40***	3090.13***	6276.98***	1734.86***				
Log likelihood	–5470.45	–5538.45	–5284.59	–5297.34				
Observations	10,929	10,929	10,929	10,929				

Note Significance levels are denoted by * $p<0.1$, ** $p<0.05$, *** $p<0.01$

Estimates in Table 7.12 imply that intergenerational mobility among advanced/middle/lower employment, excluding primary school, which is insignificant; middle school and high school in neighborhoods and villages increases intergenerational occupational mobility. The intergenerational occupational mobility of those with middle schools in their neighborhoods and villages is 5% higher than that of those with none around their communities. It is 8% higher in the

Table 7.11 Robustness check: genetic influence

Variable	Dep. var. Advanced/middle/lower employment			Dep. var.Public/non-public employment		
	(1) Logit	(2) PPOM		(3) Logit	(4) PPOM	
		Mobility = −1	Mobility = 0		Mobility = −1	Mobility = 0
Primary	–0.5377*** (–3.48)	–0.4656*** (–3.34)	–0.4656*** (–3.34)	–0.4886* (–1.79)	–0.3355 (–1.08)	–0.3355 (–1.08)
Middle	0.5430*** (3.31)	–0.2743 (–1.03)	0.4753*** (2.99)	0.6447** (2.05)	–1.5440** (–2.26)	0.4239 (1.16)
Senior–high	0.4060* (1.92)	0.4621** (2.51)	0.4621** (2.51)	0.3678 (0.97)	0.4914 (1.14)	0.4914 (1.14)
Distance-high	–0.0841 (–1.50)	–0.0086* (–1.70)	–0.0086* (–1.70)	–0.2350 (–1.50)	–1.5469*** (–3.24)	–0.0134 (–1.07)
Hospital	0.0184 (0.39)	0.0112 (0.26)	0.0112 (0.26)	0.2545*** (2.60)	–0.5453** (–2.44)	0.2144* (1.83)
Distance-hospital	–0.0203 (–1.51)	–0.0102 (–0.87)	–0.0102 (–0.87)	–0.2025* (–1.89)	–0.0729 (–1.44)	–0.0729 (–1.44)
Medicare	1.1899*** (7.38)	1.1470*** (7.92)	1.1470*** (7.92)	0.0469 (0.18)	0.5861** (2.07)	0.5861** (2.07)
Clean-toilet	0.5563*** (3.62)	0.6408*** (4.74)	0.6408*** (4.74)	0.3159 (1.14)	1.4797** (2.22)	0.0580 (1.19)
Sanitation	0.3923** (2.26)	0.3794** (2.42)	0.3794** (2.42)	0.2272 (0.71)	0.2695 (0.80)	0.2695 (0.80)
Water-supply	0.2364* (1.66)	0.3328** (2.54)	0.3328** (2.54)	0.3362 (1.25)	0.2177 (0.78)	0.2177 (0.78)
Highroad	0.2921** (2.12)	0.3468*** (2.80)	0.3468*** (2.80)	0.2798 (1.07)	0.0936 (0.34)	0.0936 (0.34)
Bus-station	0.2019 (1.42)	0.2574** (2.04)	0.2574** (2.04)	0.3606 (1.51)	0.5373** (2.05)	0.5373** (2.05)
_cons	–5.1007*** (–7.70)	4.6850*** (7.21)	–4.7340*** (–7.98)	–3.6926*** (–2.95)	15.0004 (0.04)	–5.0951*** (–3.72)
Control variables	Yes	Yes	Yes	Yes	Yes	Yes
Parents' occupational class	Yes	Yes	Yes	Yes	Yes	Yes
Province FE	Yes	Yes	Yes	Yes	Yes	Yes
Year FE	Yes	Yes	Yes	Yes	Yes	Yes
Pseudo R2	0.3190	0.4610	0.1198			
LR/Wald chi2()	843.77***	762.41***	366.46***			
Log likelihood	–900.46	–1140.86	–1346.76			
Observations	2547	2681	2681			

Note Significance levels are denoted by * $p<0.1$, ** $p<0.05$, *** $p<0.01$

case of "high school". Similarly, "medical insurance" is 31%, "clean toilets" is 12%, "sanitation" is 6%, "water supply" is 10.4%, "roads" is 7.4%, and "bus stations" is 6.2%. As for the intergenerational mobility between public/non-public employment, the influence of public goods is the same. On average, the intergenerational occupational mobility of those with public goods in their neighborhoods and villages is between 21% and 30.4% higher than those without.

Table 7.12 Impact of the heterogeneous provision of public goods on intergenerational occupational mobility

Propensity score matching	k-Nearest neighbors matching		Radius matching		Kernel matching		Average effect
	ATT	T value	ATT	T value	ATT	T value	
Advanced/middle/lower employment							
Primary	0.0196	0.57	–0.0285	–1.44	0.0110	0.59	—
Middle	0.0525***	3.08	0.0648***	4.03	0.0400***	2.81	0.0524
Senior-high	0.0963***	3.90	0.0904***	4.04	0.0623***	3.08	0.0830
Medicare	0.3140***	10.72	0.3283***	13.01	0.3008***	13.86	0.3144
Clean-toilet	0.1010***	3.51	0.1378***	6.46	0.1273***	6.35	0.1220
Sanitation	0.0559***	2.61	0.0755***	4.59	0.0518***	3.24	0.0611
Water-supply	0.1024***	4.22	0.1146***	6.94	0.0951***	5.77	0.1040
Highroad	0.0790***	4.63	0.0638***	4.35	0.0793***	5.64	0.0740
Bus-station	0.0699***	4.42	0.0529***	3.57	0.0631***	4.91	0.0620
Public/Non-public employment							
Primary	0.00093	0.05	–0.0084	–0.69	–0.00017	–0.02	—
Middle	0.0285***	2.68	0.0333***	3.31	0.0279***	3.09	0.0299
Senior-high	0.0578***	3.71	0.0488**	3.47	0.0490***	3.86	0.0519
Medicare	0.2925***	15.57	0.3325***	19.89	0.2881***	21.39	0.3044
Clean-toilet	0.0959***	5.40	0.1011***	7.46	0.0991***	7.75	0.0987
Sanitation	0.0605***	4.71	0.0590***	5.63	0.0560***	5.65	0.0585
Water-supply	0.0632***	4.46	0.0724***	6.92	0.0822***	7.80	0.0726
Highroad	0.0741***	7.02	0.0668***	7.12	0.0672***	7.52	0.0694
Bus-station	0.0256***	2.59	0.0140	1.52	0.0230***	2.79	0.0209

Notes In the k-nearest neighbors matching, k is set to 2. In the radius matching, caliper is set to 0.0001. Significance levels are denoted by * $p<0.1$, ** $p<0.05$, *** $p<0.01$

7.6 Conclusion

This study successfully argues that access to public goods within the local community is beneficial for increasing intergenerational occupational mobility among the next generation. At the macro level, equitable distribution of public goods is extremely important to reduce or eliminate rural–urban gaps which can be achieved through intergovernmental transfer or direct subsidies. However, at the same time, it is a must to enlarge the coverage of public-goods beneficiaries at the micro (i.e. household/neighborhood) level. In the stage of transition and rapid urbanization, China's provision of public goods is still unbalanced across neighborhoods and villages.

This study explores the relationship between high expenditures and low intergenerational occupational mobility. Categorizing occupations in two parts, we observe the existence of occupation-based class solidification and low upward intergenerational occupational mobility in China. Considering the heterogeneous provision of public goods by communities, rural–urban differences, and regional differences, we find that households experience higher upward intergenerational occupational mobility if they have access to middle and high school, services and facilities of public health, and public facilities in their communities and have health insurance. These results are also verified by three extensive robustness checks. Among advanced, middle, and lower employment, the intergenerational mobility of those with public goods including schools, medical insurance, clean toilets, sanitation, water supply, roads, and bus station in their neighborhood or villages is higher than that of those with none. Similarly, between public/non-public employments, the intergenerational occupational mobility of those with public goods in their neighborhoods and villages is 2.1–30.4% higher than those without.

This study does not take into account the quality of public goods. Having access to a school facilitates intergenerational occupation mobility, however, having access to a very good school will have a greater effect on intergenerational occupation mobility. CHNS data does not allow us to examine this hypothesis and we believe this should be further examined.

References

Antman FM (2012) Gender, educational attainment, and the impact of parental migration on children left behind. J Popul Econ 25(4):1187–1214

Bian Y (2002) Chinese social stratification and social mobility. Ann Rev Sociol 28(1):91–116

Bian Y, Breiger R, Davis D, Galaskiewicz J (2005) Occupation, class, and social networks in Urban China. Soc Forces 83(4):1443–1468

Björklund A, Lindahl M, Plug E (2006) The origins of intergenerational associations: lessons from Swedish adoption data. Q J Econ 121(3):999–1028

Black SE, Devereux PJ (2011) Recent Developments in Intergenerational Mobility, in Handbook of Labor Economics (Volume 4, Part B) edited by Ashenfelter, O., and D. Card, 1487–1541

Böhlmark A, Lindquist MJ (2006) Life-cycle variations in the association between current and lifetime income: replication and extension for Sweden. J Law Econ 24(4):879–896

Causa O, Dantan S, Johansson Å (2009) Intergenerational social mobility in European OECD countries. OECD Economics Department Working Papers No. 709

Chou S, Jintan L, Grossman M, Joyce T (2010) Parental education and child health: evidence from a natural experiment in Taiwan. Am Econ J Appl Econ 2(1):63–91

Dollar D (2013) The Chinese dream: will china become a land of opportunity? Social mobility memo, Brookings institute. 2013. Retrieved December 2013 http://www.brookings.edu/blogs/social-mobility-memos/posts/2013/12/20-will-china-become-land-of-opportunity-dollar

Eriksson T, Pan J, Qin X (2014) The intergenerational inequality of health in China. China Econ Rev 31:392–409

Ermisch J, Francesconi M (2002) Intergenerational mobility in Britain: new evidence from the BHPS. *Center for Household, Income, Labour, and Demographic Economics Working Paper* No. 06

Fan S, Kanbur R, Zhang X (2011) China's regional disparities: experience and policy. Rev Devel Fin 1(1):47–56

Fan Y, Yi J, Zhang J (2013) Intergenerational mobility in China: patterns and determinants. Working paper in Conference on the Studies of Inequality in China, Beijing, China. Retrieved June 2013 http://www.oal.cuhk.edu.hk/files/wun/1._Session_1_keynote_-_Junsen_Zhang.pdf

Frost MB, Forste R, Haas DW (2005) Maternal education and child nutritional status in Bolivia: finding the links. Soc Sci Med 60(2):395–407

Henderson JJ, Yi J, Human capital, economic growth, and inequality in China." Forschungsinstitut zur Zukunft der Arbeit Institute for the Study of Labor Discussion Paper No. 6550

Hu A, Li D, Liu M (2011) Research on influencing factors of welfare level about left-behind children in china rural areas. J Agric Sci 3(2):233–229

Kanbur R, Zhang X (1999) Which regional inequality? The evolution of rural-urban and inland–coastal inequality in China from 1983 to 1995. J Comp Econ 27(4):686–701

Lall R, Campbell MJ, Walters SJ, Morgan K (2002) A review of ordinal regression models applied on health-related quality of life assessments. Stat Methods Med Res 11(1):49–67

Lee M-H (2011) Migration and children's welfare in China: the schooling and health of children left behind. J Devel Areas 44(2):165–182

Liu H, Zeng J (2009) Genetic ability and intergenerational earnings mobility. J Popul Econ 22(1):75–95

Liu Y, Hsiao WC, Eggleston K (1999) Equity in health and health care: the Chinese experience. Soc Sci Med 49(10):1349–1356

Lu M, Chen Z (2006) Urbanization, urban-biased policies, and urban-rural inequality in China, 1987–2001. Chin Econ 39(3):42–63

Lu Y (2012) Education of children left behind in rural China. J Marriage Fam 74(2):328–341

Mayer SE, Lopoo LM (2008) Government spending and intergenerational mobility. J Public Econ 92(1–2):139–158

Moore JC, Welniak EJ (2000) Income measurement error in surveys: a review. J off Stat 16(4):331

OECD (2010) A family affair: intergenerational social mobility across OECD countries. Chapter 5, *Economic Policy Reforms Going for Growth*. http://www.oecd.org/eco/public-finance/chapter%205%20gfg%202010.pdf

Peterson B, Jr. Harrell FE (1990) Partial proportional odds models for ordinal response variables. J Royal Stat Soc Series C (Appl Stat) 39(2):205–217

Plug E (2004) Estimating the effect of mother's schooling on children's estimating schooling using a sample of adoptees. Am Econ Rev 94(1):358–368

Pronzato C (2012) An examination of paternal and maternal intergenerational transmission of schooling. J Popul Econ 25(2):591–608

Qian Y, Weingast BR (1996) China's transition to markets: market-preserving federalism, Chinese style. J Pol Ref 1(2):149–185

Qu J (2012) The project system: a new form of state governance. Soc Sci China 33(4):28–47

Restuccia D, Urrutia C (2004) Intergenerational persistence of earnings: the role of early and college education. Am Econ Rev 94(5):1354–1378

Sanchez TW (2007) The connection between public transit and employment. J Am Plan Assoc 65(3):284–296

Sanchez TW, Schweitzer L (2008) Assessing federal employment accessibility policy: an analysis of the JARC program. Metropolitan Policy Program at Brookings Institute. Retrieved Feb. 2008 http://www.brookings.edu/research/reports/2008/02/21-transportation-sanchez

Sanchez TW, Shen Q, Peng Z (2004) Transit mobility, jobs access and low-income labour participation in US metropolitan areas. Urban Stud 41(7):1313–1331

Sen A, Development as Freedom, Oxford New York: Oxford University Press

Solon G (1992) Intergenerational income mobility in the United States. Am Econ Rev 82(3):393–408

Tomer A (2012) Where the jobs are: employer access to labor by transit. Metropolitan Infrastructure Initiative Series and Metropolitan Opportunity Series No. 32. Retrieved July 2012 http://www.brookings.edu/research/papers/2012/07/11-transit-jobs-tomer

Tsai LL (2007) Accountability without democracy: solidary group and public goods provision in rural China. Cambridge University Press, Oxford

Williams R (2006) Generalized ordered logit/partial proportional odds models for ordinal dependent variables. Stata J 6(1):58–82

Yang DT (1999) Urban-biased policies and rising income inequality in China. Am Econ Rev 89(2):306–310

Yang DT, Cai F (2003) The political economy of China's rural-urban divide: how far across the river? in *Chinese Policy Reform at the Millennium*, edited by Hope NC, Yang DT, Li MY. Redwood City, CA: Stanford University Press, pp 389–395

Ye J, Pan L (2011) Differentiated childhoods: impacts of rural labor migration on left-behind children in China. J Peas Stud 38(2), Routledge:355–377

Zhang X, Fan S, Zhang L, Huang J (2004) Local governance and public goods provision in rural China. J Public Econ 88(12):2857–2871

Zhou X, Suhomlinova O (2001) Redistribution under state socialism: a USSR and PRC comparison. Res Soc Strat Mob 18:163–204

Rescuing the Sinking City: A Strategy to Create Sustainable Megacity Jakarta

8

Christopher Silver

Abstract

Drawing upon the historical framework in *Urban Flood Risk Management: Looking at Jakarta* (2022), coupled with new research on current flood mitigation strategies in Jakarta, it explores the implications of various approaches tried but none wholly successful to effectively deal with flooding and overall water management. Kusno's recent study, *Jakarta: City of a Thousand Dimensions* (2023), identifies a variety of initiatives proposed or tested to address flooding over the past three decades reflecting the views of experts, politicians, community advocates and development interests. Their conflicting positions have not worked and have engendered a paralysis of action. Simultaneous application of multiple mitigation and adaptation strategies aimed at reducing overreliance on subsurface water and addressing surface water pollution can create a path forward to a healthier and more equitable urban environment.

C. Silver (✉)
University of Florida, Gainesville, USA
e-mail: silver2@ufl.edu

8.1 The Challenge of Sinking Cities

There is a whole array of major cities in the world currently confronted with the prospects of being inundated by seawater within the not-too-distant future. These are commonly referred to as "sinking cities," a label that acknowledges that the climate change crisis is causing the seas to rise but also that the effects of this are compounded by a simultaneous process of urban lands that are continuously sinking. Higher water on the coasts coupled with dropping landscapes creates the flooding conditions. One might assume that the sinking is largely the product of the weight of the urban substructure pushing the land down, and that does contribute in some places. Yet as excavations of historic sites show, most cities in the process of growing actually raise their ground level as streets and other surface projects actually add to its elevation. The real culprit in almost every sinking city is over-extraction of subsurface water to meet the needs of the continuously growing population and service needs of those in the city and causing the deep aquifer to collapse.

There are no fewer than ten major cities in the world with significant sinking conditions, including Lagos, Ho Chi Minh City, Bangkok, Miami, Amsterdam, Alexandria, New Orleans, Dhaka, Houston, Venice and Jakarta. Indonesia's

A. Alakshendra et al. (eds.), *Development, Inclusion and Sustainability*,
Sustainable Development Goals Series, https://doi.org/10.1007/978-981-97-6863-9_8

capital city of Jakarta is universally accepted as number one among this challenged group. **1** The complexities of Jakarta's problems with sinking and flooding puts it in a class of its own given that it is the centerpiece of the world's second most populous megaregion and the rate at which it is sinking dwarfs all of the others in this group by a substantial margin. One of the many essays documenting Jakarta's critical conditions published in late 2023, "Sinking Megacity Has One Last Chance to Save Itself," suggested that there is a possible "longshot fix" to the megaregion's subsidence problem. **2** Yet like so many other responses to the flooding and sinking problems over the past two decades, some of which will be discussed below, this assumes that a project proposed by one of Indonesia's wealthiest individuals to take over supplying clean water directly to homes and businesses would resolve the problem. I will return to discussing in great detail not only this latest initiative but the most salient ones that have been put forth (some started, some not, and all never finished) since the crisis openly entered the public discourse in the early 2000s. It is a challenge that has cost several Jakarta governors their jobs, and served as one key rationale behind the decision by the national government to transfer the capital from its home on Java to another island.

Drawing upon the recently published monograph, *Urban Flood Risk Management: Looking at Jakarta* (Routledge 2022) **3** and supplemented by additional research over the past two years, the objective of this essay is to demonstrate that there are critical water-based improvements needed to reverse the processes of land subsidence, all of which relate specifically to changing the source of water to serve a megacity region of over 30 million inhabitants and the businesses that it houses by reducing reliance of subsurface sources. Yet even if expanded use of surface water is undertaken in a serious manner, it will take considerable time to reduce the incidence of land subsidence. That means that there also needs to be continued flood prevention measures in addition to water supply changes. Moreover, the long-term potential for a water supply from usable surface sources assumes that the highly polluted condition of virtually all of Jakarta's major surface water sources (apart from several regulated reservoirs) needs to be corrected as well. And likely expanding the utilization of surface water rather than subsurface resources will be even slower unless the basic quality of all rivers and streams serving the region is improved. In sum, the efforts to address the sinking city require simultaneous interventions on several fronts.

To understand the current conditions in Jakarta, it is necessary to briefly recount key points in the emerging sinking city saga, particularly over the past three decades and the responses by government (both in Jakarta and from the national level) as well as the role of private industry. This will focus on the most substantial interventions and how these impacted urban conditions. This will be followed by discussion of alternative interventions that should be considered to overcome the paralysis that currently characterizes the battle to stop the sinking of Jakarta (as well as so many other megacities, especially in the Global South). The point is that there is no single or simple fix, and that the commitment to addressing the sinking city phenomenon in Jakarta requires multiple simultaneous interventions that engage all of the stakeholders, government, business, and communities. But the leadership needs to begin with Indonesia's governments at the local, provincial and national levels, with both policy initiatives as well as targeted investments.

8.2 Flooding Rises to the Surface

In late January 2002, Jakarta experienced "torrential rains, resulting in serious flooding… reaching up to nearly 10 feet in some places" **4** and even higher in some of Jakarta's 20 districts as the rains continued into early February. The major source of the flood waters was the Ciliwung River, the central water body in a river system that flows from the 3000 foot high mountains to the south along a 97 km corridor through the heart of Jakarta to the Java Sea. The communities in East and West Jakarta took the brunt of the 2002 flood since the government closed the flood gates to prevent the flood water passing through the center of the city, diverting it instead to the flood canals that passed through residential areas. A less serious flooding occurred in 1992 and then again in 1996 (incidentally I was living there during both of them) and was blamed on the problem of too much rain but also led some to predict that these usually big floods were likely to be a recurring event. But there really was no openly discussed analysis of the underlying causes although academic research was already beginning to dig into the data. When the Asian fiscal crisis beginning in late 1997 toppled Indonesia's economy and eventually toppled the Suharto dictatorship in 1998, the flooding problem took a back seat to economic and political concerns.

The 2002 flood, occurring during the first decade of Indonesia's shift to a decentralized and more democratic governance system, with the accompanying tolerance of public engagement, produced the first serious generated political fallout from the annual problem of rising waters. Governor Sutioyoso's administration was criticized for failing to prepare residents and businesses for the flooding even through the Indonesian Meteorological and Geophysics Board predicted potential flooding. Although public criticism died out by the end of 2002, it had generated discussion of some possible interventions although none were initially acted upon. They provided the framework for future efforts when it was clear that mass flooding was not going away. The hardest hit areas were those occupied by the marginalized citizens and, as in the past, flooding there was treated by government as the price of illegal settlements and a condition to endure. **5**

One proposed intervention that provided the groundwork for a future initiative was to expand the number and size of the canals used to regulate water passing through Jakarta. The Japan International Cooperation Agency (JICA) funded a study that called for a 913 m diversion canal to assist in managing the Ciliwung River water volume, as well as improvements to the existing flood canals, most that had been constructed initially by the Dutch in the early twentieth century and never upgraded. There was also recognition that inadequate solid waste management contributed to the massive volume of trash in the rivers that clogged the canals and rivers and the need to create more landfills and better use of them. A Work Bank Study found that much of the trash collected intended for the city's one landfill was routinely dumped in rivers. **6** Another explanation offered to explain the flooding was the insufficient green spaces along the path of the Ciliwung and other rivers flowing through Jakarta to absorb the rainfall and prevent erosion, a problem exacerbated by the extensive suburban development.

So expanding and upgrading the canals that channeled the river water through the city was now in the government's tool bag but it would not be used until initiation of the World Bank financed Jakarta Emergency Dredging Initiative (JEDI) in 2012 that the existing canal system would be upgraded. **7** This was not just a project to increase the flow capacity of these channels but also served as a mechanism to remove hundreds of thousands of residents whose housing and community institutions had been allowed to occupy critical environment spaces along the rivers. This enabled poor residents to access water that otherwise was not available through a formal system piped into houses. We'll return to that matter below.

But an even more important contributor to the flood management strategy formulation came in the aftermath of the next and even more catastrophic flood in 2007, a flood that covered 60% of the city and displaced one-half million residents. Jakarta's new governor, Fauzi Bowo, elected following this new historic flood event, promised to take action and made his signature project completing part of the original Dutch canal system, the East Flood Canal, a project intended to reduce flooding in especially hard-hit areas in East Jakarta. When completed four years later, it provided relief to some inner city communities in East Jakarta but did nothing to address the root causes of citywide flooding.

More important than the East Flood Canal project was the post-flood assessment undertaken by two consultants associated with the Dutch water company, Deltares. Some analysts cited heavier rain owing to climate change as a factor but Jaap Brinkman and Marco Hartman maintained that the level of rainfall was not particularly unusual in the 2007 event. And certainly excessive sediments that had accumulated in rivers and canals that were not properly handled, with especially high volumes of trash, probably factored in as well. The real culprit was land subsidence that had been underway for decades and was caused largely by extensive extraction of water from the aquifers below Jakarta. By the 1970s, Jakarta residents and businesses were no longer able to utilize surface sources as they previously had because of widespread pollution and no treatment facilities to purify it. The Deltares experts calculated that 58% of Jakarta's lands, including much of North Jakarta, were already below sea level and continued land subsidence would likely put much of North Jakarta underwater by 2020. Cutting reliance on subsurface water would be necessary but difficult since there were few and inadequate sources of usable surface waters. The entire Jakarta metropolitan region had access to just one reservoir and much of that water supported agricultural and industrial uses and came from the polluted Citarum River that began in the mountains around Bandung. In the absence of sufficient piped clean water, residences and businesses relied instead on deep well water for their potable source. As development of Jakarta rapidly expanded, so too did the volume of water pulled from the aquifer. At the same time, the lack of sanitary infrastructure meant that much of wastewater not captured by septic systems simply went into the rivers making that source completely nonpotable.

There would be future efforts to create recharge areas to replenish the groundwater supply, but the new strategy emerging following the 2007 to combat land subsidence was to build floodwalls along the north coast and introduce greater pumping capacity to carry flood waters over the walls during flood events. Of course, as long as land subsidence continued, the walls themselves would sink and thus would need to be continuously expanded upward.

In 2012, these same Dutch consultants devised a bolder strategy (modeled on the Dutch approach in cities below sea level). The Jakarta Coastal Defense Strategy (JCDS) called for construction of a series of dikes on reclaimed lands, one following the coastline but two more situated between 6 and 14 km into the Java Sea where there was no evidence of land subsidence taking place. The retention basin created between these dikes would handle flood waters from land and manage it through a system of pumps. The outer dikes on reclaimed lands would not just protect Jakarta from rising seas but also provide space for development of a modern city to transform the flood barrier into a "livable dike". The project assumed the concurrent development of a comprehensive water supply system supported by upgrading the quality of surface sources to reduce the volume of deep groundwater extraction. A plan for a citywide system of sanitary sewers prepared with donor support identified how to carry this out, but it remained just an idea on paper owing to concerns about massive dislocations necessitated to do it in Jakarta's dense neighborhoods as well as the lack of a budget to even do a small part of it. Most important, the Dutch consultants envisaged the reclaimed lands on the "livable dikes" would generate the necessary revenue to finance a substantial portion of the future infrastructure deficits (cleaning the rivers and installation of sanitary sewers to help keep them clean). **8**

The JCDS was formally launched in late 2014, just prior to the governor of Jakarta, Joko Widodo, being inducted as Indonesia's next president. While officially declared underway in December 2014, JCDS immediately encountered procedural hurdles that ultimately derailed its implementation. Ongoing disconnects between central government and Jakarta stalled action except for the construction of the flood wall along the coast and expansion of the Pluit reservoir in North Jakarta to help contain water during the heaviest rains. It is also important to note that the proclaimed benefits of the project failed to persuade a vocal opposition who pointed to serious environmental concerns involved with building the livable city on the outer dikes and creating, as some noted, a "septic lagoon of trapped freshwater pumped into it from Jakarta's polluted rivers, while at the same time destroying traditional fishing villages and other working-class communities that had enjoyed settlement along the water front for decades." **9.**

Fast forward a decade to early 2024, the Giant Seawall Project, as it became popularly known, remained an idea without a sponsor and with a serious competitor for government investments, namely the new capital city of Nusantara being built in eastern Kalimantan as a legacy of Widodo's second and final term as Indonesian president. The *Jakarta Post* noted in a January 22, 2024 editorial, building the "livable seawall" that initially had been conceived as financed through private investments on the reclaimed lands, was now being advanced as a national government project using public funds. This was the position taken by a group of ministers within the Widodo government, led by the Defense Minister Prabowo, candidate for the upcoming presidential election (and who had lost two previous tries to Widodo). The *Jakarta Post* conceded that there was some possibility that the Giant Sea Wall could help, but that Thailand and Japan had taken the much less costly and more effective route of curbing groundwater extraction and thereby reducing land subsidence that way. Indonesia had done virtually nothing along these lines to address this fundamental contributor to flooding. In their view, Prabowo's proposal was merely injecting the promise of a flood wall into the political campaign without any consideration of the widespread opposition to this approach that had derailed it over the past decade. **10.**

Despite Prabowo's effort to revive the Giant Sea Wall plan, in reality the government action that sealed the fate of water city was the decision, made early in Widodo's second presidential term, to build a new "green" capital city in Kalimantan on government owned lands but with much of the investment in the built environment coming from the same private sectors interests that originally were going to fund a great portion of the "livable dikes" costs. At the same time, national government funding for infrastructure under the Widodo administration focused largely on the other great challenges facing its largest urban agglomerations, namely long overdue development of modern transit and in the case of Jakarta, coupling modern transit with expanded commercial, industrial and residential developments. These were the signature national initiatives, not water and sanitary infrastructure and flood mitigation. At the same time, all of these development priorities actually exacerbated the environmental challenges related to land subsidence.

In 2019 the government announced that contracts had been executed with the Dutch and South Koreans to support construction of a 32-km dike to begin in 2030, again pushing the grand flood mitigation strategy even further into the future. Ironically, it was in April 2019 that Jakarta's governor, Anies Baswedan had to deal with a massive flood that severely inundated 37 Jakarta neighborhoods. Knowing that flood control was a contentious political issue, he undertook several incremental initiatives to manage flood waters, such as adding some green spaces along the rivers and canals being dredged under the nearly completed JEDI project. And he initiated talks with the neighboring Bogor regency concerning construction of upstream retention areas to divert water when they exceeded the Ciliwung River's capacity. He also called for construction of infiltration wells to help recharge the aquifer even though experts pointed out that it was the deep aquifers, not those that could be reached through this method, that were the source of the problem.**11** Virtually none of these schemes yielded results.

A most recent assessment of Jakarta's response to flooding and critical issue of land subsidence (the "sinking city phenomenon") appeared in an August 2023 article in *Space 4 Water*, a United Nation's supported portal to disseminate research utilizing space-based technology. In this case, it focused on the use of interferometric synthetic aperture radar to measure the rate of land subsidence in Jakarta and also sea surface heights in order to assess the problem of sea level rise. The 26 cm of land subsidence per year (or 10 inches), based upon 2012 data, suggests that there has likely been 9 feet of decline especially along the coastal area of Jakarta over the past decade if the sinking continued at the same rate. **12** This explains why the not-unusually high levelsof rainfall have generated catastrophic floods on an almost every other year basis since the 2007 historic event. This includes major floods in 2009, 2011, 2012, and a new "historic flood" in 2013 that was especially hard hitting in the affluent Pluit area of North Jakarta. **13.**

These examples of Jakarta's vacillating and typically tepid responses to the growing incidence and intensity of flooding underscore, as Kusno puts it, Jakarta's dysfunctional systems of water management owing to the absence of a universal public good approach to key urban services, a tradition carried over to modern times from colonial precedents. The result has been infrastructure projects and policy decisions (or indecisions) intended to benefit only certain stakeholders. The one commonality is a "shared sense of uncertainty and anticipation around…*banjir*" (flooding). p.73. If one closely assesses key interventions (both those carried out as well as those proposed and remain unrealized) the past three decades of mass flooding produced mostly losers, and only a handful of winners such as those tasked with building new flood protection infrastructure. Flood walls along the coastal shores have provided some relief to new high-end developments but at the expense of communities removed to create space to build the walls. Removal of solid waste from the tributaries to reduce clogging of the flood gates and to remove sources of pollution of surface water is routine maintenance that has not been able to keep any of the waterways clear. Lack of sanitary facilities for those businesses and households along the waterways have to routinely use the surface water to dispose of their waste. In effect, there is no functioning system to clean up the rivers. The Jedi project removed a significant amount of material that enabled sections of the rivers and canals within the city to handle a greater volume of water and to prevent some flooding along their passage. But this came at the cost of mass displacement of existing residents, some few who secured replacement housing (and better living conditions) but overall most simply bore the costs of their lost community and livelihoods without any improvements in their circumstances or in the water quality of the surface waters they had been blamed for polluting. Some of the riverfront communities gained access to community water pipes but non-polluting sanitary infrastructure was virtually nonexistent within the remaining low-income housing communities. The US Agency for International Development's sponsored water and sanitation project, WASH PLUS, assisted a small number of Jakarta kampungs to access improved drinking water and sanitation facilities but largely through community septic systems. While making their domestic lives healthier, it did little to deal with the overall qualities of surface water conditions throughout the city.

The hard infrastructure constructed along Jakarta's north coast provided some protection against the effects of high waters coupled with the land subsidence that was so substantial there. The enlargement of the Pluit Reservoir to manage river waters prior to being pumped over the seawall was continually kept dredged but could not protect the surrounding communities when the heavy rains came. Management of the flood gates further upstream determined how much storm water the reservoir had to handle, but as in 2013 not enough to prevent serious flooding throughout both the poor and affluent sections of Pluit New City. Except for continued construction of the flood wall along Jakarta's seacoast, the continuing need to raise it as land subsidence continued, finishing the World Bank dredging project and the functioning of the new flood canal in East Jakarta represented the essence of the hard infrastructure additions to Jakarta's flood management system. Flooding continued, the sinking of more than 40 per cent of the city's land area continued, and even more hard impervious surfaces were added to increase runoff and minimize absorption of rains. The sinking city kept sinking.

8.3 What is Needed to Bring Meaningful Relief to Jakarta's Flooding?

To bring about meaningful improvements, two basic needs must be addressed, encompassed by the full implications of the concepts, clean and green. Jakarta must prioritize securing and maintaining clean surface water sources sufficient to substantially reduce reliance on subsurface water resources to slow down land subsidence. There are other coastal cities that have endured the problems of excessive flooding like Jakarta, including conditions created as a consequence of land subsidence. Although Bangkok is among the pantheon of "sinking cities" owing to continued overreliance on underground water extraction as well as land compaction from surface development, it also suffers from sea level rise impacts due to climate change and reduced green spaces to absorb rising rivers and rainfall. To their credit, several decades ago the government instituted higher pricing on underground water use and a prohibition on new wells. Expanded development in central Bangkok minimized the potential benefits of these interventions. A 2019 report noted that Bangkok was "sinking fast" with annual land subsidence of between 1 and 2 cm. Although the report stressed the need for further interventions, the city's early initiatives to restrict groundwater extraction probably explains why its annual subsidence is a fraction of the rate underway in Jakarta. **14** Their use of regulatory and fiscal policy enforcement to address underground water extraction offers a model that Indonesia could use to slow the land subsidence in Jakarta and its other cities.

The singular focus of local and national leadership in Indonesia on hard infrastructure for protection against the annual flooding misses two fundamental conditions—excessive reliance on underground water for consumption and utilization of surface water bodies to handle much of the waste and runoff generated by urban growth. "The situation is exacerbated by lax regulation allowing just about anyone, from individual homeowners to massive shopping mall operators, to carry out their own groundwater extractions" since the surface water is unfit for consumption except by those who have no other choice. Along a Jakarta main thoroughfare, Jalan Thamrin, an inspection of 80 buildings found that 56 had their own groundwater pumps and 33 were extracting without proper license. According to then Governor Anies Baswedan, people should be able to extract groundwater so long as they replace it through recharge efforts. He favored the biopori method that involves digging a hole in the ground, 10cm in diameter and 100cm deep, to allow water to be reabsorbed into the soil. **15** This is not a viable way to recharge groundwater and curb land subsidence since that problem is associated with the collapse of the deep aquifers that this method cannot affect. At best, it can help to reduce standing water during rain episodes.

Also, the problem of water pollution in Jakarta is not just a consequence of action by those residing along the waterways within the urbanized area but are directly connected to upstream conditions involving runoff, residential, agricultural and industrial discharges, as well as the volume of water contributed not only by the main arteries but also its tributaries. This suggests the need for a regional approach to managing the waterways rather than disjointed interventions resulting from multiple and competing jurisdictions. Under Governor Fauzi Bowo in 2009, there was a plan to construct a water treatment facility in cooperation with the West Java Province in connection with the Jatiluhur reservoir on the Citarum River thar supplies Jakarta with raw water from a surface source. It never happened. Instead, the Jakarta water company continued to draw its raw surface supply from the highly polluted West Tarum Canal and from subsurface sources. **16.**

Since the 1990s, leaders in Jakarta and the national government have sought to draw attention to the deteriorated quality of the rivers running through the city. The initial programmatic manifestation of this movement, the Clean River Program (*Program Kali Bersih*), was launched during implementation of the 1985–1995 master plan. It then became a centerpiece of the subsequent 2010 master plan. Whereas the river cleanup movement in 1990s was spearheaded by non-governmental groups with the support of Emil Salim, an economist and Minister for the Environment, the 2010 plan made river clean-up and reconstruction as well as expanding green areas a centerpiece of the city's "vision" and linked to a proposed Waterfront City scheme initiated by the national government in the closing years of the Suharto regime. Coupled with the river clean up was increased attention to solid waste management. Still the 2010 plan and subsequent implementation documents included no specific programs to improve water quality but, as occurred in subsequent administrations, merely cited broad policy "ideals" such as to transform the "polluted places hidden in the 'back yard' into scenic avenues located in the 'front yard'…[by] clearing the riverbanks of illegal and temporary structures." **17** Removing kampungs along the rivers was carried out during the administrations of governors Sutiyoso, Bowo, Widodo and Purnama from 2000 to 2019 but no change in the water quality was observed. The one exception was when President Widodo took on the task of tackling pollution in Indonesia's contribution to one of the world's dirtiest rivers, the 290-km Citarum. The Citarum is a major source of water for over 25 million users in Bandung, Jakarta and Bekasi as well as replenishment of three reservoirs. In the case of Jakarta, it accounts for approximately 80 percent of the surface water used in its distribution system. With 2000 textile factories contributing to the 20,000 tons of liquid and solid waste deposited daily, the clean-up challenges of the Citarum are on a megascale. In 2018, Widodo (in cooperation with West Java Province (and with donor funding) initiated a program, Citarum Harum (Fragrant Citarum) to make its waters potable by 2025. This effort was showcased at the 10th World Water Forum held in Bali in June 2024. By the day after the forum, however, images in the *Jakarta Globe* confirmed that the solid waste pollution was just as intense as always.**18.**

The Citarum case underscores several key takeaways concerning how Indonesian cities need to approach challenges of flooding, land subsidence and ensuring a healthy environment. One is that interventions simply focused on Jakarta alone are inadequate. Water management, including provision of clean water and management of waste requires regional and even national interventions. A second takeaway is that these interventions must engage all stakeholders, not just the providers but also those affected, accompanied by a vigorous approach to education on the necessity of intervention and enforcement of regulations to prevent discharge of waste. A water management program funded under the UN Development Program began in 2023 to improve "responsible business practice" along the Citarum. underscores that merely a corp of workers to clean-up of debris is not enough but must also be reinforced through continuous monitoring of those businesses as well as community residents that have traditionally used surface waterways as waste receptacles. The UNDP program in intended to gain adherents from business to voluntarily stop contributing to the problem but rather to the solution.

This also assumes that there is in place systems to capture and appropriately dispose of waste before it enters the waterway. Another component of the Citarum Harum program involved planting thousands of new trees to support a green corridor along the river. Here is a strategy that city and regional authorities through regulation of development can ensure increases in open lands to help absorb rainfall, reduce runoff (with its polluting impacts) and contribute to groundwater replenishment.

In the case of Jakarta, ensuring an adequate clean water supply without over-utilizing underground sources while systematically reducing the pollution effects of human and non-human impacts on surface and non-surface water, these are prerequisite to slowing land subsidence and reducing the intensity of flooding. As previously noted, over the past three decades, necessary modernization of the water and waste management systems has been sidelined by policies and practices to elevate infrastructure investments for transportation in support of economic development, as well as hard infrastructure investments to address flooding conditions, rather than dealing with core environmental deficits. Throughout this period, urban development continued to pave over green spaces within Jakarta. The result is that green spaces currently represent less than ten percent of the city's land area (and that includes counting sparsely populated offshore islands officially within the city limits). While clean rivers and streams, greater green spaces, and sanitary infrastructure are fundamental conditions for flood control and slowing (if not reversing) the problems of the sinking city, focusing just on Jakarta proper is unlikely to do the job. Without comprehensive and integrated regional water and waste management systems, long-term effectiveness of all of the interventions to date will be limited and likely compromised in the near future. **19.**

With the Indonesian government focused on completing construction and moving the national capital to the new "green city" of Nusantara in East Kalimantan, public investment in modern water infrastructure in Jakarta and other needy Indonesian cities seems highly unlikely. In the case of Jakarta, a ban on new private groundwater extraction wells by owners of buildings of 54,000 square feet or greater, enacted in 2021 by outgoing governor Baswedan, is set to take effect in 2023. Its effect on the problem will depend upon the new leadership to bring about enforcement but also to make the investment necessary to remove current illegal private deep wells. **20.**

There is a plan, however, to help with the transition to greater reliance on surface water sources and like so much of the development in Indonesia it relies on attracting private capital into the water management system. A deal has been struck between the government water company, PAM Jaya, and a private firm, PT Air Bersih Jakarta to increase the supply of potable water for the government water company. PT Air Bersih, the creation of corporate mogul Antoni Salim, bought up the five largest water treatment plants serving Jakarta and is building another to expand the supply of potable water from surface sources. Their financing of the water treatment plan will be reimbursed over time by PAM Jaya while at the same earning revenues from the water utility that intends to expand piping into homes and businesses with the expanded supply of clean water. For Salim, heir to and current head of the Salim Group of investments, helping Jakarta reduce reliance on private licensed and illegal wells is a winning business arrangement since his company will be reimbursed not only for the cost of construction of the new treatment plant but also secure revenue from selling clean water to PAM Jaya to distribute through a system that will be expanded and upgraded. PT Air gets paid for the water it supplies to PAM Jaya even if the current estimate of a 40% loss of the water passing through is poorly maintained pipes is not corrected by the utility. For PAM Jaya, and with support from the Jakarta government, getting consumers to transition from private wells extracting subsurface to this new sources is intended to take place over the next several decades. PAM Jaya's challenge will be to convince (or have the government compel) existing businesses and households relying on private (and often illegal wells) to switch to the piped system.**21** It seems that the plan of relying on a private company to increase access to clean water rather than having the full burden on a public enterprise is the only feasible means for Jakarta to reduce its reliance on groundwater and to take the necessary first big step to a modern water provision system.

Based on the challenges of keeping the Citarum River clean enough to support the critical downstream role that it plays as a source of potential potable water, cleaning up the system of polluted rivers that flow through the densely developed Jakarta megacity offers a far more daunting challenge, and one for which there is not yet a workable plan. No more than 50 years ago, many of Jakarta's rivers were considered usable. With only about 5% of the area of Jakarta supported by sewer service, addressing river pollution remains the critical missing component of environmental challenge. Bringing modern sanitary infrastructure to urban kampungs to eliminate the need for open defecation, expanding and enforcing solid waste collection, and creating a culture of treating Jakarta's waterways as precious commodities to protect are all within the means of the government to implement if given the priority it deserves. **22** Only then will Jakarta and its surrounding municipalities be able to ensure a healthy and sustainable environment.

References

Adi Renaldi (2022) Indonesia's Giant Capital City is Sinking. Can the Government's Plan Save It?" *National Geographic*. https://www.nationalgeographic.com/environment/indonesias-giant-capital-city-is-sinking-can-the-governments-plan-save-it

Algi M Gifari (2024) After world water forum display: Citarum river returns to Trash-Ridden state. Jakarta Globe.Silver, 2022, pp 180–181

Earthbound (2021) Ten Sinking Cities on the Verge of Disaster (video). http://www.bing.com/videos/riverview/sinking+cities+in+the+world/

Gabrielle Smith (2023) From Jakarta to Nusantara: Land Subsidence and Other Pressing Water Challenges in a Sinking Mega City, *Space4Water.* https://www.space4water.org/news/Jakarta-nusantara-land-subsidence-and-other-pressing-water-challenges-in-sinking-mega-city

Jakarta Indonesia Flooding:CWS/ERP Situation Report 31 Jan 2002 (2020). http://reliefweb.int/report/indonesia/jakarta-indonesia-flooding-cwserp-situation-report-21-jan-2002

Kusno A (2023) Jakarta: the city of a thousand dimensions. National University of Singapore Press, Singapore, p 98

Lin and Hidayat, op. cit

Mayuri Mei Lin, Rafki Hidayat (2018) Jakarta, Fastest Sinking City in the World, BBC Indonesia

Ng AH, Ge L, Li X, Abidin HZ, Andreas H, Zhang K (2012) Mapping land subsidence in Jakarta, Indonesia Using Persistent Scatterer Interferometry (PSI) Technique with ALOS PALSAR. Intern J Appl Earth Observ Geoinform 18:232–242. https://doi.org/10.1016/j.jag.2012.01.018

Phen-wej N, Giao PH, Nutalaya P (2006) Land subsidence in Bangkok, Thailand. Eng Geol 82(4):187–201. https://doi.org/10.1016/j.enggeo.2005. Bankok is Sinking Fast," *The Asean Post*, December 31, 2019; Ron Gluckman, "Bangkok: The Sinking City Faces Severe Climate Challenge," *Urban Land*, June 10, 2019

Republic of Indonesia (2004) Special assistance for project implementation for Ciliwung-Cisadane river flood control project (I) IP–496. Final Report. Jakarta: Japan Bank for International Cooperation

Rethinking the Seawall (2024) editorial, *Jakarta Post*

Sagita Adesywi (2023) A River Runds Through It: How Responsible Business Practices Can Clean Up Indonesia's Waterways, United Nations Development Program Blog. https://www.undp.org/indonesia/blog/river-runs-through-how-responsible-business-practice-can-clean-indonesias-waterways

Siabutar NV, Soesilo DM, Buhi TE, Hutapea RC (2017) The quality of raw water for drinking water unit in Jakarta-Indonesia. *International Conference on Chemistry, Chemical Process and Engineering (IC3PE)*, in AIP Conference Proceedings 1823:020067–9; https://doi.org/10.1063/1.4978140.

Sheryl Tian Tong Lee, Grace Schomburg (2023) Sinking megacity has one last change to save itself, *Bloomberg*. www.bloomberg.com/features/2023-jakarta-sinking-salim/

Sheryl Tian Tong Lee, Grace Sihombing (2023) The World's Fastest Sinking Megacity Has One Last Chance to Save Itself, *Bloomberg*. https://www.bloomberg.com/features/2023-jakarta-sinking-salim/

Sherwell Philip (2016) "40bn to Save Jakarta: The Story of the Great Garuda," *The Guardian* November 22.

Silver C (2011) Planning the megacity: Jakarta in the Twentieth Century. Routledge, Abingdon, UK, pp 207–208

Silver C, Management UFR (2022) Looking at Jakarta. Routledge, London New York

Silver (2022) p 120, 179

Triple-A Team Consultants (2019) *Jakarta Coastal Defense Strategy* (September 1, 2010-July 31, 2012) http//triple-a-team.com/index.php?id=icwrmip-rcmu

World Bank (2019) *Implementation Completion and Results Report,* IBRD-8210, Jakarta Emergency Dredging Initiative

Urban Inequality and Infrastructural Violence: Socio-spatial Dimensions in Delhi and Patna

9

Bhim Reddy and Manoj Bandan Balsamanta

This chapter deals with the structure of urban inequality and infrastructural violence along the spatial and social dimensions in the cities of Patna and Delhi. More specifically, it analyses inequality in housing and living conditions and access to basic amenities on spatial and social lines. Drawing on the notion of 'infrastructural violence' (Rodgers and O'Neill 2012) we illustrate and argue that deprivation of urban infrastructure in Indian cities manifests in both forms of violence, viz. structural violence and direct/physical violence (Galtung 1969; Farmer 1996; Winter 2012). Urban inequality, poverty and violence manifest in multiple deprivations in intersecting social and spatial categories (Balsamanta and Reddy 2018).

First, the chapter attempts to analyse the nature of growth of these cities in terms of change in the composition of social classes of population using proxies that represent such classes. This in turn has implications for the nature of contemporary urbanization and how inclusive or exclusionary this process has been in the recent decades. One of the crude proxies for urban poverty or urban poor has been slums and slum population respectively (Eswaramma 2006; Bhan and Jana 2013) Here, we consider (the number of) dwelling rooms as a proxy for identifying different socio-economic classes to analyse the nature of growth in cities.

In the second part, using data from household survey conducted as part of the Safe and Inclusive Cities (SAIC) research project (Reddy and Balsamanta 2016), it analyses spatial and social inequality in these cities. So far, other studies using secondary data, such as Census, have analysed urban spatial inequality at two levels: one, at the level of wards i.e. inter-ward inequality and two, inequality between slums and non-slums (or slums vs. the city) (Bhan and Jana 2015; Sidhwani 2015). Ward, being a large administrative unit of large size, consists of heterogeneous residential settlements and social classes; or in other words, it represents socio-spatial diversity with varying levels of infrastructure, public service delivery and living conditions. Analysis of inequality at this level cannot sufficiently capture the complex socio-spatial geographies of the cities that house settlements ranging from informal slum and non-slum settlements, the old villages incorporated with the expansion of the cities, resettlement areas to planned and authorized colonies. This study attempts to fill this gap by analysing inequality between these varied types of residential settlements through the levels of access to social infrastructure, basic amenities and quality

B. Reddy (✉)
Senior Fellow, Institute for Human Development, New Delhi, India
e-mail: bhim@ihdindia.org

M. B. Balsamanta
Institute for Human Development, New Delhi, India

A. Alakshendra et al. (eds.), *Development, Inclusion and Sustainability*, Sustainable Development Goals Series, https://doi.org/10.1007/978-981-97-6863-9_9

of housing. This is a unique attempt to capture spatial inequality and spatio-residential patterns of different social groups and the corresponding inequalities. Inequality could be very stark in terms of individual household wealth and income, but it is expected to be least in terms of basic public goods (civic amenities) where the state plays an important role in their provision. This inequality reflects on the role of state, planning and governance in reducing or perpetuating such inequality which in turn have implications for the well-being of various sections of the urban population.

In the third section, we explore the notion of infrastructural violence and how it manifests in the cities of Delhi and Patna.

9.1 Inclusive or Exclusive Urbanisation?

One of the parameters, among material conditions, to assess the wellbeing or impoverishment of the urban population is housing condition. Housing condition is depicted through the available physical space including the number of rooms, the material used for the house construction and having or absence of a separate enclosed kitchen, bathroom and a latrine. It is obvious that there is a great degree of variation in the quality and the size of what is defined as a dwelling room or an enclosed kitchen or bathroom between households. But a crude measure of urban inequality is presented by the number of rooms households possess, while an absence of enclosed kitchen and/or bathroom indicates a simplest measure of deprivation.

A room with a minimum length, breadth and height of 2, 1.5 and 2 m respectively for residence purpose only is considered as a dwelling room. It is significant to note that there is marginal increase between 2001 and 2011 in both the cities of the percentage of households *without* an exclusive room. In absolute terms there has been an 80% increase in this category of households. In 2011, 1.3% of households in Delhi and 1.6%in Patna did not live in an exclusive dwelling room. Going by the Census of India manual, these households could be living in a single room that is used for other purposes as well like shop or other economic/livelihood activity, or multiple households could be living in a single room. Beside this category of households, there is an extremely pauperized population without a home, termed as houseless households by the Census. The homeless population, known to be under-enumerated, has increased in urban India with NCT of Delhi (0.28% of total population) figuring in the top five states/union territories of India in 2011.

The composition of households based on the number of dwelling rooms reveals marked inequality in the quality of housing accessed in the cities. Importantly, at the bottom of this dwelling compositional structure nearly one third (31.1%) of the households in Delhi and one fifth (21.1%) in Patna live only in one room according to Census 2011. This one room in a large number of cases could be a multipurpose residential space used for domestic activities like cooking, washing and bathing besides being used for sleeping, sitting and storing. This is because when a household has only one room that is used for multiple purposes like cooking as well as sleeping then the lone room is counted as a dwelling room, although kitchen, bathroom, storeroom are not counted as dwelling rooms when they are enclosed or separate from the living room. The share of households living in a single room is a crude indicator of the most deprived section of the population living in the cities which is remarkably large (except sections like students living in rented single rooms). On the other end, those living in three rooms (equivalent to 2BHK in popular parlance) and more rooms together constitute about 38% in Delhi and 46% in Patna. In the lower middle of this structure, around 30% live in two rooms in both the cities.

Dwelling based household population composition in the cities has shifted towards households having more than one room in both the cities during the last decade. The percentage of households dwelling in a single room has significantly declined and at the same time the shares of households living in more than one room have increased. Significant to note that not only the percentage share but even in absolute terms the households living in one room have marginally declined if one considers MCD excluding urban villages during the previous Census decade. We separated urban villages within MCD to look at the compositional and inequality changes within the core areas which may significantly vary from the urban peripheries and outgrowths. Whereas both the absolute numbers as well as their percentage shares of households living in two, three, four and more number of rooms have increased substantially. Looking at urban villages separately, although the absolute number of households living in one room has increased their percentage share has declined indicating a higher growth rate of households living in more than one room (Tables 9.1 and 9.2).

Does the decline in growth of the section of households living in single room in both these cities and a relative increase in the growth rate of sections living in more than one room an indication of reducing urban inequality in the quality of housing? Or, are the cities, particularly the core areas, becoming relatively less conducive to the expansion of housing for the under-privileged sections and, therefore, exclusionary? The limited secondary evidence along with primary qualitative fieldwork suggests reduction of inequality to a small extent and the process of exclusion to a large extent. In Delhi, the decadal growth of population has declined from about 47% in 1991–2001 to 21% in 2001 2011.[1] Such decline in population growth in metropolitan cities in India is suspected to be a result of 'exclusionary urbanisation' wherein cities are turning out to be "less welcoming" to poor migrants (Kundu and Saraswati 2012). Spatial patterns of housing condition, according to the household survey of this study (discussed later), suggests higher concentration of single room tenements in slums and also, to a lesser extent, in urban villages and unauthorized colonies. Conditions for expansion of slum settlements became unfavorable and thereby growth of single room tenements stagnated. Strict regulation in restricting new slum settlements from coming up and evictions of slum settlements[2] from the core areas of the cities along with a lack of affordable housing have curtailed in-migration of deprived sections into these areas. Ironically, the increase in census houses in NCT of Delhi was much higher (36%) during this decade than the population growth (21%), leaving 0.51 million census houses in 2011 vacant.[3] Yet, as the increase in the number of households without an exclusive room suggests, the surplus dwellings are not affordable for the low income groups. The Technical Group on Urban Housing Shortage (2012–17) also observed that the "shortage of housing is mostly at the lower economic strata of the population" (Govt. of India 2012). The insights from field research, particularly in slums both in Delhi and Patna, suggests the following. The slums that exist in the core areas of these cities have emerged more than twenty years ago and some date back to more than fifty years. Absence of recently evolved slums in the middle of the cities is a result of administration's role in checking against fresh encroachments on public lands. This is also reflected in the relative decline in the growth of slum population in Delhi, whose share in total population reduced from 14.65% in 2001 to 10.65% in 2011, notwithstanding underestimation of such population owing to definitional considerations.[4] In the existing slums, some of the longstanding

[1] See Government of NCT of Delhi (2012): *Statistical abstract of Delhi 2012.*

[2] Based on the official sources Dupont (2011) estimated the number of evictions of slums (JJ clusters) was 217 between 1990 and 2007; during the run up for 2010 Common Wealth Games Delhi witnessed a major slum clearance drive, and as a result, a study by Housing and Land Rights Network (2011) suggests, about 200,000 people were reportedly forcibly evicted since 2004.

[3] See Government of NCT of Delhi (2012): *Statistical abstract of Delhi 2012.*

[4] NSSO's 69th round(2012) estimated around 30% of households in urban areas of NCT of Delhi living in slums (Govt. of NCT of Delhi (2015): *Urban Slums in Delhi*); this is because NSSO considers slum as a compact settlement of at least 20 households whereas the Census puts a bar of minimum of 60–70 houeholds or a population of 300 to include an area as a slum and thus exclusing many dispersed slum like settlements with less population.

Table 9.1 Percentage of households based on number of dwelling rooms

City	2011								2001							
	Total number of households	No exclusive room	One room	Two rooms	Three rooms	Four rooms	Five rooms	Six rooms and above	Total number of households	No exclusive room	One room	Two rooms	Three rooms	Four rooms	Five rooms	Six rooms and above
Delhi (with urban village & MCD)	100	1.3	31.1	29.6	20.7	10.8	3	3.6	100	0.9	37.6	26.7	18.5	9.1	2.9	4.3
MCD	100	1.2	30.1	28.8	21.5	11.4	3.1	3.9	100	0.9	36.9	26.1	18.9	9.5	3	4.6
Urban village	100	1.3	37	34.2	16.1	7.3	2.1	2	100	0.7	44.4	32.1	13.9	5.6	1.6	1.7
Patna	100	1.6	21.1	31.1	22.1	12.4	4.2	7.5	100	1.1	24.3	28.1	21.9	11.6	4.6	8.5

Source Census of India

Table 9.2 Number of households according to number of dwelling rooms they live in

City	2011								2001							
	Total number of households	Households having number of dwelling rooms							Total number of households	Households having number of dwelling rooms						
		No exclu-sive room	One room	Two rooms	Three rooms	Four rooms	Five rooms	Six rooms and above		No exclusi-veroom	One room	Two rooms	Three rooms	Four rooms	Five rooms	Six rooms and above
Delhi (with urban village & MCD)	2,643,391	33,307	821,864	781,604	548,027	285,589	78,402	94,686	2,026,707	18,507	762,565	540,572	374,223	185,298	58,602	86,940
MCD	22,51,616	28,137	6,76,753	6,47,792	4,85,060	2,57,038	70,063	86,773	1,843,243	17,301	681,049	481,696	348,781	174,939	55,595	83,882
Urban village	391,775	5170	145,111	133,812	62,967	28,551	8339	7913	183,464	1206	81,516	58,876	25,442	10,359	3007	3058
Patna	2,81,986	4,509	59,541	87,562	62,432	34,876	11,946	21,120	2,16,985	2,316	52,755	60,865	47,448	25,177	10,058	18,366

Source Census of India

residents claiming possession rights have increased the number of dwelling rooms mostly by building double storied structures. However, there is no scope for new in-migrants to access housing in the existing slums even on rent due to overcrowding. In some of the slums, we observed that households live in semi-permanent room (either rented or with informal possession rights) on the first floor of double storied tenements using wooden ladders with no permanent staircase. New encroachments within these slums are also resisted by the old residents.

Census of India categorises the condition of houses occupied by the households into three groups: good, livable and dilapidated. Between Census 2001 and 2011, there is about 7 percent increase in the share of 'good' quality houses in Delhi and a relative decline in the share of 'livable' and 'dilapidated' houses. During this period, Patna shows only marginal change in the composition of households in terms of quality of housing. There was only a marginal increase (1.4%) in the share of 'good' houses and a marginal decline in both 'livable' and 'dilapidated' houses. The figures depict an improvement in the quality of housing, particularly in Delhi, seen in these terms. Yet more than 30 percent of households in both the cities do not live in 'good' houses. Nearly three out of every hundred households in Delhi and five out of every hundred in Patna live in dilapidated tenements.

Other parameters that indicate quality of housing and the household economic situation that the Census of India provides information about are availability of separate/enclosed kitchen, bathroom and latrine. With regard to separate kitchen within or outside housing premises the situation seems to have improved in Delhi but remained almost the same in Patna. But still both in Delhi and Patna a large section of households did not have a separate enclosed kitchen in 2011 and some still cook outside their homes in the open. The situation with regard to bathroom facility and latrines have also improved. However, inadequate public infrastructure in terms of public toilets is still glaring through the figures on open defecation in both the cities and especially in Patna (Tables 9.3, 9.4, 9.5 and 9.6).

9.2 Spatial and Social Inequality

This study through a household survey in both the cities of Delhi and Patna sought to map the urban spatial and social patterns of inequality on which the secondary sources of data offer very limited insights. Census data enables one to analyse spatial inequality in the cities only at the level of urban administrative wards. While such an analysis throws light on inter-ward or ward-level inequality, it has many limitations owing to the large size of wards with heterogeneous residential spaces with varying levels of housing quality, public infrastructure and access to services. Census data on slums provides insights on conditions of slum households and scope for a comparison of slum conditions with the cities at large. This apart, there are no sufficient sources of data that capture the complex socio-spatial geographies of the cities that house settlements ranging from informal (non-slum) settlements known as un-authorised colonies, villages that were incorporated into growing cities and urban administrative structures known as urban villages, resettlement areas, planned and authorized colonies beside recognized and unrecognised slums. This study attempts to fill this gap while analysing inequality between these varied types of residential settlements through the levels of access to social infrastructure, basic amenities and quality of housing besides other indicators like employment/unemployment. This is a unique attempt to capture spatial inequality and spatio-residential patterns of different social groups and the corresponding inequalities. But the problem in analyzing spatial inequality is in identifying a relevant spatial unit at the level of which one could meaningfully analyse inequality. As noted above, the available sources of data allowed analysing urban inequality at two levels: one, at the level of administrative wards and two, between slums and non-slums (rather, slums vs city as a whole). The limitations these two levels of analysis have are that the first is too large a unit with heterogeneous spatio-residential patterns with varying degrees of levels of access and infrastructure and the second does not throw light on the deprivation levels and urban poverty

Table 9.3 Households by the condition of census houses occupied by them

City	2011												2001											
	Total				Residence				Residence-cum-other use				Total				Residence				Residence-cum-other use			
	Total	Good	Livable	Dilapidated	Total	Good	Livable	Dilapidated	Total	Good	Livable	Dilapidated	Total	Good	Livable	Dilapidated	Total	Good	Livable	Dilapidated	Total	Good	Livable	Dilapidated
Delhi (with urban village & MCD)	100	67.6	29.7	2.7	96.0	65.2	28.2	2.6	4.0	2.4	1.5	0.1	100	60.1	34.4	5.5	94.6	56.8	32.5	5.3	5.4	3.3	1.9	0.2
MCD	100	69.3	28.0	2.8	96.0	66.9	26.5	2.6	4.0	2.4	1.5	0.1	100	61.7	32.7	5.6	94.6	58.3	30.9	5.4	5.4	3.4	1.8	0.2
Urban village	100	58.2	39.7	2.2	95.4	55.5	37.9	2.1	4.6	2.6	1.8	0.1	100	43.9	51.7	4.4	94.8	41.5	49.1	4.3	5.2	2.4	2.6	0.2
Patna	100	68.6	26.9	4.5	96.0	66.1	25.6	4.3	4.0	2.5	1.3	0.2	100	67.2	27.4	5.4	95.4	64.1	26.1	5.2	4.6	3.0	1.3	0.2

Source Census of India

Table 9.4 Availability of separate kitchen

City	2001					2011							
	Total	Available	Not available	Cooking in open	No cooking	**Total**	**Cooking inside house**	Has kitchen	Does not have kitchen	**Cooking outside house**	Has kitchen	Does not have kitchen	**No cooking**
Patna (M Corp.+OG)	100	69.6	23.1	6.9	0.4	100	97.5	74.8	22.7	2.2	0.7	1.6	0.2
Delhi	100	66.5	27.8	5.2	0.5	100	98.4	79.1	19.3	1.3	0.3	1.0	0.3
DMC (U)	100	69.5	25.1	4.8	0.5	100	98.4	81.1	17.4	1.2	0.3	1.0	0.3
Urban village	100	54.3	38.5	6.8	0.4	100	98.3	74.5	23.8	1.5	0.3	1.2	0.2

Source Census of India

Table 9.5 HH-9: Households by availability of bathing facility and type of drainage connectivity for waste water outlet

City	2011							2001								
	Total number of households	Number of households having bathing facility within the premises			Waste water outlet connected to			Total number of households	Number of households having bathroom facility within the house	Type of latrine within the house				Type of connectivity for waste water outlet		
		Yes		No	Closed drainage	Open drainage	No drainage									
		Bathroom	Enclosure without roof							Pit latrine	Water closet	Other latrine	No latrine	Closed drainage	Open drainage	No drainage
Delhi (with urban village & MCD)	100.0	87.4	3.9	8.7	59.8	36.2	3.9	100.0	71.6	15.7	46.9	16.5	20.9	50.7	40.3	9.1
MCD	100.0	87.7	3.1	9.2	72.5	25.1	2.3	100.0	73.8	9.9	52.4	17.5	20.2	59.0	35.1	5.8
Urban village	100.0	86.6	6.0	7.5	29.6	62.6	7.8	100.0	62.7	39.3	24.1	12.5	24.1	16.6	61.2	22.1
Patna	100.0	79.4	10.7	9.9	67.0	25.5	7.5	100.0	72.0	8.1	62.5	20.4	9.0	55.2	32.7	12.1

Source Census of India

Table 9.6 HH-8: Households by availability of type of latrine facility

City	2011													2001				
	Total number of households	Number of households having latrine facility within the premises	Type of latrine facility within the premises								Number of households not having latrine facility within the premises	No latrine within premises		Total number of households	Type of latrine within the house			
			Flush/pour flush latrine connected to			Pit latrine		Night soil disposed into open drain	Service Latrine			Alternative source			Pit latrine	Water close	Other latrine	No latrine
			Piped sewer system	Septic tank	Other system	With slab/ ventilated improved pit	Without slab/ open pit		Night soil removed by human	Night soil serviced by animal		Public latrine	Open					
Delhi (with urban village & MCD)	100	90.0	60.0	25.2	0.9	1.6	0.2	2.1	0.0	0.0	10.0	7.0	3.0	100	15.7	46.9	16.5	20.9
MCD	100	89.2	73.8	11.4	0.8	0.7	0.1	2.3	0.0	0.0	10.8	8.2	2.7	100	9.9	52.4	17.5	20.2
Urban village	100	92.0	27.2	58.0	1.2	3.6	0.4	1.6	0.0	0.0	8.0	4.2	3.9	100	39.3	24.1	12.5	24.1
Patna	100	92.3	21.6	65.6	1.8	1.7	0.5	1.0	0.0	0.2	7.7	2.8	4.8	100	8.1	62.5	20.4	9.0

Source Census of India

Table 9.7 Distribution of social groups (official caste/class/Scheduled Tribe population) by locality

		Type of locality					
		Slum	Authorized colony	Unauthorised colony	Urban village	Resettlement colony	Total
Patna	Schedule caste/tribe	42	58	5	24	0	129
	Other backward Class-I	2	83	2	24	0	111
	Other backward Class-II	3	234	14	101	0	352
	General	1	395	26	14	0	436
	Total	48	770	47	163	0	1028
Delhi	Schedule caste/tribe	76	185	56	107	39	463
	Other backward class	85	150	92	105	14	446
	General	97	690	152	181	8	1128
	Total	258	1025	300	393	61	2037

Source IHD Household Survey 2015

outside the category of slums. Not only slums are known to be under-recognised officially but official statistics like Census too under-count them owing to definitional parameters that identify them. As slums are identified by Census in terms of a cluster of households numbering at least 60–70 households or a population of 300 living in poor living conditions, similar or poorer clusters with less than this number of households or poor households dispersed spatially do not figure in slum population. However, the existence of substantial number of households living in houses categorised by Census as 'semi-permanent', 'temporary' and 'non-serviceable' houses clearly indicates poverty in non-slum areas of the cities (see Bhan and Jana 2011). This study categorized the surveyed households in Delhi (Muncipal Corporation of Delhi (MCD) area) using the official broad sub-systems of residential settlements such as 'authorised' colonies, 'unauthorised' colonies, 'resettlement' colonies, 'urban villages' and 'slums' (notified slums and JJ clusters). These reflect different levels of urban infrastructure, service delivery and living conditions in the process of urbanization and urban planning notwithstanding the possibility of certain variation between the same categories. But similar or some other official categorization of residential settlements is not found for the city of Patna. Only recognised slums and urbanized villages are identifiable separately. There is no clear distinction of authorized/planned and unauthorized colonies here. Most of the residential settlements, though unplanned, were authorized/regularized. For convenience and to separate households living in settlements other than urbanized villages and slums, the terms 'authorised' and 'unauthorised' are used for Patna as well. But most of the households fall in the 'authorised' category only and the category of 'resettlement' colonies does not exist in Patna, at least within the Patna Municipal Corporation (PMC) area which this study covers. Table 9.7 shows spatial distribution of sample households according to the type of residential settlements in both the cities. In Patna, the major section, i.e. three-fourths, of the households live in authorized colonies and of the remaining about 4.5% live in slums, 4.5% in unauthorized colonies and 16% live in urbanized villages. In Delhi, about half of the surveyed sample households live in authorized colonies, about 13% live in slums, 15% live in unauthorized colonies, 19% live in urban villages and nearly 3% live in resettlement colonies.

Table 9.7 also presents spatial distribution of social groups clubbed according to official class/caste status categories in Patna and Delhi. Although all social groups are spread across different spatial types in the cities, the socially backward groups, particularly the Scheduled Castes and Scheduled Tribes (SC and ST), are concentrated disproportionately high in slums and urban villages in Delhi and in slums in Patna. In Patna there are 12.5% ST& ST households in a total of 1028 sample households. But nearly one-third of them are found to be residing in slums. Owing to issues of representativeness of the sample households, the table shows that more than 80% of the slum population in Patna is composed of SC& ST households. But qualitative fieldwork in six slums in Patna suggests that the slums' population is largely constituted by both SCs and OBCs. However, as noted earlier the slum population is very low in Patna and the poorer households are dispersed and live in clusters numbering less than 60 households along the railway lines, roads, sewerage canals etc. Some of the SC households live in such clusters which may fall under authorized colonies and some indeed are better off and live in authorized colonies as a result of their access to government regular salaried jobs. On the other hand, the upper caste groups, the General category, constitute about 42% of the sample households but more than half of them reside in authorized colonies which represent better living and infrastructural conditions (which will be shown below). Similarly in Delhi, SC &ST households constitute about 22% of the sample households but are present disproportionately higher in slums (about 30%), urban villages (27%) and resettlement colonies (more than 60%), which represent poor living conditions, and are lower in authorized colonies (18%). Whereas, the general category population which forms about 55% of the total 2037 sample households is disproportionately higher in authorized colonies (67%) but lower in slums (37%) and resettlement colonies (13%) and nearly half in unauthorized colonies and urban villages (47%). On the whole, the general category (upper caste) population is higher than other groups in these cities living disproportionately higher in numbers in better off areas while the contrary is true for the socially backward groups.

9.2.1 Inequality in Housing Condition

Spatial inequality measured in terms of housing condition and access to basic amenities is quite stark between slums and non-slums but all the following indicators suggest presence of poor households in varying extents in non-slums too, particularly in urban villages, unauthorized and resettlement colonies. More than half of the slum households still live in *kutcha* houses in Patna while such households constitute about 7% in the city at large (see Table 9.8). One the contrary, about 95% of the households live in *pucca* houses in authorized colonies. But it should be noted that a significant 4.4% households in authorized colonies and 8% in urban villages too live in kutcha households. In Delhi the share of households living in *kutcha* houses is low (1.5%) compared to Patna, but Delhi's slums still have a significant number (7.4%) of them. A small percentage of households are also found to be living in *kutcha* houses in unauthorized colonies (1.7%), urban villages (0.8%) and resettlement colonies (0.6%).

Table 9.8 Type of house by locality

City	Caste category	Pucca	Kutcha	Total
Patna	Slum	47.9	52.1	100
	Authorized colony	95.6	4.4	100
	Unauthorised colony	95.7	4.3	100
	Urban village	92.0	8.0	100
	Total	92.8	7.2	100
Delhi	Slum	92.6	7.4	100
	Authorized colony	99.7	0.3	100
	Unauthorised colony	98.3	1.7	100
	Urban village	99.2	0.8	100
	Resettlement colony	98.4	1.6	100
	Total	98.5	1.5	100

Source IHD Household Survey 2015

Table 9.9 Type of house by caste

City	Caste category	Pucca	Kutcha	Total
Patna	Schedule caste/tribe	70.5	29.5	100
	Other backward class-I	89.2	10.8	100
	Other backward class-II	94.9	5.1	100
	General	98.6	1.4	100
	Total	92.8	7.2	100
Delhi	Schedule caste/tribe	97.8	2.2	100
	Other backward class	98.0	2.0	100
	General	98.9	1.1	100
	Total	98.5	1.5	100

Source IHD household survey 2015

Between the social categories, the highest share of households live in kutcha households among the SC & ST group both in Patna and Delhi (see Table 9.9). It is very high in Patna at nearly 30% and it is 2.2% in Delhi, followed by OBCs and General category. A comparison of inequalities between spatial and social terms with regard to the type of house suggests that spatial inequality is higher than social inequality in both these cities.

In the type of house categorized as pucca and kutcha, the section of households living in the latter represent the most deprived section of the population after the homeless. The next level of deprivation can be inferred from the number of rooms a household lives in with those living in a single room representing the bottom most section of the urban population. Spatially, the pattern of inequality in terms of households' access to number of rooms is shown in Table 9.10. In Patna, a very high percent of the households in slums (58%) live in one room dwellings while such households are relatively low, but significant in numbers, in authorized colonies (10.5%), and urban villages (14.7%). In Delhi, about half of the slum households and a quarter of households in unauthorized colonies, urban villages and resettlement colonies live in single room houses. The share of such households is lowest in authorized colonies but still a substantial percent of nearly 17% live in single rooms. The difference in the share of households living in single rooms between social groups follows similar pattern to that of households in *kutcha* houses but the gap between social groups is less in this case, particularly in Delhi (see Table 9.11). Patna shows a clear contrast with more than 30% of SC&ST households on one side and only 6.2% of General category households on the other living in single rooms. The difference is less in Delhi with similar share of households among SC& ST and OBC, with about 29 and 28% respectively, and lower share among General category households (20%) living in single rooms. The trend reverses as the number

Table 9.10 Distribution of number of rooms in a household's dwelling by locality

City	Locality category	1 Room	2 Room	3 Room	4 Room	5+Room	Total
Patna	Slum	58.3	22.9	10.4	4.2	4.2	100
	Authorized colony	10.5	30.0	26.9	13.9	18.7	100
	Unauthorised colony	8.5	23.4	34.0	2.1	31.9	100
	Urban village	14.7	30.1	22.1	19.6	13.5	100
	Total	13.3	29.4	25.7	13.8	17.8	100
Delhi	Slum	49.2	34.5	9.7	5.8	0.8	100
	Authorized colony	16.7	42.5	21.7	11.0	8.1	100
	Unauthorised colony	25.3	33.3	22.0	10.3	9.0	100
	Urban village	25.2	37.9	19.1	7.6	10.2	100
	Resettlement colony	24.6	57.4	11.5	3.3	3.3	100
	Total	24.0	39.7	19.4	9.4	7.6	100

Source IHD household survey 2015

Table 9.11 Distribution of number of rooms in a household's dwelling by caste group

City	Caste category	1 Room	2 Room	3 Room	4 Room	5+Room	Total
Patna	Schedule caste/tribe	31.8	27.1	17.1	12.4	11.6	100
	Other backward class-I	20.7	43.2	18.0	8.1	9.9	100
	Other backward class-II	13.1	29.5	25.9	14.2	17.3	100
	General	6.2	26.4	30.0	15.4	22.0	100
	Total	13.3	29.4	25.7	13.8	17.8	100
Delhi	Schedule caste/tribe	28.9	42.3	16.6	5.0	7.1	100
	Other backward class	28.3	40.4	16.1	8.5	6.7	100
	General	20.2	38.4	21.8	11.5	8.1	100
	Total	24.0	39.7	19.4	9.4	7.6	100

Source IHD Household Survey 2015

of dwelling rooms increases beyond two, with the share of general category households living in more number of rooms being the highest in both the cities.

Tenure conditions of the residential dwellings differ across localities in both Patna and Delhi, especially with many households in slum settlements having no legal rights on the occupied residential land. In Patna, 75% of the surveyed slums households claimed informal possession of dwelling land without ownership rights, and a small section in urban villages (3.1%) and unauthorized colonies (4.3%) too have no legal rights on the occupied land. Such settlements have come up on public land under different government bodies like Municipal corporation and railways. In Delhi too, 36% percent of slum households claimed informal possession without legal titles to the residential land. A small section both in authorized (0.5%) and unauthorized colonies (0.7%) also informally occupied dwelling space. It may be noted that even though most of the households in the survey from unauthorized colonies in Delhi responded as having ownership rights, legality of property in unauthorized colonies in Delhi is an ambiguous issue without transferable rights on the land until they are regularized. The Government of NCT of Delhi claimed about 25 percent of the city's population living in unauthorized colonies. Attempts in the recent years to regularize unauthorized colonies prompted applications seeking regularization from 1639 unauthorised colonies. Of which some of the colonies have come up on agricultural land and private lands not conforming to urban land regulations and reportedly 583 such colonies emerged partly or wholly on government lands (Sheikh and Banda 2014). The share of households living in houses on rent is lowest in slums in both the cities compared to authorized and unauthorized colonies. But it also points to the instances of renting out by the locally powerful individuals in slums encroaching on public lands and renting out to the poor households. Between social groups, the most significant difference in tenure conditions of residence is seen in the share of households renting-in dwellings. In both the cities the share of SC&ST households living in rental housing is remarkably lower than other social groups (see Table 9.13). This aspect requires further exploration and the dimensions that could have important role are affordable/low-cost rental housing and social discrimination in house rent markets (Table 9.12).

Housing condition is also reflected in the presence or absence of separate enclosed kitchen, private latrine and space for bathing and washing within premises. We analyze household status based on these parameters both spatially and socially. Table 9.14 present the status of kitchen locality wise. Here, we look at availability of enclosed kitchen separately irrespective of whether it is inside the house or outside the house, those cooking inside the living room without an enclosed space and those cooking

Table 9.12 Nature of tenure of households' residence by locality

City	Caste category	Self owned with legal rights	Informal possession/no legal rights (government land)	Rented	Provided by employer (pay rent)	Provided by employer (without rent)	Others	Total
Patna	Slum	18.8	75.0	2.1	0.0	0.0	4.2	100
	Authorized colony	71.6	0.5	23.5	0.9	2.9	0.6	100
	Unauthorised colony	61.7	4.3	34.0	0.0	0.0	0.0	100
	Urban village	82.2	3.1	14.1	0.0	0.6	0.0	100
	Total	70.3	4.6	21.5	0.7	2.2	0.7	100
Delhi	Slum	55.0	36.0	8.5	0.0	0.0	0.4	100
	Authorized colony	77.6	0.5	14.3	6.2	1.0	0.4	100
	Unauthorised colony	81.0	0.7	16.0	1.0	0.0	1.3	100
	Urban village	78.4	0.0	19.8	1.0	0.0	0.8	100
	Resettlement colony	90.2	0.0	9.8	0.0	0.0	0.0	100
	Total	75.7	4.9	14.8	3.5	0.5	0.6	100

Source IHD household survey 2015

Table 9.13 Nature of tenure of households' residential space by social groups

City	Caste category	Self owned with legal rights	Informal possession/ no legal rights (government land)	Rented	Provided by employer(pay rent)	Provided by employer (without rent)	Others	Total
Patna	Schedule caste/tribe	53.5	27.1	15.5	0.0	2.3	1.6	100
	Other backward class-I	72.1	1.8	21.6	0.0	4.5	0.0	100
	Other backward class-II	76.4	2.6	18.5	0.0	2.3	0.3	100
	General	70.0	0.2	25.7	1.6	1.6	0.9	100
	Total	70.3	4.6	21.5	0.7	2.2	0.7	100
Delhi	Schedule caste/tribe	78.0	6.9	9.9	3.7	0.6	0.9	100
	Other backward class	71.7	8.5	15.5	2.9	0.4	0.9	100
	General	76.4	2.7	16.5	3.6	0.4	0.4	100
	Total	75.7	4.9	14.8	3.5	0.5	0.6	100

Source IHD household survey 2015

Table 9.14 Status of kitchen of households by locality

City	Caste category	Separate kitchen (enclosed space)	Inside living room without enclosed space	Outside in the open	Do not cook	Others	Total
Patna	Slum	18.8	56.3	25.0	0.0	0.0	100
	Authorized colony	83.4	15.5	0.9	0.0	0.3	100
	Unauthorised colony	87.2	8.5	4.3	0.0	0.0	100
	Urban village	56.4	38.7	4.3	0.0	0.6	100
	Total	76.3	20.7	2.7	0.0	0.3	100
Delhi	Slum	21.7	77.9	0.4	0.0	0.0	100
	Authorized colony	89.6	10.2	0.1	0.1	0.0	100
	Unauthorised colony	71.0	28.7	0.3	0.0	0.0	100
	Urban village	80.2	19.8	0.0	0.0	0.0	100
	Resettlement colony	85.2	14.8	0.0	0.0	0.0	100
	Total	76.3	23.5	0.1	0.0	0.0	100

Source IHD household survey 2015

outside in the open. Locality wise in Patna, more than half of the households in slums cook inside their living space without an enclosed kitchen and a quarter in slums cook outside in the open. Significantly, other localities which show better living conditions as well house households without enclosed kitchens: about 39% in urban villages and 15% in authorized colonies cook in living rooms and about 4% each in unauthorized colonies and urban villages cook in the open. Delhi has even higher share of households cooking in living rooms compared to Patna but has very negligible number of households cooking in the open. Here, more than three-fourths of the households in slums, nearly 29% in unauthorized colonies, 20% in urban villages, 15% in resettlement colonies and 10% in authorized colonies cook in living rooms without an enclosed kitchen.

Social groups wise in Patna, both SC&ST and OBC-I have a high share of households (37 and 39% respectively) that cook food inside living space without a separate kitchen, while a higher share (13%) of SC%ST households cook outside in the open compared to the OBCs (see Table 9.15). On the other hand, only 6% of general category households cook inside living rooms and the remaining 94% cook in enclosed kitchens. In Delhi, the difference between caste groups with regard to having an enclosed kitchen or not is less compared to Patna. 29% among SC%STs, 33% among OBCs and 18% among the General category households cook inside living rooms with no enclosed kitchens.

Open defecation is still widespread in Patna with majority of the population in slums not accessing either private or public toilets. There seems no significant reduction in the share of such population between the period Census 2011 (5%) and this household survey in 2015 (4.8% households as shown in Table 9.16) in this city. This survey recorded toilet facility for men, women and children separately, but the difference between men and women is very small. About 60% of the households in slums, 1% in authorized colonies, 2.2% in unauthorized colonies and 3.7% in urban villages reported that women in the households resorted to open defecation. This was higher for men in urban villages at 6.2%. The shortage or absolute lack of

Table 9.15 Status of kitchen of households by caste groups

City	Caste category	Separate kitchen (enclosed space)	Inside living room without enclosed space	Outside in the open	Do not cook	Others	Total
Patna	Schedule caste/tribe	48.1	37.2	13.2	0.0	1.6	100
	Other backward class-I	58.6	38.7	2.7	0.0	0.0	100
	Other backward class-II	70.2	27.3	2.3	0.0	0.3	100
	General	94.0	6.0	0.0	0.0	0.0	100
	Total	76.3	20.7	2.7	0.0	0.3	100
Delhi	Schedule caste/tribe	70.6	28.9	0.2	0.2	0.0	100
	Other backward class	67.0	32.7	0.2	0.0	0.0	100
	General	82.3	17.6	0.1	0.0	0.0	100
	Total	76.3	23.5	0.1	0.0	0.0	100

Source IHD household survey 2015

public toilets is clearly evident in Patna's slums through this survey as well as through the qualitative field research in six slums in Patna. Here, only about 19% accessed private toilets, 17% public toilets and 4% shared toilets (more than one household using the same private toilets). In Delhi, the share of households defecating in the open seems to have further reduced after the year 2011 till 2015 from 3 to 0.5%, and the share of households using public toilets has marginally increased from 7 to 7.5%. But the share of such population is still significant in slums with 4.3 and 3.2% households reporting men and women respectively defecating in the open.

In terms of social groups, the share of households among ST&ST using public toilets (8.5%) and defecating in the open (25%) is not only very high but much higher than other caste groups in Patna (see Table 9.17). In Delhi, the share of households using public toilets among both SC&ST and OBC is very high, 12% and 11% respectively, but the difference between these groups on the one side and the general category (4.5%) on the other is high. Open defecation is overall low in Delhi; yet the share was more than 1% among SC&ST households, lower among OBC and much lower among the general category. Even with regard to toilets/open defecation inequality is more pronounced in spatial than social terms. But social inequality and deprivation is quite sharp in Patna than Delhi as in case of most other parameters.

The survey also covered aspects related to washing space in housing condition. These include washing of utensils and clothes and bathing space for females within or outside housing premises. Women take bath in temporary structures outside housing premises in half of the slum households in Patna and another 20% take bath near the water source in the open. Such households were negligible in authorised colonies, while it was about 3% in urban villages. Similar is the spatial pattern in the share of households washing of utensils and clothes either outside the house or near the water source. In Delhi, only slums still have a small Sect. (3.5%) of households without bathing space inside the house for women. Also a significant share of 8 and 10% households respectively wash utensils and clothes outside their house premise in slums; unauthorized colonies as well have 1% of such households. Among the social groups, households without washing space and bathing space inside house are high

Table 9.16 Share of households (gender wise) by locality based on type of facility for defecation

City	Locality	Community toilets			Household toilets			Open defecation			Shared toilets			Others			Total
		Male	Female	Children	Male	Female	Children	Male	Female	Children	Male	Female	Children	Male	Female	Children	
Patna	Slum	16.7	16.7	16.7	18.8	18.8	18.8	60.4	60.4	58.3	4.2	4.2	4.2	0.0	0.0	2.1	100
	Autho-rized colony	1.6	1.6	1.5	93.4	93.4	93.2	1.2	1.0	1.2	3.9	4.0	4.1	0.0	0.0	0.0	100
	Unau-thorised colony	2.1	2.2	2.3	93.6	93.5	93.2	2.1	2.2	2.3	2.1	2.2	2.3	0.0	0.0	0.0	100
	Urban village	1.2	1.2	1.2	86.4	89.0	87.0	6.2	3.7	5.6	5.6	5.5	5.6	0.6	0.6	0.6	100
	Total	2.2	2.2	2.2	88.8	89.2	88.6	4.8	4.3	4.8	4.1	4.2	4.3	0.1	0.1	0.2	100
Delhi	Slum	56.0	55.7	59.2	37.4	38.7	34.0	4.3	3.2	3.9	1.6	1.6	1.9	0.8	0.8	1.0	100
	Autho-rized colony	0.4	0.4	0.3	98.6	98.5	98.4	0.0	0.1	0.0	0.9	0.9	1.1	0.1	0.1	0.2	100
	Unau-thorised colony	1.7	1.7	1.3	97.3	97.3	97.5	0.0	0.0	0.0	1.0	1.0	1.3	0.0	0.0	0.0	100
	Urban village	0.0	0.0	0.0	99.0	99.2	98.9	0.0	0.0	0.0	1.0	0.8	1.1	0.0	0.0	0.0	100
	Resett-lement colony	0.0	0.0	0.0	100.0	100.0	100.0	0.0	0.0	0.0	0.0	0.0	0.0	0.0	0.0	0.0	100
	Total	7.6	7.5	9.1	90.8	90.9	88.9	0.5	0.5	0.6	1.0	1.0	1.2	0.1	0.2	0.2	100

Source IHD household survey 2015

Table 9.17 Share of households (gender wise) by social groups based on type of facility for defecation

City	Caste category	Community toilets			Household toilets			Open defecation			Shared toilets			Others			Total
		Male	Female	Children	Male	Female	Children	Male	Female	Children	Male	Female	Children	Male	Female	Children	
Patna	Schedule caste/tribe	8.5	8.5	8.8	60.5	62.0	59.2	24.8	24.0	25.6	6.2	5.4	5.6	0.0	0.0	0.8	100
	Other backward class-I	3.6	3.6	2.8	89.2	89.2	89.8	2.7	2.7	2.8	4.5	4.5	4.6	0.0	0.0	0.0	100
	Other backward class-II	2.0	2.0	2.1	88.0	88.6	87.9	4.0	2.8	3.5	5.7	6.3	6.2	0.3	0.3	0.3	100
	General	0.2	0.2	0.2	97.7	97.7	97.6	0.0	0.0	0.0	2.1	2.1	2.2	0.0	0.0	0.0	100
	Total	2.2	2.2	2.2	88.8	89.2	88.6	4.8	4.3	4.8	4.1	4.2	4.3	0.1	0.1	0.2	100
Delhi	Schedule caste/tribe	12.4	11.9	12.9	85.4	86.1	84.5	1.3	1.1	1.4	0.7	0.7	0.9	0.2	0.2	0.3	100
	Other backward class	10.9	10.7	11.5	87.3	87.2	86.1	0.5	0.7	0.6	1.1	1.1	1.5	0.2	0.2	0.3	100
	General	4.3	4.5	6.0	94.3	94.3	92.4	0.3	0.1	0.1	1.1	1.0	1.3	0.1	0.1	0.1	100
	Total	7.6	7.5	9.1	90.8	90.9	88.9	0.5	0.5	0.6	1.0	1.0	1.2	0.1	0.2	0.2	100

Source IHD household survey 2015

among SC&ST in Patna with stark difference from other groups. In Delhi, such households are on the whole very low across social groups including SC&STs and difference between the groups is also less.

9.3 Socio-spatial Disparity in Access to Basic Amenities

This section deals with access to basic infrastructure and services in the cities and maps how it varies between socio-spatial categories. Access to toilets was included as part of the above section on housing condition, which has both private and public good components. The latter includes sewerage system, with piped sewerage system representing a better public infrastructure that connects even private latrines and public toilets (which was already discussed). Other aspects this section deals with are access to water, sanitation and cooking fuel. In Patna, about 9% of the households do not access water within housing premises (see Table 9.18). Locality wise, the share of such households is highest in slums followed by urban villages—more than two-thirds in slums, 10% in urban villages, 6% in unauthorized colonies and 5% in authorized colonies. Delhi has a high share of population (17% households) without the main source of water located within housing premises. Spatial disparity in this regard is very high with about 70% of slum households with no water source within house, followed by more than 40% households in unauthorized colonies, 8% in resettlement colonies, about 2–3% in urban villages and authorized colonies. As mentioned above, the emergence of unauthorized colonies not conforming to urban plans and regulations for land development in Delhi and their ambiguous legal status has resulted in lack or inadequate public services in many of these colonies. Lack of piped water is a major issue in these areas (discussed below).

Social disparity in access to water within housing premises is more significant in Patna than in Delhi. Nearly 29% of SC&ST households have no access within house while, on the other extreme, only about 2% lack main source of water within house among the general category. In Delhi, about 22 and 23% households among SC&ST and OBC do not have main drinking water source in house premises, while it is 12% among the general category.

Table 9.18 Locality wise share of households having main source of drinking water within and away from house premise

City	Locality	Within the premises	Within 50 m radius	Within 500 m radius	More than 500 m radius	Total
Patna	Slum	37.5	52.1	8.3	2.1	100
	Authorized colony	94.9	4.3	0.6	0.1	100
	Unauthorised colony	93.6	4.3	2.1	0.0	100
	Urban village	89.6	7.4	3.1	0.0	100
	Total	91.3	7.0	1.5	0.2	100
Delhi	Slum	29.1	51.9	19.0	0.0	100
	Authorized colony	97.4	2.0	0.7	0.0	100
	Unauthorised colony	58.0	34.7	7.3	0.0	100
	Urban village	97.7	2.3	0.0	0.0	100
	Resettlement colony	91.8	8.2	0.0	0.0	100
	Total	82.8	13.4	3.8	0.0	100

Source IHD household survey 2015

The socio-spatial pattern with regard to the main source of drinking water is complex in Patna with more of privately developed groundwater source though borewells and tubewells among the better off localities and sections. Public piped water supply is highly insufficient here, with no system of public supply to many localities, despite the privileged location of the city on the banks of river Ganga. The main source of water for more than half of the households both in slums and urban villages is piped water, which in case of slums, as shown, is largely away from the house. Of the remaining slum households, the main source for 21% households is public and private hand pumps and private borewells and tubewells for 17% households. Whereas, in authorized and unauthorized colonies about 50% households depend on private borewells and tubewells and the remaining largely depend on piped water as the main drinking water source. The major reason for more than half of the households in authorized colonies not using piped water as the main source of drinking water is that the infrastructure for public piped supply has not reached many new colonies. The scenario is contrasting in Delhi with planned/authorized colonies having much better access to public supplied piped water than unauthorized colonies and slums. Here the authorised colonies have 94% of households accessing piped water as the main drinking source while it is only 57% in slums and 54% in unauthorized colonies. Significant to note, 17% of households in slums and 25% in unauthorized colonies depend on public (DJB) and private water tankers and nearly 2% in slums and 4% in unauthorized colonies buy water in 20 L cans. Urban villages in Delhi also have better access to public piped supply which is the main drinking water source for 93% households.

Access to piped water according to social groups presents a reverse trend in disparity in Patna with piped water being the main source of drinking water for a higher share of SC&ST (57%) and OBCs (64%) than the General category households (43%), as shown in Table 9.21. Whereas the share of households depending on hand pumps (both public and private) as the main source of water is high among SC&ST at 11%, 4.5% among OBC-I, 6.5% among OBC-II and 1.4% among the General category. (This, as noted above, is a result of insufficient public supply of water in the city where better off sections and localities depend on private borewells/tubewells located within housing premises while the poorer localities and sections depend on public piped water accessing mostly from public stand posts away from the houses). In Delhi, the trend in access to piped water by social groups is similar to the disparities in other basic indicators with the backward social groups in a more disadvantageous position than the general category population. Piped water is the main drinking water source for a higher share of general category households (86%) than the SC&STs (80%), while it is lowest among OBCs (77%) (Tables 9.18, 9.19 and 9.20).

There is a marked spatial disparity in drainage systems in both the cities. Drainage system is severely lacking in general in Patna and there is no proper storm water drainage system (City Development Plan (2010–30), Patna); there is, however, significant spatial variation in what is available. The waste water outlets of the houses of majority of households living in better off localities are connected either to closed drains or open but concrete drainage (see Table 9.22). Less than 20% of houses (equal to households, here) in slums and only about 42% in urban villages are connected to closed drainages. Most of the remaining houses in slums do not have their water outlets connected to concrete drainages. In Delhi, drainage system and connectivity is on the whole better but is still deficient, especially, in slums and unauthorized colonies. Only 16% of the houses in slums and 21% in unauthorized colonies have their waste water outlets connected to closed drainage lines and of the remaining houses 29 and 19% respectively in these locales are not connected to concrete drainage lines. Relatively, except a small section, houses in authorized colonies and also urban villages are connected to concrete drainage system.

Table 9.19 Social groups' wise share of households having main source of drinking water within and away from house premise

City	Caste category	Within the premises	Within 50 m radius	Within 500 m radius	More than 500 m radius	Total
Patna	Schedule caste/tribe	71.3	23.3	4.7	0.8	100
	Other backward class-I	86.5	12.6	0.9	0.0	100
	Other backward class-II	91.8	6.0	2.0	0.3	100
	General	98.2	1.6	0.2	0.0	100
	Total	91.3	7.0	1.5	0.2	100
Delhi	Schedule caste/tribe	77.1	18.1	4.8	0.0	100
	Other backward class	76.0	18.4	5.6	0.0	100
	General	87.9	9.4	2.7	0.0	100
	Total	82.8	13.4	3.8	0.0	100

Source IHD household survey 2015

Disparity between caste categories with regard to the type of drainage facility the houses are connected to is marked in Patna than in Delhi like in other parameters (see Table 9.23). More than three-fourths (78%) of houses of the households among the general category are connected to closed drainage lines, while it is only 34% among SC&ST, 41% among OBC-I and 53% among OBC-II. In Delhi, although there is no significant difference between social categories regarding the availability of concrete/*pucca* drainage, there is considerable disparity with regard to closed drainage. 39% and 37% respectively of the houses occupied by the SC&ST and OBC households are connected to closed drains, while it is 52% among the general category. In this parameter, more OBC households are at a disadvantageous position than SC&ST with about 13% having no concrete drainage lines.

Similarly, there is significant disparity in access to cooking fuel in both these cities as shown in Table 9.24. In Patna, still more than 11% of the households do not access LPG for cooking; only one-third of the households in slums access this fuel. The remaining households use firewood, cow dung and kerosene for cooking. Of those who use LPG, 95 percent access it through official connections and the remaining but it in the black market in the form of large cylinders (14 or 15 L) of small (2to 4 L) in black market. In Delhi, more than 99% of the households use LPG as cooking fuel but only about 92% have access to official source (see Table 9.25). More than 8% of them procure it through unofficial means by paying higher price. The share of such households is high (20%) in slums in Delhi.

One of the major issues that poor households in Patna face is lack of access to subsidized food grains through entitlements issued through ration cards. Many households in slums expressed this grievance owing to government's non-recognition of these households under the category of below poverty line households. Some of those who earlier had access to subsidized grains through Public Distribution System were later eliminated from the list of BPL households. A large number of households (82%) in Patna do not have ration cards altogether whether below or above poverty line (see Table 9.26). 75% of the households living in Patna slums and 12% in Delhi slums do not possess ration cards. Absence of a ration card not only denies subsidized food grains but availing other facilities such as LPG connection, telephone connection, driving license etc. because it is often sought as an official proof of one's

Table 9.20 Locality wise main source of drinking water for households

City	Locality	Piped water supply	Private bcrewells	Public well	Private well	Public hand pump	Private hand pump	Public tube well	Private tube well	DJB/ PJB tanker/ private tanker	Buy water cans (20ltr bottles)	Other	Total
Patna	Slum	52.1	4.2	0.0	4.2	8.3	12.5	2.1	12.5	0.0	0.0	4.2	100
	Authorized colony	42.7	46.0	0.3	0.8	1.8	1.7	0.1	5.7	0.0	0.0	0.9	100
	Unauthorised colony	44.7	40.4	0.0	0.0	0.0	2.1	0.0	10.6	0.0	0.0	2.1	100
	Urban village	55.8	23.9	1.2	1.8	2.5	3.7	0.0	9.2	0.0	0.0	1.8	100
	Total	45.3	40.3	0.4	1.1	2.1	2.5	0.2	6.8	0.0	0.0	1.3	100
Delhi	Slum	57.4	3.5	0.0	0.0	1.2	0.8	11.6	0.4	16.7	1.6	7.0	100
	Authorized colony	94.0	3.9	0.0	0.0	0.2	0.3	0.0	0.2	0.0	0.9	0.6	100
	Unauthorised colony	51.7	9.7	0.0	0.0	1.3	1.3	3.7	0.7	25.3	4.0	2.3	100
	Urban village	93.1	4.3	0.0	0.0	0.3	0.3	1.0	0.0	0.0	0.5	0.5	100
	Resettlement colony	83.6	6.6	0.0	0.0	0.0	0.0	1.6	0.0	3.3	0.0	4.9	100
	Total	82.6	4.9	0.0	0.0	0.5	0.5	2.3	0.2	5.9	1.3	1.8	100

Source IHD household survey 2015

Table 9.21 Main source of drinking water for households by social groups

City	Caste category	Piped water supply	Private borewells	Public well	Private well	Public hand pump	Private hand pump	Public tubewell	Private tubewell	Buying water, DJB/ PJB tanker/private tanker	Buy water cans (20ltr bottles)	Other	Total
Patna	Schedule caste/tribe	57.4	17.1	1.6	1.6	3.9	7.0	0.8	9.3	0.0	0.0	1.6	100
	Other back-ward class-I	64.0	20.7	0.0	1.8	1.8	2.7	0.0	5.4	0.0	0.0	3.6	100
	Other back-ward class-II	46.0	35.2	0.3	2.0	3.4	3.1	0.3	7.7	0.0	0.0	2.0	100
	General	36.5	56.2	0.2	0.0	0.7	0.7	0.0	5.7	0.0	0.0	0.0	100
	Total	45.3	40.3	0.4	1.1	2.1	2.5	0.2	6.8	0.0	0.0	1.3	100
Delhi	Schedule caste/tribe	79.9	3.0	0.0	0.0	0.4	0.6	4.1	0.4	8.2	1.3	1.9	100
	Other back-ward class	77.4	8.3	0.0	0.0	0.2	0.7	2.0	0.4	7.0	2.2	1.8	100
	General	85.8	4.3	0.0	0.0	0.6	0.4	1.6	0.1	4.6	1.0	1.7	100
	Total	82.6	4.9	0.0	0.0	0.5	0.5	2.3	0.2	5.9	1.3	1.8	100

Source IHD household survey 2015

Table 9.22 Locality wise share of houses (households) connected to the type of drainage facility

City	Locality	Pucca (Covered)	Pucca (Uncovered)	Pucca- Kutcha/ Mixed	Informal nali	Others	Total
Patna	Slum	18.8	6.3	16.7	56.3	2.1	100
	Authorized colony	65.5	23.2	5.7	4.4	1.2	100
	Unauthorised colony	72.3	14.9	8.5	2.1	2.1	100
	Urban village	41.7	31.3	14.1	12.3	0.6	100
	Total	59.8	23.3	7.7	8.0	1.2	100
Delhi	Slum	16.3	54.7	3.5	25.6	0.0	100
	Authorized colony	55.8	34.3	0.6	2.0	7.2	100
	Unauthorised colony	21.0	49.3	8.0	11.3	10.3	100
	Urban village	55.7	32.8	1.5	3.6	6.4	100
	Resettlement colony	60.7	39.3	0.0	0.0	0.0	100
	Total	45.8	39.0	2.2	6.6	6.4	100

Source IHD household survey 2015

Table 9.23 Type of drainage facility by social groups

City	Caste category	Pucca (Covered)	Pucca (Uncovered)	Pucca- Kutcha/ Mixed	Informal nali	Others	Total
Patna	Schedule caste/ tribe	34.1	20.9	16.3	26.4	2.3	100
	Other backward class-I	41.4	33.3	14.4	9.9	0.9	100
	Other backward class-II	52.6	27.3	9.7	8.8	1.7	100
	General	78.0	18.3	1.8	1.4	0.5	100
	Total	59.8	23.3	7.7	8.0	1.2	100
Delhi	Schedule caste/ tribe	39.1	44.7	3.7	6.0	6.5	100
	Other backward class	37.0	43.5	1.8	11.0	6.7	100
	General	52.0	34.8	1.8	5.1	6.2	100
	Total	45.8	39.0	2.2	6.6	6.4	100

Source IHD household survey 2015

Table 9.24 Share of households using different types of cooking fuel by locality

City	Locality category	Firewood	Kerosene	LPG	Cow dung	Coal	Not applicable (if food not cooked at home)	Other	Total
Patna	Slum	35.4	8.3	33.3	8.3	12.5	0.0	2.1	100
	Authorized colony	1.4	1.0	93.8	3.5	0.3	0.0	0.0	100
	Unauthorised colony	4.3	0.0	95.7	0.0	0.0	0.0	0.0	100
	Urban village	9.2	1.8	79.1	8.0	1.8	0.0	0.0	100
	Total	4.4	1.5	88.7	4.3	1.1	0.0	0.1	100
Delhi	Slum	1.2	0.8	98.1	0.0	0.0	0.0	0.0	100
	Authorized colony	0.0	0.0	99.6	0.0	0.0	0.2	0.2	100
	Unauthorised colony	0.0	0.3	99.7	0.0	0.0	0.0	0.0	100
	Urban village	0.3	0.0	99.2	0.0	0.0	0.3	0.3	100
	Resettlement colony	0.0	0.0	100.0	0.0	0.0	0.0	0.0	100
	Total	0.2	0.1	99.4	0.0	0.0	0.1	0.1	100

Source IHD Household Survey 2015

Table 9.25 Source of procurement of LPG cooking fuel

City	Caste category	Through official LPG connection	Buy 14/15 kg cylinder paying extra money	Buy small cylinder in black market	Total
Patna	Slum	50.0	18.8	31.3	100
	Authorized colony	97.1	0.8	2.1	100
	Unauthorised colony	100.0	0.0	0.0	100
	Urban village	86.8	6.2	7.0	100
	Total	95.0	1.9	3.2	100
Delhi	Slum	79.8	7.1	13.0	100
	Authorized colony	96.3	1.9	1.9	100
	Unauthorised colony	86.3	7.4	6.4	100
	Urban village	90.8	2.8	6.4	100
	Resettlement colony	98.4	0.0	1.6	100
	Total	91.7	3.5	4.8	100

Source IHD household survey 2015

residence and identity. Thus, lack of a ration card denies many citizenship rights and one's urban identity.

9.4 Urban Inequality and Infrastructural Violence

The notion of violence adopted here is extended beyond the shallow confines of liberal notions of direct violence identified with an agent and intent to 'structural' or 'invisible' violence (Winter 2012). Following Galtung (1969), scholars increasingly talk about structural violence wherein poor is the 'chief victim' (Farmer 1996: 280). They study the nature and scale of the disadvantages that poor embody by virtue of an ingrained unequal distribution of culture and materiality. However, structural violence is too vague and broad a term for the comfort of empirical research. Consequently, as part of a much-needed simplification, scholars often

Table 9.26 Household possessing a ration card and the type of card

City	Locality	% of HHs with a ration card	APL (White/Saada card)	BPL (Red/ Laal card)	Antyodaya (Yellow/ Peela card)	Other	Total
Patna	Slum	25.0	25.0	66.7	8.3	0.0	100
	Authorized colony	13.2	43.1	35.3	17.6	3.9	100
	Unauthorised colony	17.0	75.0	25.0	0.0	0.0	100
	Urban village	37.4	36.1	63.9	0.0	0.0	100
	Total	17.8	41.0	46.4	10.4	2.2	100
Delhi	Slum	88.0	27.8	68.3	1.8	2.2	100
	Authorized colony	73.8	91.4	5.8	1.3	1.5	100
	Unauthorised colony	75.3	76.5	16.4	1.3	5.8	100
	Urban village	65.1	85.2	10.9	1.6	2.3	100
	Resettlement colony	93.4	84.2	14.0	1.8	0.0	100
	Total	74.7	78.4	17.9	1.4	2.3	100

Source IHD household survey 2015

focus on the institutional arrangements and mode of functioning to unpack the extent of structurally induced invisible violence. A case in point is the role of state as a chief perpetrator of such forms of violence. State impacts on the lives of poor severely through unequal policy implications of governance (Gupta 2012). Infrastructure is one of the ways of how state organises society (Scott, 19,998.) Further, infrastructure is the 'material channel' of systemic or structural violence which situate the responsibility of individual suffering at the scale of social (Rodgers and O'Neill 2012: 404–05).

As, of late, increasingly emphasized, infrastructure is a key site which greatly conveys the discriminatory practices of state. Urban scholars presently talk about 'infrastructural violence', by which they seek to demonstrate the way poor population and localities are infrastructurally discriminated against in contrast to the infrastructure-privileges that non-poor effortlessly experience (Rodgers and O'Neill 2012). Discussion in the previous sections on the inequalities in basic infrastructure issues like water, toilets, drainage and garbage in poor localities clearly showed spatial and social infrastructural discrepancies in the cities. Also, infrastructure is critical in urban research from a different dimension. Acute shortage of infrastructures lead to routine disputes and conflicts in poor localities. Here, we also seriously consider the conflict/violence generative dimension of infrastructural inadequacy. Thus, we argue that infrastructural violence manifests both in the sense of structural/systemic violence as well as creates conditions for direct violence.

Infrastructure, and thus, by implication, state plays an implicit role in marginalizing poor in a double sense. First, poor experience, and live with, infrastructural inadequacy in their everyday life and suffer the consequences of deprivation from material resources. Second, they are forced to compete and fight among themselves to avail the limited material infrastructures they are provided with. As a result, disputes and conflicts frequently emerge among poor, particularly around accessing water in public places, public toilets, drainage, and issues of water clogging, open defecation, garbage etc., as we have shown elsewhere (see Reddy and Balsamanta 2016; Balsamanta and Reddy 2018). Further, these individual infrastructural disputes sometimes acquire collective form, as people identify themselves on the lines of caste, religion, region and neighbourhood and so on. Thus, as emerges in the field, poverty is enmeshed with infrastructural inadequacy, embedded in state apathy, to produce disputes and conflicts which are routine, pervasive and immensely harmful for the poor in cities.

9.5 Conclusion

This chapter considers spatial and social locations on the one hand and infrastructure on the other as important vectors or dimensions in understanding urban inequality in India. Spatial inequality in terms of access to basic amenities or, in other words, public material infrastructural resources is stark in the cities of Delhi and Patna. Spatial and social locations largely overlap and intersect in cities and manifest in multiple deprivations and produce horizontal inequalities. State plays a vital role in the reproduction of such inequalities through the provision of unequal material infrastructure and its implications, especially the negative consequences of the deprivations of basic social infrastructure to the poor. Such suffering, as scholars call it 'infrastructural violence' inflicted by the state manifests not only in a systemic manner in structural violence, but is also direct-conflict/-violence generative. Urban infrastructural deprivation creates a fertile ground for direct conflict or violence.

More often, such direct violence emanates within or among the poor while competing to access and share scare infrastructural resources. Further, we have also shown that the contemporary nature of urbanisation appears to be more exclusive with limited scope for the residential spaces for the poor. Other scholars have observed such tendencies based on the decline of slum populations. We have shown this exclusionary nature especially in case of Delhi based on the share of single room dwellings which are taken as a proxy for the poor or lower economic classes and their share in the city population.

References

Balsamanta MB, Reddy B (2018) 'State, violence and everydayness: some insights from Delhi' in social theories of urban violence in the global south: Towards safe and inclusive cities, edited by Jennifer Erin Salahub, John de Boer, and Markus Gottsbacher. Routledge, London

Banda, Sheikh (2014) The Case of Sonia Gandhi Camp: The Process of Eviction and Demolition in Delhi's JhuggiJhopri Clusters. A report of the Cities of Delhi project, Centre for Policy Research, New Delhi

Bergen M (2013) Urban planning's role in making India safer for women. *The Atlantic Cities*. http://www.theatlanticcities.com/neighborhoods/2013/01/urban-plannings-role-making-india-safer-women/4456/. *Accessed 7 October 2014*

Bhan, Jana (2013) Of Slums or Poverty, Economic and Political Weekly 48(18)

Bhan, Jana (2015) Reading spatial inequality in urban India. Econ Pol Week 50(22)

Bhan G (2009) "This is no longer the city I once knew": evictions, the urban poor and the right to the city in millennial Delhi. Environ Urban 21(1):127–142

Bhan G (2014) The impoverishment of poverty: reflections on urban citizenship and inequality in contemporary Delhi. Environ Urban 26(2):547–560

Delhi Human Development Report 2013: Improving Lives Promoting Inclusion, Institute for Human Development and Academic Foundation, New Delhi

Dupont V (2008) Slum demolition in Delhi since the 1990s: an appraisal. Econ Pol Wkly 43(28):79–87

Dupont V (2011) The dream of delhi as a global city. Int J Urban Reg Res 35(3):533–554

Essex S, Chalkey B (1998) Olympic games: catalyst of urban change. Leis Stud 17(3):187–206

Eswaramma P et al (2006) Socio-economic characteristics of the Slum Households in Tirupati Town. Indian J Reg Sci XXXVIII 1:65–72

Farmer P (1996) On suffering and structural violence: A view from below. Daedalus 125(1):261–283

Galtung J (1969) Violence, peace, and peace research. J Peace Res 6(3):167–191

Gupta A (2012) Red tape: Bureaucracy, structural violence, and poverty in India. Duke UniversityPress, Durham

Housing, Land Rights Network (2011) Planned Dispossession: Forced Evictions and the 2010 Commonwealth Games, Housing and Land Rights Network, Habitat International Coalition, New Delhi

Kalyan R (2014) The magicians' ghetto: moving slums and everyday life in a postcolonial city. Theory Cult Soc 31(1):49–73

Kundu A, Saraswati L (2012) Migration and exclusionary urbanisation in India. Econ Polit 47(26–27)

Muggah R (2012) Researching the Urban Dilemma: Urbanisation, Poverty and Violence. http://www.idrc.ca/EN/PublishingImages/Researching-the-Urban-Dilemma-Baseline-study.pdf.IDRC.Canada

Reddy B, Balsamanta MB (2016) Poverty, inequality and violence in Indian cities: A study in Delhi and Patna—a report of the safe and Inclusive cities project funded by IDRC, Canada and DFID, UK. Institute for Human Development, New Delhi

Read CJ (2014) Un-settlement: demolition, home remaking, and the everyday politics of citizenship in a low-income delhi neighborhood. Home Cultures. 11(2):197–218

Removal of Boundary Walls and Reclaiming of Setbacks into Public Realm (2010) *UTTIPEC* . http://uttipec.nic.in/writereaddata/linkimages/9614247194.pdf [Accessed 31st May 2013]

Rodgers Dennis, Bruce O'Neill 2012 Infrastructural violence: Introduction to the Special Issue. Ethnography 134:401–412

Sudworth J (2006) Slum dispute over commonwealth games, *BBC News.* http://news.bbc.co.uk/2/hi/south_asia/5325034.stm *[Accessed 26th May 2013]*

TRUST LAW POLL (2011) http://www.trust.org/item/20110615000000-na1y8/?source=spotlight

UNHABITAT (2008) State of the World Cities 2010–2011: Bridging the Urban Divide, Earthscan, London and Sterling, VA

Winter Y (2012) Violence and visibility. New Polit Sci 34(2):195–202

Social and Community Mapping of Inclusivity: A Spatial Analysis of Patna, Bihar

10

Jesse Anderson and Abhinav Alakshendra

Abstract

Urbanisation in India has been rapid and largely unplanned. Utilising the GIS and the Census 2011, this paper attempts to visually map the level of urban services and amenities available to the residents of Patna, Bihar. We focus on the key services such as sanitation, drinking water access, health, and housing quality to understand the level of stress on the urban service delivery system. This article also ranks administrative wards in Patna on the basis of availability of urban services using interval ranking scheme. We find huge inequality in the access and availability of services in peripheral areas of the city compared to centrally located wards. We find that areas with low density and higher presence of informal settlements are worse off. Finally, we discuss strategies to reduce the existing inequalities in the urban service delivery system to achieve inclusive development.

Keywords

Social mapping · Community mapping · Urban services · Inequality · Inclusive urbanisation

10.1 Introduction

India has only 31% urban population as per 2011 census, but the rate of urbanisation is expected to significantly grow during the next two to three decades. It is widely acknowledged that the process of urbanisation has been marked with haphazard development in India and there has been hardly major policy initiatives until recently for urban development. However, during recent years some significant policy measures such as smart city, AMRUT and Swachh Bharat (Clean India) have been initiated for inclusive urban development. Addressing the unplanned growth of urban areas and improving the service delivery and environment are the most important objectives of the recent policy initiatives.

As urbanization and population growth related pressures continue to increase, there is growing realisation that methods of sustaining this

'Some Issues of Urban Planning in Bihar and Patna' that informs the reader that "This section draws from the dissertation of one of the authors and the complete document can be found at https://ufdcimages.uflib.ufl.edu/UF/E0/05/64/08/00001/Anderson_J.pdf."

J. Anderson
Senior Planning Project Manager, Poulos & Bennett, Orlando, USA
e-mail: jessed.anderson@icloud.com

A. Alakshendra (✉)
Department of Urban and Regional Planning, University of Florida, Gainesville, USA
e-mail: alakshendra@dcp.ufl.edu

A. Alakshendra et al. (eds.), *Development, Inclusion and Sustainability*,
Sustainable Development Goals Series, https://doi.org/10.1007/978-981-97-6863-9_10

phenomenon must be approached in a proactive manner, as to better prepare an urban environment for the future. The *Swachh Bharat* scheme can be categorized as both reactive and proactive in terms of planning measures due to simultaneous launch of the associated e-learning portal. It promotes a furthering in the efficiency in which sanitation and solid waste management duties are carried out, resulting in maximizing the utility behind these government services. Proactive development strategies in India are providing the potential for proper sustainable urban development through the parameters of SMART Cities and inclusive development. Yet in this pursuit, culture and heritage should not be sacrificed. Accordingly, the Heritage City Development and Augmentation of Yojana (HRIDAY) scheme has been conceived with the intent to "rejuvenate the soul of urban India" by focusing on the holistic development of heritage cities through preservation and revitalization efforts. These steps are very important, but more should be due to address urban social and economic inequalities.

The most important challenge of current urban planning and policy regime is to address the social and economic inequalities in the urban areas. These inequalities and exclusions relate to provision of basic infrastructure and services that could act as a driver of conflict, poverty and socio-economic inequality within the city. An important method of evaluating the level of inclusiveness in the pursuit of sustainable development of an urban area is spatial analysis by means of Geographical Information Systems (Yeh and Li 1997). Spatial analysis is performed using the combination of two techniques: social mapping and community resource mapping, which are viewed as two of the most efficient and effective methods of evaluation (World Bank 2005b).

Social mapping is defined as a visual method of showing the social structure and organization of an area along with the location and distribution of households and people within the same environment. In contrast, community resource mapping shows the occurrence, distribution, access to and use of said resources (World Bank 2005b). Although uses of these tools, in the case of analysing inclusive development, they are used to identify locations of different sub-groups of the city, for example ward populations, and their relation to housing quality, as well as their accessibility to basic infrastructure and services provisioned by the municipal government. Social and community resource mapping can also be used to determine a relative ranking of the level of social development of areas, regarding specific indicators chosen to produce assessments of categorical and overall standings. Both approaches of social and community mapping are present within this summary of conducted spatial analyses. In this regard, social and community mapping enables a user to evaluate locations in a social context with various focuses, such as evaluating government initiatives or identifying specific issues of an urban area.

The Swacch Bharat (clean India) national government program is a prime example of a government initiative that can be evaluated through social mapping. Through the premise of Swacch Bharat being a complete sanitation scheme with specified objectives to be reached for all cities within the nation, a spatial assessment can be conducted to show which areas of an urban environment are complying with and meeting the standards of the initiatives, in addition to identifying locations suffering from a backwards stance regarding the level of sanitation infrastructure and service provisions present. In order to achieve a balanced urban growth, it is important to have healthy communities (Anttiroiko 2018).

This paper provides a breakdown of social and community mapping done at a localized level with intentions to evaluate the current level of urban inclusiveness in the city of Patna, which is the capital city of Bihar, with a population of more than two million as per the 2011 census. Following this introductory section, the next Sect. (10.1.1) of the paper discusses the emerging issues and concerns relating to urban planning in Bihar, particularly in Patna. Section 10.1.2 provides the methodology of the study. Sections 10.1.3 and 10.1.4 give the major findings based on qualitative and quantitative approaches respectively. The last part (Sect. 10.2) provides some broad conclusions and recommendations.

10.1.1 Some Issues of Urban Planning in Bihar and Patna

Bihar is one of India's least urbanized states and is positioned to experience a significant rise in the urban–rural population ratio in the near future. Consequently, further development impediments are emerging, which pose increasing concerns as to how the State will be able to "catch up" in terms of social development (Pangotra and Govil 2008). Only a select few towns in the state have modern sewerage and the State has historically failed to comply with the Ministry of Development's solid waste management rules (Singh and Stern 2014). Furthermore, lack of safe drinking water and poor sanitation plague the area, leading to increased health concerns (Alakshendra 2019). To adequately address the location specific deficiencies, Urban Local Bodies must act as functioning independent planning agencies. Bihar serves as a primary example of a state that has not decentralized to the intended extent. While Bihar has abided by the 12th schedule of the constitution and has transferred development functions to Urban Local Bodies, it has done so without fiscally decentralizing funds. Furthermore, rather than allowing for Urban Local Bodies to perform municipal planning functions, the State formed district-planning committees tasked with the creation of urban locality development plans, which subversively absorbs the administratively decentralized responsibilities of Urban Local Bodies (Singh and Stern 2014).

The lack of decentralised power and fiscal responsibilities has directly impeded attempts of Urban Local Bodies to attain sustainable development practices. Fiscal control remains one of the biggest hurdles for the development of Bihar's Urban Local Bodies. This is due to their heavy reliance on state transfers, as Urban Local Bodies in the state are unable to produce enough revenue to meet necessary expenditures (Pantgotra and Govil 2008). Even in Bihar's capital region, Patna Municipal Corporation's spending habits have only worsened the situation as expenditures on establishments exceed revenue generation (Pantgotra and Govil 2008). To combat this, in 2013, the State Government approved the tax and non-tax revenue collection rules under the 2007 Bihar Municipalities Act, permitting Urban Local Bodies to issue collection notices of tax and non-tax revenue (Government of Bihar 2014).

Progressive actions, such as this, are enabling Urban Local Bodies to develop internally. However, fragmented administrative structuring at the local level is hampering these efforts (World Bank 2005a). This is also the case in Patna Municipal Corporation, as the urban management structure of departments is overly complex and fragmented, leading to overlapping responsibilities and inefficient governance (Pantgotra and Govil 2008). It is particularly problematic in the case of the capital region, as Patna is already the most densely populated municipality in the State and is the only district of Bihar to have recorded an urbanization rate surpassing that of the national average (Kumar and Punia 2014). Overlapping administrative functions disassociates single bodies from responsibility, which is exemplified in Patna, as both Patna Municipal Corporation and Bihar Rajya Jal Parishad are responsible for the maintenance of the water supply system, while the Patna Municipal Corporation remains detached from related functions of upgradation and the lying of the network (Pantgotra and Govil 2008).

In order to address this growing concern, Patna Municipal Corporation has openly advocated further devolution of power and responsibilities within its own administration (PMC 2007). These proposed efforts are intended to establish additional subdivisions of governance, building on the current level of decentralization such a movecan help PMC provide efficient provision of services along with a better monitoring system (Fenton and Gustafsson 2017). The Patna Municipal Corporation is presently constructed with a vertically aligned administration, containing 72 wards of division at its base. Each of these wards falls under the executive control of one of the four city circles, depending on the geographical location. Wards are represented by ward councillors who are intended to report to their respective circle's executive officer. Through the envisioned efforts of further decentralizing the planning process, a present requirement to establish additional circles has been proposed (PMC 2007). This alone will not solve the issues erected through a fragmented administration. As purported by decentralization advocates, it is the more intensive scrutiny by constituents of lower levels of government that catalyses increases in efficiency and directs efforts more towards the best interests of the public (Yang 2014).

Although further devolution of powers and responsibilities enables this to occur, it can only come to fruition when the administrative alignment is publicly understood and single body responsibilities are clearly mandated. While the latter of these two requirements has been acknowledged and is intended to be addressed through the creation of additional administrative sub-units, the former has been a much less recognized impediment to efficient governance. Complexity behind administrative structuring has created a disconnection between efforts of the administration and needs of the public (Pantgotra and Govil 2008). Residents of the Patna Municipal Corporation are severely uninformed in regards to their government's organization, which became overwhelmingly apparent throughout the spatial analysis process of the municipality. The general public is largely unaware of the subdivisions of administration and consequently which ward and administrative circle they reside in.

This lack of awareness greatly disrupts public involvement, as the typical resident is uninformed as to which administrative unit is responsible for the development activities of their residential area, thus disconnecting concentrated public scrutiny from the associated government officials. Although the passing of a public disclosure law at the State level was anticipated to resolve such general unawareness, low literacy rates and the absence of a public participation law have largely negated this resolution (Singh and Stern 2014). Consequently, public voices are suppressed without mandated participation in the development process and a lack of public awareness amongst citizens. As a result, Patna Municipal Corporation residents, as a whole, have greatly suffered in the context of municipal development, with increasing economic and social disparities leading to dangerous levels of inequality. Tacking inequality and exclusion is important as quite often they are difficult to reverse (McGranahan et al. 2016).

Moreover, without public participation and the presence of a strong administrative alignment, development in the city has become spatially disproportionate, with the city core facing issues of overcrowding. Natural growth and expansion of the municipality is directed towards the west, while the old portion of the city lying in the east and the overcrowding centre continue to suffer from inefficient and inadequate infrastructure. The spatially disproportional growth adequately depicts the discontinuity present within the sublevels of governance. With each of the four circles representing a geographical section of the city (East, West, North-Central, South-Central), a clear inequality of development amongst the circles becomes extremely apparent through the spatial concentration, expansion and categorization purported in the growth process of the municipality. This is particularly problematic in terms of the national efforts towards inclusive development. As the capital of Bihar, the Patna Municipal Corporation is expected to progressively make strides towards more inclusive and equitable development.

Although social and economic inequality runs rampant through the city, its relation to administrative presence and cohesion has sparsely been evaluated in a spatial context. Taking into account the factor of administrative efforts and responses is imperative when assessing the inclusivity of a municipality's development. Inequality and exclusion can and often does manifest through fragmented provisioning of public services and infrastructure (Mahadevia 2011). Fragmented provisioning generally stems from improper fiscal expenditure or a lack of spatial growth awareness. The

Patna Municipal Corporation has been susceptible to both with the former cited as a cause for concern. This is due to Patna's extreme level of under spending on urban services, which reportedly is only 5.6% of the normative requirement (Bandyopadhyay 2012; Rao and Bird 2011, 2010).

In order to avoid or abate such fragmentation, spatial distribution and extension patterns must be identified. Therefore, spatial analysis is a necessary component of the planning process and is the ideal method of assessing a municipality's level of inclusiveness. It is this need that validates and necessitates the analysis conducted in this paper, which identifies social inequalities and exclusion present in the Patna Municipal Corporation, concerning the provision of basic infrastructure and services.

10.1.2 Methodology

Throughout the 72 wards of Patna Municipal Corporation there are different levels of accessibility to numerous forms of public amenities and services. Because of this, various forms of inequality have become present throughout the city. This has largely been due to the overwhelming influx of population that has not been accounted for through proper development practices, mainly caused by the lacking of a cohesive urban planning structure. In the following sections the aforementioned inequalities are deliberated upon through various maps and their associated interpretations, representing the diverse forms of inequality that are overwhelmingly present within the Patna Municipal Corporation. The GIS data for ward 21 is not available and therefore not used in the analysis. Recently, Patna Municipal Corporation has increased the number of wards from 72 to 75 and has divided 4 administrative circles into 6. However, the GIS shape file is yet not available and therefore we could not map 75 wards.

Social and community resource mapping is the primary method of analysis done in this study. Use of the Geographic Information System, ArcMap, enabled the data in question to be analysed in a spatial context with visual representations. However, the manner in which social and community resource mapping was used as an analytical tool has been consistently evolving which eventually led to incorporating both qualitative and quantitative based analyses. By conducting spatial analysis, it was possible to examine the inequalities present between the levels of social development of Patna Municipal Corporation's wards. Qualitative based analysis was used to visually represent provisional extents of infrastructure and services, in addition to depicting household quality concentrations and patterns. Quantitative based analysis was used to rank wards according to categorical and overall social development in order to compare inter-ward levels of social development.

Although both methods do allow for consistencies to be uncovered, each of these analyses provides additional analytical evaluations of Patna Municipal Corporation's wards in this respect. The qualitative approach allows for more in depth interpretation of service provisioning, clustering and extension tendencies, while the quantitative approach gives validity to the concerns unearthed through the qualitative tactics by means of a structured numerical model. Essentially, the qualitative approach concentrated on identifying the extreme cases of inter-ward inequalities, lacking the ability to distinguish between variations of inequality. For example, the qualitative sanitation map portrays layers that identify wards enduring extreme stress from the absence of sanitation infrastructure and services. However, the entire Patna Municipal Corporation, in general, suffers from poor sanitation, ranking as the second most polluted city in India and being labelled as the dirtiest capital in the nation (Bihar Times 2014). Thus, it must be noted that the qualitative visual represents the extremes within the municipality. This is not intended to discount the sanitary issues plaguing wards that are portrayed as "unstressed", rather it is intended to highlight the most "stressed" or worst performing wards to acknowledge the disparity amongst the inter-ward levels of social development. As this approach only identifies the presence of inequality at the ward level in the municipality, it excludes the evaluation of non-extreme cases. This created the need for an additional method of assessment that could be applied to all wards, which resulted in the creation of a numerical inter-ward ranking scheme. Although the ranking scheme distinguished between each wards' level of development, it did not convey the severe disparities between the extreme and the more typical cases. An example of this being that the quantitative sanitation map portrays the rank of each ward yet does not convey that the while the top ranked ward has the best sanitary services connectivity, the worst ranked ward has been excluded from specific sanitary services connections, altogether. Therefore, it was ultimately necessary to include both approaches in the overarching analysis.

To further exploit patterns of the present inequalities, wards were further associated with the Circle level of administration. The municipality is divided into four circles that are composed of different wards. Together the four circles; New Capital Circle (1–28 and 37), Kankarbagh Circle (29–35, 44–46 and 55), Bankipore Circle (36–43 and 47–51) and Patna City circle (52–54 and 56–72), account for all 72 wards, which is represented in Fig. 10.1. Through ward level analysis in combination with circle level association, spatial analysis was strategically conducted. The two methods of spatial analysis, implementing social and community resource mapping, and the associated datasets are described in the next two sections.

The data analysed for this study was derived from the 2011 India Census and the 2009 baseline survey from the Support Program for Urban Reforms (SPUR) of Bihar. The census information provided population and household data associated with the ward level of administrative divisions. The data from SPUR was derived from the ward level of administration as well and included locations of variables such as parks, hospitals and slums, in addition to providing the location, connection and extent of variables such as sewerage and drainage systems.

10.1.3 Qualitative Analysis

The first level of analysis was of a qualitative base with an emphasis on visual analysis and interpretation of social and community resource mapping schemes. Variables from both the 2011 Census data and 2009 Support Program for Urban Reforms data sets were entered into the Geographical Information System and presented as different data layers that could be depicted over the base map of the city, which was divided into its respective wards. Data layers such as slum locations and park locations could be visually represented on the base map. However, to enable the production of a more in-depth analysis, data transformations were performed as the data was manipulated in a way to present different, more indicative variables that were not given through Census or the Support Program for Urban Reforms data sets. Examples of these data transformations, which were also used in the second method of spatial analysis, are producing a slum household ratio from slum data and total ward household data, as well as producing ward level population densities from ward population and area data. Basic maps were then produced to show the extent of the provision of public infrastructure and services, in addition to giving the first look at potential inequalities and exclusion. Layers such as the drainage and sewerage systems were displayed to show the extent of connectivity and to identify excluded or under-connected wards.

Once all the necessary data and the accompanied transformations were organized into data layers, the qualitative method of spatial analysis was done. In this process a tool of spatial analysis termed overlay was implemented using various layers of data. Performing an overlay enables Geographic Information System technicians to identify areas that are influenced by multiple variables and begin to distinguish spatial patterns that may be present within the data sets. As the focus of this study is to examine the levels of social development of each ward and identify any present inequalities, the overlay tool has been used to uncover patterns of development regarding sub-divisional categories in a social context. Sub-divisional categories of social development that were of focus included: sanitation, clean drinking water access, unplanned housing, and household quality. Each of these categories was comprised of related indicative variables that represented the current level of associated development. Said indicators were determined through a discussion led by experts with the following factors in consideration: most prevailing issues in the Patna Municipal Corporation, current government policy initiatives and progressive planning techniques for social development. Once the decision was made of which social categories should be examined and which indicative variables appropriately represented a ward's associated level of development, the overlay tool was implemented to show concentrations of provisions and areas of exclusion.

By overlaying the indicative variables, a map of each category of social development was produced in which spatial patterns presented themselves amongst the varying levels of ward development. The following is a breakdown of each category, followed by the conclusions reached through this qualitative map interpretation.

Sanitation: The Swacch Bharat initiative of India places a high level of importance on sanitation of urban areas. The community resource mapping of ward level sanitation indicators provides a platform of analysis concerning the current status of sanitation and in regards to the government programme's goals. Through identifying areas stressed by lack of sanitation service provisions, a consistency can be drawn regarding the administrative response to city wide sanitary conditions. Figure 10.1 portrays layers that identify wards enduring extreme stress from the absence of sanitation infrastructure and services. In this case the peripheral wards in the southwest and southeast experienced the most stress through exclusive provisioning. Consequently, these stresses can also be associated with low population densities, in addition to peripheral location. In contrast, central wards, primarily situated within the Bankipore and Kankarbagh Circles, generally were unaffected by extreme sanitary stressors. This was also the case for the more centrally located wards of the administrative area of New Capital Circle.

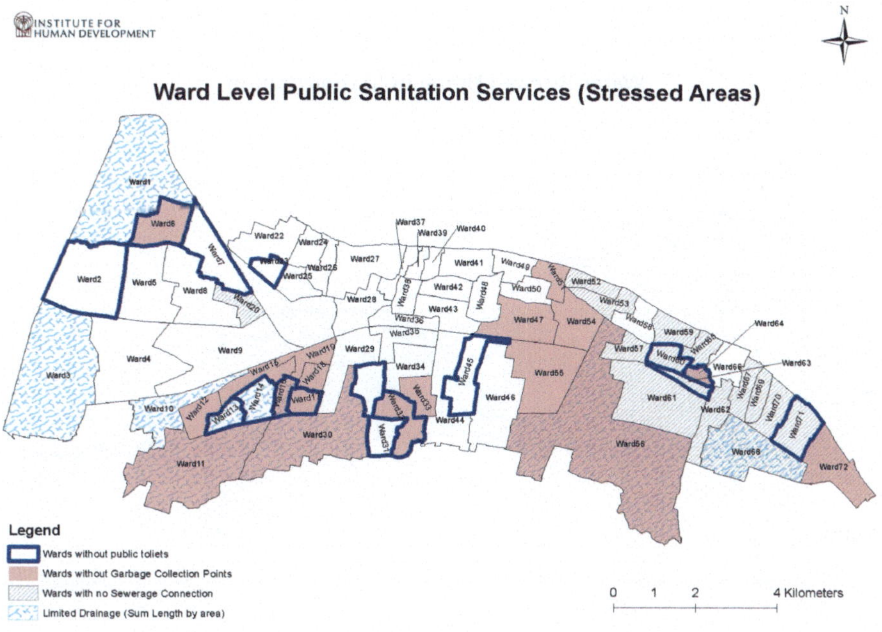

Fig. 10.1 Ward level sanitation

Drinking water access: Drinking water is one of the basic necessities and an important proxy of level of development. Figure 10.2 represents just how stressed the city is when it comes to the access to clean drinking water. While a large portion of wards severely lack household access to treated drinking water, patterns of stressed areas present themselves in this overlay analysis. Wards located along the bank or within close vicinity of the bank of the river, in the north, have better access to drinking water. In contrast, wards situated on the eastern and western peripheries of the municipality showed greater stress in access to clean drinking water.

Unplanned housing: Fig. 10.3 depicts the level of residential planning in the city. Informal housing is a factor that plagues many of the cities in India as the rate of urbanization continues to grow. The mapping of unplanned housing represents areas that are subjected to extensive informal settlements, which results in lower levels of infrastructure and service provisions. By identifying slums and other informal housing in a spatial context, one can begin to understand the absence of such provisions regardless of their geographical location. This is most notable for wards that would be considered anomalies in their respective geographical situations. Through the mapping of informal housing in the Patna Municipal Corporation, a few wards were noted as anomalies in this instance. Two wards in the New Capital Circle, wards 9 and 19, are riddled with the presence of informal housing; where over 40 per cent of all households in each ward are considered slums. Due to New Capital Circle being the administrative sector of the city, poor quality housing and lack of infrastructure and service provisions would seem surprising. However, with the knowledge of the informal housing infestation, one can surmise what has led to low levels of provision and housing quality. Moreover, by comparing the informal housing map with a density map, one can further deduce that large slum concentrations appear to occur in low-density, well-off areas; exaggerating the influences informal housing has on the overall quality of social development in a ward.

Housing quality: While other social development categories have been more exclusively analysed, the decision to incorporate housing quality, in addition to unplanned housing was two-fold: The first reasoning behind

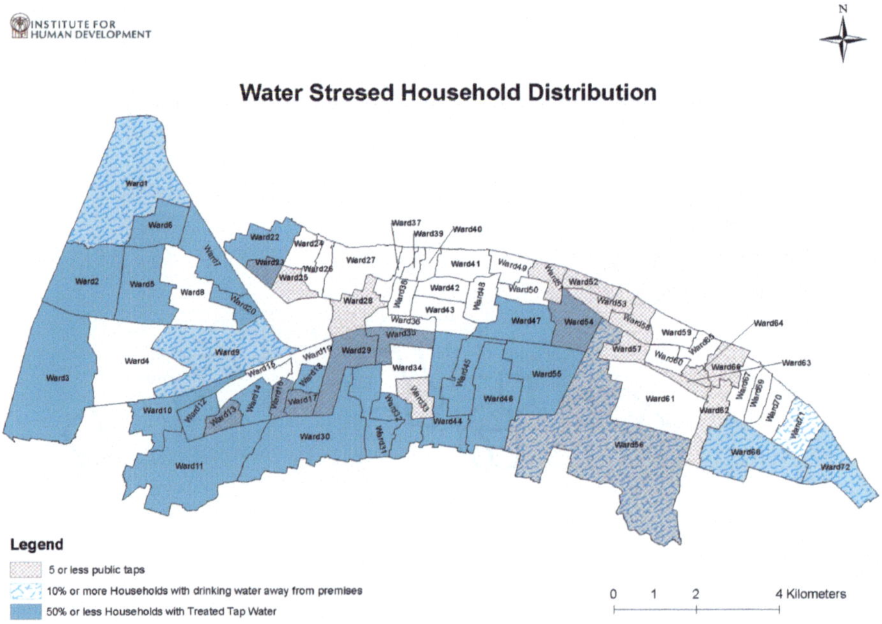

Fig. 10.2 Water distribution

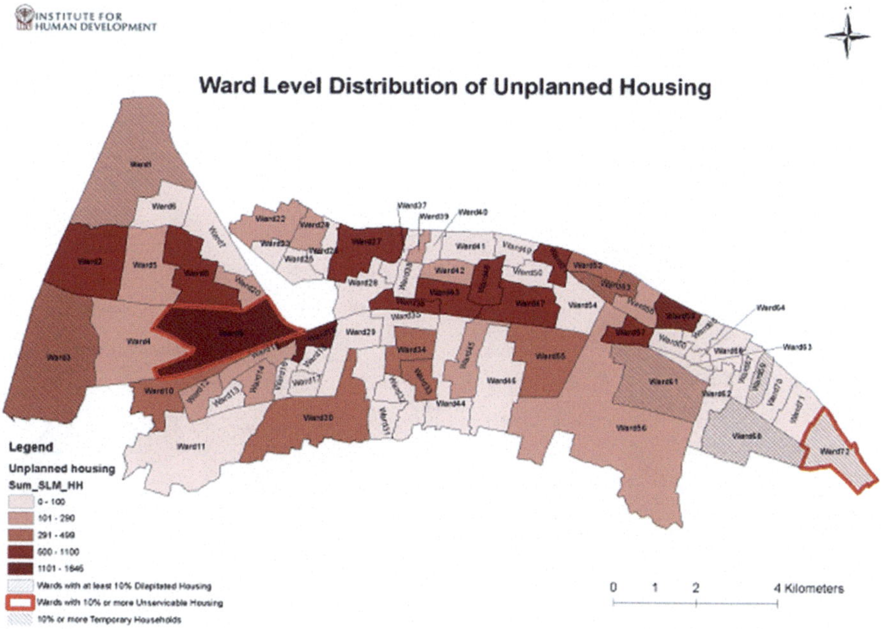

Fig. 10.3 Unplanned housing

this inclusion was to represent the stark contrast between housing quality and unplanned housing, which enabled us to show that while unplanned housing can explain housing quality to an extent, there is still a level of exclusivity that differentiates the two. It also allows for connections to be made regarding how to explain the reasons for the concentration of poor quality housing in certain wards. Figure 10.4 attempts to visualize and explain this concentration. While wards in more central or administrative related locations have their housing qualities largely influenced by an extensive presence of slums, housing qualities of peripheral wards, specifically in the east, do not appear as impacted due to the lack of slum presence. Instead, the housing quality of these wards appears to be influenced more through peripheral location in low-density areas. Moreover, with further knowledge of the city and its employment divisions, it is also suggested that the agricultural economic concentration of these wards could be a prevailing factor in the presence of lower quality households.

As noted previously, patterns emerged through the qualitative spatial analysis. Slum presence was an obvious factor in impacting some of the more central wards in terms of their unplanned housing stance, as well as their overall housing qualities and provisions of services. While slum presence was a primary factor of exclusion in the centrally stressed wards, low population density and peri-urban settings were the pivotal factors regarding wards outside of the city centre. Due to the consistency of low-density wards presiding in the peripheral parts of the city, it was not feasible to attempt to dissociate the two factors. Thus, both comparative low density and peripheral locations were viewed with equal influence in regards to provisions of both basic infrastructure and services.

10.1.4 Quantitative Analysis

In the quantitative method of spatial analysis, wards were ranked based on their level of social development, which was defined through predetermined variables discussed in the qualitative section. As planning often requires public participation to be successful, public participation was incorporated in the format of expert opinion

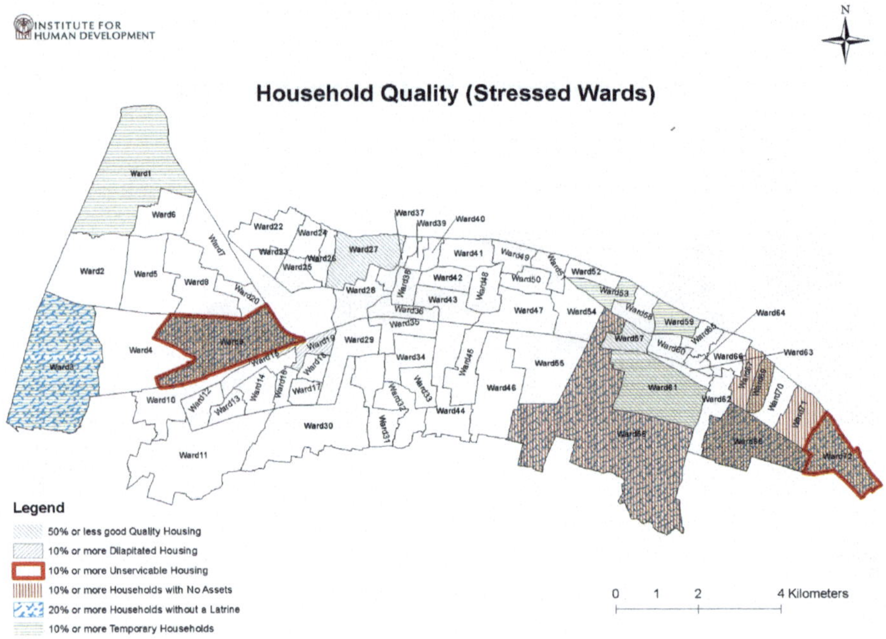

Fig. 10.4 Housing quality

to determine which representative variables to use in the overall assessment of how inclusive the social development of the city is. Through a focus group discussion and interviews with local residents and experts a list of 17 variables were chosen covering the glaring issues perceived regarding the current level of social development in the city.

While the method of applying the ranking system was, itself, quantitative, the approach to how the variables were assigned weights was qualitative in nature. A Priori weighting scheme was constructed upon the conclusion of a discussion led by experts with the following factors in consideration: most prevailing issues in Patna, government policy initiatives and planning for social development. The determined weights of each variable are displayed in Table 10.1. Once the variable weights were assigned, a quartile multiplier was implemented. The quartile multiplier is identified as the quartile in which a ward's indicator falls in comparison to all other indicators within the data set. Once the quartile of a wards variable is identified, the weighted value of the said indicator is multiplied by the

Table 10.1 Variable weights

Variables	Weight
Garbage	20
Main source of drinking water	19
Sewerage system	18
Drainage system	17
Literacy rate	16
Households with latrine	15
Slum household ratio	14
Good quality household	13
Location of drinking water	12
Households without sewage disposal	11
Household structure type	10
Households without drainage	9
Hospital accessibility	8
Persons per household	7
Employment rate	6
Parks	5
Availability of assets	4

Authors Calculations

Variable names and assigned weights determined through a focus group discussion and interviews with local residents and experts ward indicator's quartile, which produced what is called a quartile value in this scheme. The quartile value once again is the product of multiplying a variable value with the quartile in which that ward's variable falls. Due to the use of both positive and negative variables of social development, negative variables received their natural quartiles as their multipliers, while positive variables received the reciprocal quartiles as their multipliers. Table 10.2 provides examples of the quartile value calculations.

After all the variables' quartile values were determined for each ward, the variables were divided into corresponding subdivisions much like the subdivisions analysed in the qualitative portion of spatial analysis. The subdivisions chosen for the quantitative portion of the spatial analysis were health, sanitation, clean drinking water access and residential planning. Rankings for each subdivision were then computed for all of the wards by summing the associated variables' quartile values. The summed quartile values of a subdivision were ranked with the lowest sum being the best and highest sum being the worst of wards. Totals of all quartile values for each ward were then computed and assigned a rank in the same manner as their subdivisions to represent an overall crude ward level social development index. Next the crude indices of the subdivisions and overall rankings were portrayed over the base map to assess the spatial organization of the rankings. Due to the visually unappealing nature of displayed rankings and the inability to easily uncover patterns, a tiered ranking system was implemented. The tiered ranking system broke down the indices into six tiers of interval rankings. The intervals of each tier had an interval of 12 ranks, where tier one contained rankings 1–12 and tier six contained rankings 61–72.

The interval-ranking scheme was applied in order to reveal spatially related ward level consistencies regarding access to and exclusion of services, infrastructure and the quality of living situation. By associating the tiered rankings of wards with the city circles, patterns further presented themselves. The following is an interpretation of the spatial organization of each tiered subdivision index, as well as the overall tiered index, in association to the administrative circles, followed by the conclusions reached through quantitative map interpretations.

Clean drinking water access rankings: Fig. 10.5 shows the drinking water access rankings. The patterns present within the administrative circle associated ward level tiered ranking

Table 10.2 Examples of quartile value calculations

Drain_Q	Drain_V	Drain_QV	PpHH_Q	PpHH_V	PpHH_QV
3	17	51	4	7	28
3	17	51	1	7	7
4	17	68	2	7	14
4	17	68	1	7	7
3	17	51	4	7	28
2	17	34	3	7	21
3	17	51	2	7	14
2	17	34	3	7	21
3	17	51	1	7	7
1	17	17	2	7	14

Authors Calculations

Drain_Q represents the quartile multipler for the drainage system. Drain_V represents the value associated with the variable drainage system. Drain_QV represents the computed quartile value. PpHH_Q represents the quartile multipler for the persons per household. PpHH_V represents the value associated with the variable persons per household. PpHH_QV represents the computed quartile value

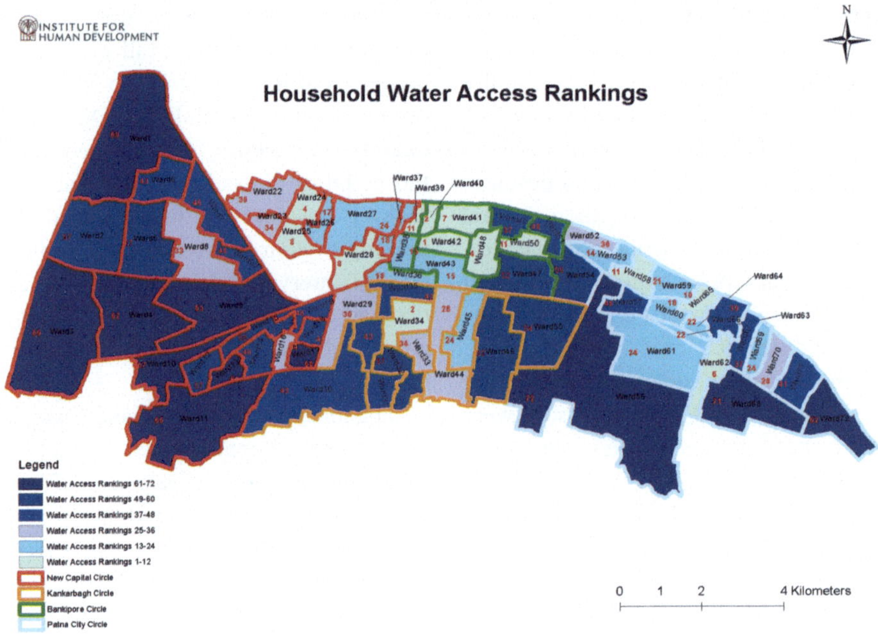

Fig. 10.5 Household water access rankings

scheme were very similar to the qualitative analysis of clean drinking water access. Again, location near the Ganges River appears to play a crucial role in clean drinking water access, as most high-ranking wards were situated along the river or in close proximity to its banks. Acknowledging the importance of proximity from the river is imperative, as the majority of wards in Patna City Circle rank considerably higher than expected, especially in comparison to other social development categories. Furthermore, periphery exclusion was more prominently represented as wards located in the western and south-eastern most parts fell within the worst of the tiers, ranking between 63 and 72 out of all wards.

Health rankings: Fig. 10.6 ranks wards based on their health outcomes. Indicators of health showed stresses in slum dominated central wards. Non-central wards that experienced health stresses were primarily situated in the Patna City Circle region, showing a clear and prominent exclusion of related services and infrastructure to that particular segment of the city. Once again, the central portion of the city, namely Bankipore Circle and Kankarbagh Circle faired well in their rankings with low variability amongst their respective wards. Additionally, New Capital Circle, the administrative section of the city ranked very well in terms of health rankings, aside from the few slum dominated areas. In this visual, peripheral location did not appear to influence ward rankings, as significantly, rather it was the municipality's westward-directed growth and development.

Residential planning rankings: Fig. 10.7. Residential planning ranking patterns closely resembled the patterns identified in the qualitative unplanned housing analysis. The wards located in the low density and extreme periphery areas in the western portion of the city, in addition to the peripheral wards of the eastern portion of the city ranked poorly in regards to residential planning. This could once, again, be associated with the concentration of agriculturally oriented economic activity and agricultural land use patterns. Moreover, centrally located stressed wards again were very much influenced

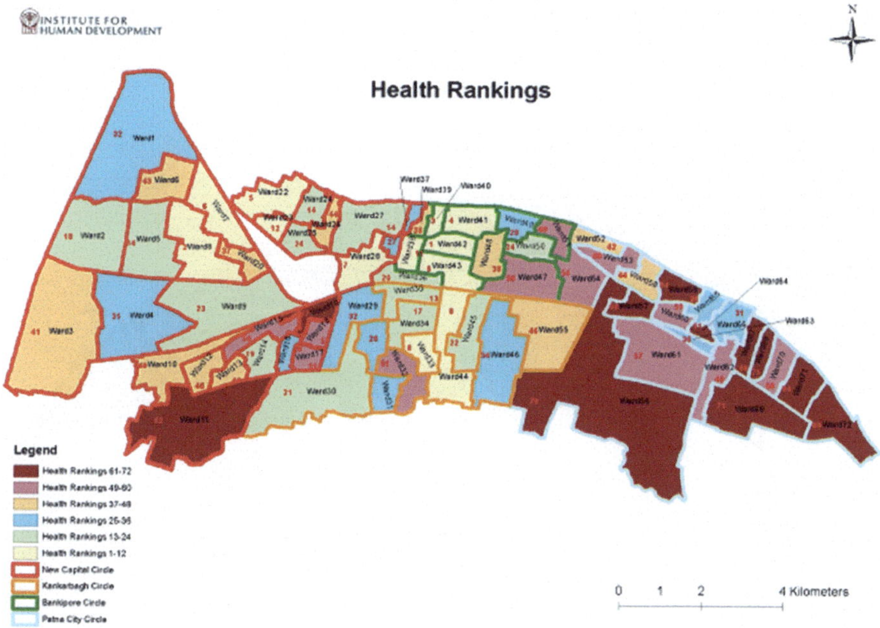

Fig. 10.6 Health rankings

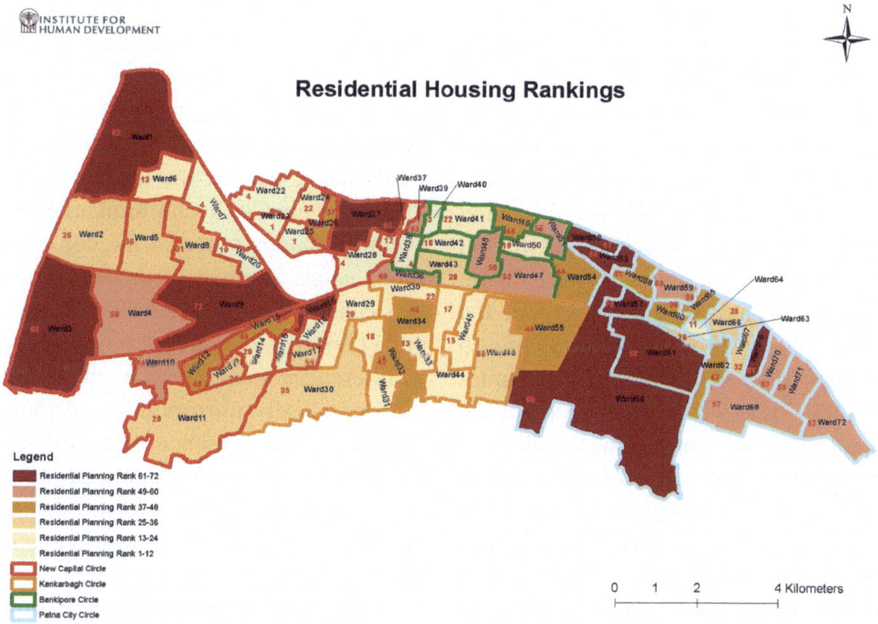

Fig. 10.7 Residential housing quality rankings

by the slum concentrations present within their boundaries. Non-slum stressed centrally located wards ranked very well in terms of residential planning, especially in Kankarbagh Circle, where development practices and dense population continue to concentrate.

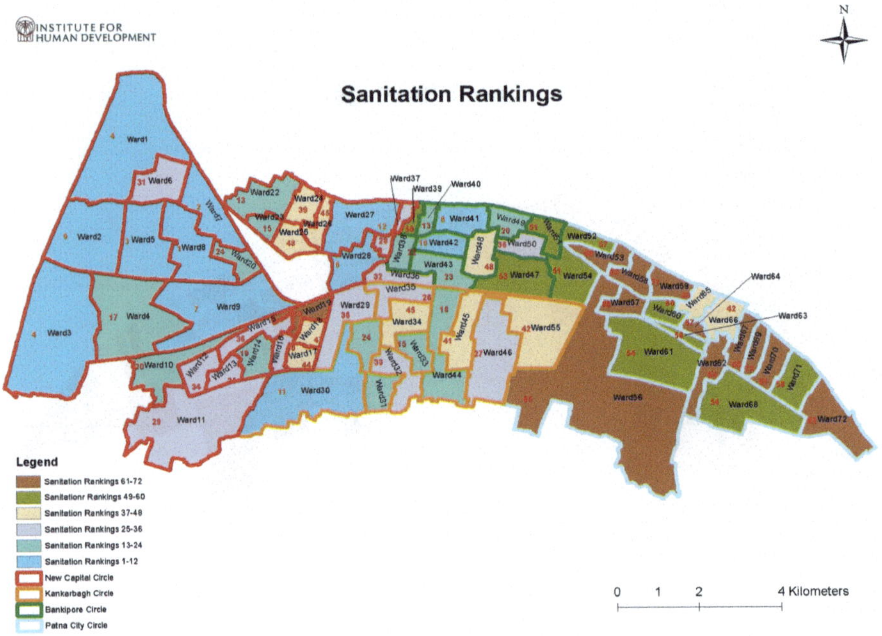

Fig. 10.8 Sanitation rankings

Sanitation rankings: As seen in Fig. 10.8, results are consistent with qualitative rankings, centrally located wards of the city ranked well concerning their provision of sanitation services and infrastructure. Furthermore, as expected, the administrative area of New Capital Circle contained the highest ranked wards and presented itself as far and above the most well off region of the city regarding sanitation. This was in stark contrast to the Patna City Circle. While the administrative region was primarily composed of first and second tier wards in terms of sanitation, Patna City Circle was primarily composed of the fifth and sixth tier wards. Once again, this demonstrated a clear exclusion of basic services and infrastructure to a large section of the city. It also gives validity to the purported direction of expansion and growth in the municipality, which is directed towards the west and away from the old sections of the city.

Overall rankings: Fig. 10.9 shows the overall ranking of the Patna Municipal Corporation wards. In the overall social development rankings, New Capital Circle was composed of mixed rankings with most of its centrally located wards ranking in the top three tiers, while its periphery and slum infested wards ranked in the two worst tiers. Patna City Circle was almost entirely composed of the bottom two tiers of wards with only one of the bottom tier wards being outside the circle's boundaries. As for Bankipore Circle and Kankarbagh Circle, the overall rankings reflected the consistency of the subdivisions as most of the wards within these two centrally located circles ranked in the top three tiers with only a few anomalies.

Inferences made in the quantitative spatial analysis closely mirrors those of the qualitative analysis. Slum presence remained a prominent influence in the quality degradation of centrally or administratively located wards in terms of their rankings. There was also a continuation of the growing connection between the influences that low population density and periphery location had on the provision of basic infrastructure and services of a ward. However, what was more evident in the quantitative spatial analysis was the incessant exclusion of the Patna City Circle division of municipality. The wards comprising this circle overwhelmingly lagged behind in

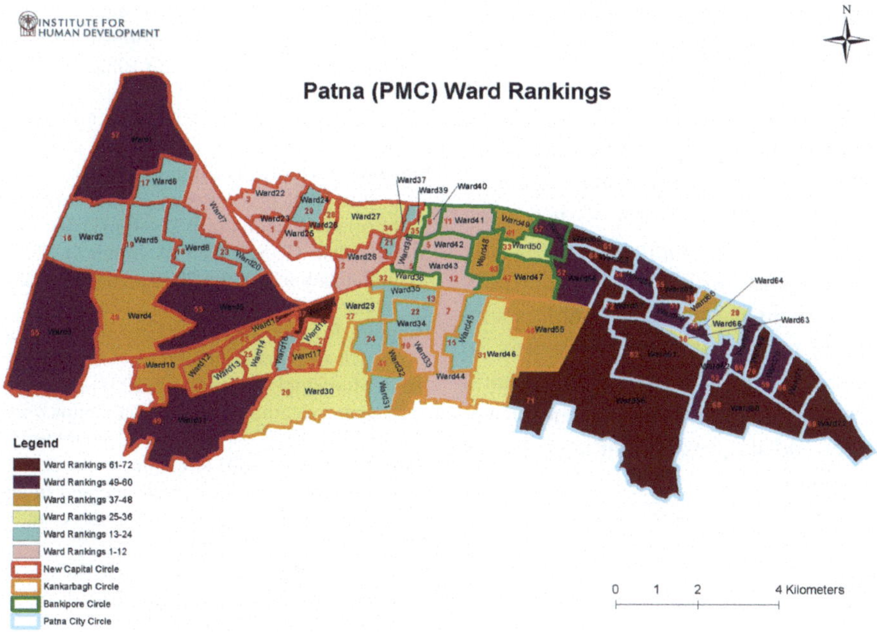

Fig. 10.9 Overall ward rankings

the majority of social development indicators. Thus, it is evident that the current development approach of Patna Municipal Corporation cannot be considered entirely inclusive. While three out of the four city circles have experienced relatively similar levels of social development, Patna City Circle has been left behind and the level of inequality is drastically increasing.

10.2 Conclusion

The analyses conducted have led to several recommendations deemed necessary to further establish a sustainable and inclusive development model throughout the Patna Municipal Corporation area. First and foremost, prioritizing basic government service provision and establishing a more responsive administrative approach must occur. The central government is playing its role by inaugurating an e-learning portal to instil best practices regarding the Swacch Bharat program. It is now necessary for the Municipal Corporation to extend the same effort in pursuits of a more inclusive development scheme. A clearer management structure is imperative to bettering administrative response. Reducing overlaps in delegated responsibilities will improve efficiency, enable constructive scrutiny and serve in the best interests of the public. Furthermore, Circle level administration must be held more accountable for the development of their circle in relation to its counterparts. It is the Circle level of administration that needs to induce an inclusive development model, as there are obvious patterns of inequality emerging through this level. To further enable the Circle level towards pursuing equitable development, the municipal government must be active in assisting a lagging Circle by reassessing allocation and spatial distribution of resources, infrastructure and services. Decentralised infrastructure provision has worked in South Africa and Latin America (Bardhan 2002; Serageldin et al. 2000) and can be adopted for the city of Patna too.

Also, there is a need for increased monitoring regarding the clustering of informal housing, which greatly reduces overall quality of living conditions, inclusiveness and land values in several wards. A cluster analysis may be necessary to better understand the clustering effect and the needs of these informal areas in order to improve their overall quality and ward wise impact. And finally it is proposed that an initiative to further raise citizen awareness is established, as it is essential in efforts of increasing the inclusiveness of development. Increasing public awareness and participation is the only way to achieve inclusive and sustainable social development. Moreover, it is the only way to attain the administratively initiated behavioural change necessary to attain the drastic improvements this city and its government seek for its people.

References

Alakshendra A (2019) City profile: Patna, India. Environment and Urbanization ASIA, 0975425319859132

Anttiroiko AV (2018) Towards inclusive wellness City. In Wellness City (pp. 139–147). Palgrave Pivot, Cham

Bandyopadhyay S (2012) Performance evaluation of urban local governments: a case for Indian cities. International Center for Public Policy Working Paper Series. Paper 70

Bardhan P (2002) Decentralization of governance and development. J Econ Persp 16(4):185–205

Bihar Times (2014) Patna second most polluted city in India after Delhi. Bihar Times. http://www.bihartimes.in/Newsbihar/2014/May/newsbihar10May3.html

Fenton P, Gustafsson S (2017) Moving from high-level words to local action—governance for urban sustainability in municipalities. Curr Opin Environ Sustain 26:129–133

Government of Bihar (2014) Bihar municipal budget manual. Urban Development and Housing Department Government of Bihar

Kumar R, Punia M (2014) Urbanization in Bihar: Need To Break Stigma. J Asian Netw GIS-Based Hist Stud 32:42

Mahadevia D (2011) Branded and renewed? Policies, politics and processes of urban development in the reform era. Econ Pol Wkly 46(31):56–64

McGranahan G, Schensul D, Singh G (2016) Inclusive urbanization: can the 2030 Agenda be delivered without it? Environ Urban 28(1):13–34

Pangotra P, Govil A (2008) Urban development strategy for Bihar: a management perspective

Patna Municipal Corporation (2007) Brief History. Patna Nagar Nigam. Patna, India. https://www.nytimes.com/2010/04/11/world/asia/11bihar.html

Rao MG, Bird R (2011) Coping with change: the need to restructure urban governance and finance in India. IMFG papers on municipal finance and governance (4)

Rao M G, Bird RM (2010) Urban governance and finance in India. Rotman School of Management Working Paper No. 1568858

Serageldin M, Kim S, Wahba S (2000) Decentralization and urban infrastructure management capacity. Cities in a Globalizing World, Global Report on Human Settlements 2001

Singh NK, Stern N (2014) The new Bihar-rekindling governance and development. Harper Collins India

World Bank (2005a) Bihar: towards a development strategy. A World Bank report. Washington, DC: World Bank

World Bank (2005b) Poverty and social impact analysis sourcebook. http://go.worldbank.org/ZGZHJEDBZ0

Yang J, Li G (2014) Fiscal and spatial characteristics of metropolitan government and planning in China: understanding centralization trends in a decentralization context. Habitat Int 41:77–84

Yeh AGO, Li X (1997) An integrated remote sensing and GIS approach in the monitoring and evaluation of rapid urban growth for sustainable development in the Pearl River Delta, China. Intern Plan Stud 2(2):193–210

11 Challenges to Sustainable Urbanization in India—Migration, Poverty and Urban Crimes

S. R. Hashim

Abstract

Urbanization is a strong indicator of productivity and growth. The world is getting increasingly more urbanized. India has been rather slow to urbanize. However, an aspiring India which has in recent decades demonstrated its potential for high growth cannot remain far behind the rest of the world. Even at the present low level of urbanization, the stresses and strains of urban expansion are being felt acutely. This paper raises some of the issues that pose a challenge to sustainable urbanization. India already has some of the largest urban sprawls in the world, while small and medium category towns are neglected, and newly emerging towns remain unrecognized and uncared for. Urban finances are in poor shape and urban governance is weak. Water supply in the towns is inadequate, sanitation poor and proper urban drainage system in bad shape in big and small cities alike. Urban environment is at great risk. Among many issues of sustainability, this paper particularly focuses on migration, poverty and urban crimes. Urban growth is largely supported by rural to urban migrants. Majority of these migrants are poor. Migrants face many difficulties in the cities. The most important is the problems of housing. They have to live in unhygienic conditions. Migrants face a variety of biases and discriminations. The pressure of limited provision of urban amenities is often blamed on migrants. They are blamed even for creating crime prone environment. Migrants loose family and community support when they move to the city, and quite often they have to leave behind even their entitlements for subsidized food etc. There is no job security in the cities, and in times of crises, they are the worst sufferers. They constitute the majority of the urban poor. The food-centric concepts of poverty are not quite relevant for measuring urban poverty and for identifying the urban poor. Residential, occupational and social vulnerabilities are the defining features of urban poverty. Crime rates are generally higher is urban areas, and crimes rates have been rising over time. Residential mobility, poverty, increasing inequalities, family disruption, lack of informal social control, and population heterogeneity are some of the factors that are suggested as contributing to the urban crime prone environment. Also, the failure of justice system and ineffective policing may be responsible for not keeping the crime rates under check. The situation is India is far from satisfactory in respect of the adequacy of police personnel, their training and

S. R. Hashim (✉)
Formerly, Government of India, Union Public Service Commission, New Delhi, India
e-mail: drsrhashim@hotmail.com

A. Alakshendra et al. (eds.), *Development, Inclusion and Sustainability*,
Sustainable Development Goals Series, https://doi.org/10.1007/978-981-97-6863-9_11

their working conditions. While abatement of sociological factors behind crimes requires a sustained social and political movement over a considerable period of time, the quality of governance and policing can go a long way is alleviating the crime situation in the near future.

11.1 Introduction

The world is becoming increasingly more urbanized. The world which was two-thirds rural in 1960 became half urban by the year 2007 and nearly 55% urban by 2017. Urbanization in the developed world is over 80%. The wealth creation in the world today is at its highest ever is history, and about 80 to 90% of this wealth is being created in urban settings. Urbanization is a strong indicator of productivity growth over the long run (Annez and Buckley 2009). The future of the world is even more urbanization.

India has been slow to urbanize. The level of urbanization was a little over 31% in 2011 and might be just around 36% at present. India is even less urbanized than a number of similarly placed countries in terms of economic development and industrialization. However, an aspiring India, which has demonstrated, in recent decades, its ability to move fast forward in terms of economic growth, cannot remain far behind in terms of urbanization. The process of urbanization poses many challenges which need to be focused, managed and resolved in the interest of a sustainable future.

In spite of relatively low level of urbanization, India has been developing some of the largest urban sprawls/agglomerations, in the world. According to Demographia World Urban Areas (2019), Delhi has become the third largest agglomeration in the world after Tokyo and Jakarta. Mumbai, Kolkata, Bangalore and Chennai are also listed among the big urban agglomerations in the world. These agglomerations have developed in countries irrespective of their overall levels of urbanization or the levels of development. There are powerful economic forces behind these agglomerations. But the nature of these forces changes over time. At one time, transport cost used to be a dominant factor behind agglomerations, but the modern technological developments have made it less important. The rising land costs are pushing human settlements and activities away from the core city, but not to a point where they get disconnected from the advantages of clustering and that is how the agglomerations turn into urban sprawls. These urban sprawls generate growth impulses, but also give rise to new set of problems. The core and the satellite settlements are all under different municipal administrations. In case of Delhi the satellite settlements are spread over different states as well. The movement of the people, goods and services between the settlements requires coordination in respect of transport, taxes and a number of other rules and regulations in the interest of smooth working of the entire agglomeration as one economic unit. These and other related issues need to be focused and better framework of coordinated development evolved.

Groups of firms with similar products, sharing common markets, technologies, workers' skills and linked by buyer seller relationships tend to cluster around together in space. Being together generates many advantages like access to a pool of skilled workers and knowledge spillovers resulting ultimately in a knowledge multiplier. The newly emerging clusters of knowledge-based industries share a tendency to become very large creating mega cities and large agglomerations. Bengaluru and Hyderabad in India are outstanding examples of such clusters. Clusters become the drivers of growth. The urban sprawls cannot be wished away because of the strong economic forces which propel them. The issues arising out of these sprawls need to be focused and better framework of coordinated development evolved.

Small and medium size cities have not been given the attention that they deserve. They are poorly governed and lack even the minimum necessary infrastructure and amenities in most cases. The so called 'census towns', i.e., the newly emerging towns, have not even been given statutory recognition as towns which places them at great handicap in further development. The most important role of small and medium towns/cities has been to bring the urban reach to the vast rural hinterland. They have the potential as growth leaders of the hinterland.

Urban governance has been a big issue, particularly for the medium and small size towns. The 74th constitutional Amendment (1992) recognized urban local bodies as the third tier of governance. The state governments were mandated to transfer (through appropriate legislative reforms) to urban local bodies a set of specified functions under 12th Schedule. These were: urban planning including town planning, regulation of land use and construction, buildings of roads and bridges, provision of water, public health services, sanitation, solid waste management, etc. The states were also mandated, through the state finance commissions, to spell out a framework of sharing funds with the urban local bodies in order to enable them to discharge their assigned responsibilities and also to building capacities/expertise necessary for discharging those functions. Even after 28 years (since the Constitutional Amendment) many states have not transferred the stipulated responsibilities, and where some responsibilities have been transferred, adequate funds have not been provided. As a result, local bodies suffer from acute financial crunch. Capabilities of planning and implementation are weak, and conditions in respect of water supply, drainage, sanitation and waste disposal are very unsatisfactory.

Universal access to urban water supply and sanitation still remains a critical issue in India. Big and small cities suffer alike from this problem. A large number of households do not have easy access to water and availability is often inadequate. Sometimes, quality is questionable. The problem is more acute in slums and unauthorized colonies. Resort is taken to unauthorized ground water exploitation. As a result, the aquifers are getting depleted and the quality of underground water is deteriorating.

Sources of public water supply have not expanded in keeping with urban expansion. The situation is going to be even more critical in future. Large scale planning is necessary for enhancing water supply by creating large storages, surface as well as ground water storages. As a result of climate change, the rainfall pattern is becoming eccentric—stormy rainfall followed by long dry spells. Creation of water reservoirs is the only long term solution. Water harvesting needs to be paid special attention. A related problem is storm water drainage. Older cities and towns had natural storm water drains. With increasing pressure on land in cities, storm water drains have been mostly encroached upon. As a result, they become narrow or completely blocked. This causes dangerous flooding and water logging in the cities whenever there is heavy rain.

Urban growth and the modern urban life style put the urban environment at great risk and make the urban air toxic, causing health hazards. The potential fixers of the problems are: reduce reliance on coal based energy, pay attention to ways in which households can conserve energy, promote use of solar energy, reduce vehicular pollution and upgrade agricultural practices with particular emphasis on banning burning of agricultural wastes. Widely available and efficient public transportation systems (like metro trains) can go a long way in reducing the use of private vehicles. Clean fuel and fuel efficient cars need to be promoted. City planning and architecture have a large role in creating an environment friendly city which conserves energy. Cities need to be planned with optimal density—neither too thinly spread nor too congested. Very tall buildings use up more energy (per household) and create heat islands. The materials used in buildings and insulation practices also have impact on energy conservation.

Rural to urban migration is a major source of contributing to the growth of urban population. Majority of the migrants are rural poor with low levels of education and skills. The migrants do find work in the city, but housing is a big problem for them contributing to the growth of slums. Migrants face a variety of discriminations and biases and suffer from insecurity and other vulnerabilities.

Poverty, in terms of conventionally defined measures may appear to be less in the cities, but poverty that pushes life to the edge appears in different forms and is often more acute in the cities.

Crime rates in cities are higher than in rural areas, and the rates of crimes have been increasing over time. Poverty, extreme inequalities and vast differences in lifestyles of different sections of urban inhabitants generate a feeling of alienation among the deprived ones. Poor housing, low education and family disruption reinforce that feeling of alienation. There is absence of a shared sense of community and mutual commitment. The community structure in the cities is too weak to exercise effective social control on individuals. Ethnic heterogeneity weakens the community structure. Opportunities that the city offers also create temptation for crimes. The new technology encourages new ways of committing crimes. It is not only the poor but the better off city dwellers also get involved in crimes. The quality of governance and law enforcement can of course make a difference.

But there is hope. The State of the World Population Report says, 'the current concentration of poverty, slum growth and social disruption in cities does point a threatening picture; yet no country in the industrial age has ever achieved significant economic growth without urbanization. Cities concentrate poverty, but they also represent the best hope of escaping it ' (UNFPA 2007, Introduction).

In what follows, we give a brief narrative on urban growth in India and then take up migration, poverty and urban crime for a little more elaborate treatment. The sections on urban growth, migration and poverty draw largely upon a recent paper by the author (Hashim 2020).

11.2 Urbanization in India

The level of urbanization is associated with the development and prosperity in the country. India experienced severe ups and towns in the level of prosperity in the past few centuries, and so did its level of urbanization. At just a little over 10% in 1911, urbanization in India was at its lowest of the last few countries (Hashim 2020), whereas the world then was about 15% urbanized. Since then, urbanization in India has progressed, but at a considerably slower rate than the urbanization in the world. Urbanization in India reached a level of 17.3% in 1951, 27.3 percent in 2001 and 31.2% in 2011. At 31.2% urban in 2011, India was still one of the least urbanized countries of the world. Since India has emerged as an economy with high potential for growth, and growth inevitably leads to higher levels of urbanization, one could expect an accelerated growth of urbanization in the future.

An important feature of urbanization in India is that there is a wide variation in the rates of urbanization in different states. If we exclude NCT of Delhi, which is a city state with a relatively small territory as compared to other states, the following four states emerge as highly urbanized (listed here in descending order of their level of urbanization): Tamil Nadu, Kerala, Maharashtra and Gujarat. The percentage of urban population in these states ranges between 48.4 and 42.6. These four states have 4 of the 8 Indian metro-cities of 2 million plus population.

Kerala particularly stands out as one of the two most urbanized states of India with not a single city larger than population size of 1 million in 2011. Three of the four most urbanized states, i.e., Tamil Nadu, Maharashtra and Gujarat are also the most industrialized states of India. Kerala is an example of an economy with a very high level of consumption multiplier boosted by income earned abroad. All these four states rank high among the states in terms of per capita income. The other states which have urbanization level higher than the all-India average are: Karnataka, Punjab, Haryana, Andhra Pradesh and West Bengal. The first four of these are prosperous states with high agricultural as well as industrial base. West Bengal's urban scene is dominated by the city of Kolkata. Kolkata emerged as the most important city of India during colonial times in the nineteenth century and maintained that status for a long time in the twentieth century. It was a leading industrial and commercial city of India along with Mumbai. West Bengal also has the highest number of smaller towns. All the other states have levels of urbanization below all-India average. In terms of economic development also the performance of these states has been much below average, Orissa. Assam, Bihar and Himachal Pradesh are still at a level of urbanization which was the average for India as a whole before independence and in early part of the twentieth century.

The overall picture that emerges from the foregoing review of the pattern of urbanization in India is that the level and the pattern of urbanization are closely related with the level of economic development and industrialization. The four metro-cities of Kolkata, Mumbai, Chennai and Delhi which dominated the urban scene during the 19th and much of the twentieth century and are still the four largest cities of India owed much of their primacy to the colonial history. They make the Indian urban scenario look like top heavy. Otherwise, the pattern of urbanization induced by economic and industrial development during the last 3–4 decades is much more even. Bangalore, Hyderabad, Ahmedabad, Pune and Surat are the newly emerging metro-cities. Indications are that by 2021. Bangalore and Hyderabad may overtake Chennai and Kolkata in size. Tamil Nadu is industrializing fast, but industries are concentrating more in middle-size towns. Bangalore and Hyderabad have developed big clusters of ICT industries. To the extent that the regional pattern of urbanization is uneven it is because the regional pattern of economic development is uneven.

Distribution of cities of population size 100 thousand and above is shown in Table 11.1. The total number of those cities in 2011 was 496 as against 394 in 2001. The total number of towns and cities, including the census towns, was 7935 in 2011 as against 4370 in 2001. Thus, between the two census years the growth in the number of bigger cities is about 26%, whereas the growth in the total number of cities and towns is about 81%. The number of smaller towns, including census towns, alone has gone up by 87%. This is quite a healthy trend. The size distribution of the 496 (100 thousand plus) cities also appears to be balanced with 450 cities in the size group of 0.1 to 1.0 million and only 8 cities in 4 million plus group.

When the economic activities in a village diversify and the village acquires characteristics to be called a town in terms of Census definition, the main feature of which is that the non-agricultural activities become dominant among the male agricultural workers, the Census of India recognizes that village as a town and designates them as 'census town'. These are not statutorily recognized by the state governments as towns. The census towns have grown at the fastest rate during the decade 2001–2011.

Some scholars believe that the pattern of urbanization in India is top heavy on the ground that 23.5% of the urban population in India lives in 5 million plus cities as compared to 7.5% in Europe, 9% in Africa and 15.5% in the world on an average (Kundu 2014). Comparison with highly urbanized regions of the world (Europe has 80 plus percent of urban population) appears to be somewhat misplaced. At low level of urbanization, when the urban population as a proportion of total population is very small and the number of cities is fewer, it is natural that the population which lived in big cities will bear a large proportion of the small urban population. At higher level of urbanization, a much larger proportion of population will live in smaller cities. Africa also is more urbanized than India, and the world on the whole has crossed urbanization level of 50%. The big metro cities in India performed a lead role in attracting technology and industry and in promotion of science and education even when overwhelmingly large part of the country was rural.

11.3 Rural to Urban Migration

Migrants from rural areas have been moving to urban centres over a long period of time. Some migrants settle down permanently in the city. Some stay in the city for a longer period but maintain connection with their places of origin. Those

Table 11.1 Level of urbanization and size distribution of cities for states with population more than 5 million in 2011

States	Population	Urban	No. of cities by population size (million)			
	(Million)	%	0.1–1.0	1.0–2.0	2.0–4.0	4.0+
NCT of Delhi	16.8	97.5	14	–	–	1
Tamil Nadu	72.1	48.4	29	2	–	1
Kerala	33.4	47.7	7	–	–	–
Maharashtra	112.4	45.2	34	8	2	1
Gujarat	60.4	42.6	25	2	–	2
Karnataka	61.4	38.7	25	–	–	1
Punjab	27.1	37.5	15	2	–	–
Haryana	25.3	34.9	19	1	–	–
Andhra Pradesh	84.5	33.5	39	2	–	1
West Bengal	91.2	31.9	59	1	–	1
Uttarakhand	10.1	30.2	6	–	–	–
Madhya Pradesh	72.6	27.6	28	4	–	–
Jammu and Kashmir	12.5	27.4	2	1	–	–
Rajasthan	68.5	24.9	26	2	1	–
Jharkhand	33.0	24.1	8	2	–	–
Chhattisgarh	25.5	23.2	8	1	–	–
Uttar Pradesh	199.8	22.3	57	4	2	–
Orissa	42.0	16.7	10	–	–	–
Assam	31.1	14.1	4	–	–	–
Bihar	104.1	11.3	25	1	–	–
Himachal Pradesh	6.9	10.0	1	–	–	–
India	1210.0	31.2	450	33	5	8

Source Constructed from Provisional Tables, Census of India 2011

who have stayed for shorter duration and those who commute for jobs on a daily basis are not counted as migrants in the major sources of data. Only those who have stayed in the place of their destination for at least six months are counted as migrants. However, the shorter duration migrants and the commuters also put considerable strains on the limited infrastructure and amenities of the city.

As per census of India, 2011, the total number of migrants in India was 454 million on the basis of place of birth criterion. This is a cumulative figure over time and could be described as the number of people on the move which includes rural to rural, rural to urban, urban to rural and urban to urban permanent and temporary migrants. It is estimated that there were 100 million rural to urban migrants (Working Group 2017). Srivastava (2020) estimates that short term and circular migrants in the informal economy (including both rural to rural and rural to urban migrants could number 60–65 million and with family members their number would not be less than 100 million. He further notes that 40% of these migrants work in construction sector and 15% in agriculture.

The numbers involved are staggering, and this is with slow pace of urbanization and reluctant migrations. It has been often argued that there have been political attempts to discourage migration and is this context a variety of rural schemes like MGNREGA, Indira Awas Yojana and other state and community specific programmes are citied as attempts to retain people within the villages. Non portability of entitlements are the source of a lot of deprivations to the poor migrants. People who are registered to claim access to a number of legal and other entitlements at the place of their residence, loose access to these benefits upon migrating to other places. This is particularly aggravated in the case of inter-state migration and is further complicated by barriers of language, etc., (working Group 2017). Migration to urban areas is slow also because the cities have not created the job opportunities enough to attract more migrants. Economically more developed states like Punjab, Haryana, Gujarat, Maharashtra, Karnataka and Tamil Nadu have been major recipients of migrants, and other states were the net senders of migrants in 2011 (Mistri 2015). During the decade 2001–2011, Tamil Nadu received the highest proportion of migrants (Mistri 2011). Tamil Nadu has emerged as the most advanced industrializing state during the last two decades. Migrants from as far as UP, Bihar, Odisha and North Eastern states are migrating to Tamil Nadu breaking the barriers of distance and language. It is also true that the Indian migrant does not want to flock to the city losing the security of living with the family and the community in the village, even though in poverty, without assessing the opportunity of work in the destination through well-established network of pioneering migrants from the community.

Living on the edge literally is how the life of a migrant worker could be described. A little nudge, and he falls from the edge. This could have never been demonstrated so vividly as during the crisis created by Lockdown in the wake of Covid 19 pandemic (March to May 2020). A very large number of workers lost their jobs and with that the means of their very existence. They lost their shelter and their hope. The lockdown had closed all the means of transport. Having no alternatives and afraid of dying of hunger, many alone or with family, started walking to their home destination, hundred or sometimes thousands of miles away. A number of them died on the road. Some of them managed to hire unauthorized private vehicles, mostly trucks, at exorbitant cost throwing away all their savings. They suffered from police actions and unwelcome reception back in the home village/town. A vivid description of their plight is available in Wikipedia, 2020, COVID-19 Pandemic in India. According to the Wikipedia, those who stayed back suffered discrimination and assault, even from neighbours. Upon return to their home destination, they were treated with either fear or class bias. Migrants travelling by special trains (after end of May 2020) reported that food and water were either not provided or dumped at the entrance by the trains leaving them crowding and fighting with each other for their share. All this happened in spite of the efforts on the part of the governments to help them. The crisis was simply too big to be managed with the given capabilities of the administration. There is no exact assessment of numbers involved as yet. According to some estimates about 10 million workers might have been involved. But the numbers were significant enough to induce a panic among the employers that they would not be able to resume their business and industrial work after the crisis melted down. It is reported that some workers have started returning. But it will take some more time to have an assessment of the full impact of the crises. Not all the workers might return.

11.4 Urban Poverty

Traditionally, the concept of poverty in India, has been food-centric. The concept evolved in the background of acute food shortages and hunger in India during the 19th and more than half of the twentieth century. It was in this background that Dandekar and Rath (1971) defined the poverty line as the expenditure level at which the average calorie intake met the norms of 2250 cal per capita per day. The recommended poverty line was Rs. 22.50 and Rs. 15.00 per capita per month respectively for urban and rural areas on the basis of the consumer expenditure data for the year 1960–61. This poverty line was considerably lower than a national minimum recommended by a Distinguished Group set up by the Planning commission. Without any reference to the caloric norm, the Group had, for the year 1962, recommended Rs. 25 per capita per month (pcpm) for urban and implicitly Rs. 18.6 pcpm for rural areas, and this was supposed to be in addition to state-sponsored free education and health services and state subsidized housing for the poor in urban areas (Lakdawala Committee 1993). A Planning Commission Task Force (Alagh Task force 1979) raised the calorie norms to 2400 and 2100 cal pcpd for rural and urban areas respectively. Using the NSS data on consumer expenditure for the year 1973–74 the Task Force recommended a poverty line of Rs. 49.1 (rural) and Rs. 56.6 (urban) for the year 1977–78. This poverty line became the basis of poverty estimates and with some adjustments still remains at the base of official poverty estimates. Lakdawala Expert Group (1993) made recommendations mainly for using state specific prices and their state specific indexation for updating the poverty line, and also for estimating poverty first at the level of states and then aggregating to get national estimates. Tendulkar Expert Group (2009) recommended that the urban poverty line consumption basket be taken as the reference also for the rural areas. In spite of all these revisions, a basket of consumption in 1973–74 and a food-centric approach still remains at the base of poverty estimates. Tendulkar Expert Group's methodology has been the basis of official poverty estimates from the year 2004–05 onwards. The estimates are given below (Table 11.2).

These estimates suggest a sharp decline in urban poverty from 25.7% in 2004–05 to 13.7% in 2011–12. These estimates do not appear to reflect the urban reality at all. Food is no longer the most important of deprivations in urban India. Availability of food has improved considerably and with rising incomes share of food in the consumption basket is going down. Share of food in the consumption basket was 54.7% in 1993–94. It came down to 38.5% in 2011–12 (Rangarajan Expert Group 2014). Housing has become the most acute problem in cities. It is not so difficult to find work in a city if only one has a foothold, i.e., an affordable place to live at a commutable distance from work. Larger the city, more is the work opportunity but also more difficult is the housing problem. Most of the workers start living in slums, but many of them cannot afford a shelter even in a slum. 17.2% of the urban households lived in slums (GOI 2019). Among the larger cities of India the proportion of slum dwellers is much higher.

Table 11.2 Percentage of population below poverty line

	Urban	Rural	All
2004–05	25.7	41.8	37.2
2009–10	20.9	33.8	29.8
2011–12	13.7	25.7	21.9

Source Press note on poverty estimates, 2011–12, Planning Commission, GOI

Cost of housing has pushed up the poverty level even in the developed countries. The Economist (Sept. 28, 2019, p.5) reported about California (USA) that 'despite the progressive states' more generous safety net, the out of control housing costs push more people into destitution than elsewhere.

Apart from the housing costs, there are other severe vulnerabilities from which the urban poor suffer. Working as self-employed in some of the lowest earning work or as casual wage labour, the poor is always living on the edge. Vulnerability is compounded by non-existence of social security and non-availability of family and community support and because of social biases for the socially low status workers. This was amply demonstrated during COVID-19 lockdown. It pushed not less than a million urban workers on the road to nowhere.

The cost of health services and education is another element in poverty. Urban poverty has to be perceived in terms of these deprivations and vulnerabilities rather than just the availability of food. Capturing these vulnerabilities and defining poverty in terms of these also requires relevant data to be generated.

An Expert Group (Hashim Committee 2010) was set up by the Planning Commission to evolve a methodology to identify the urban poor and measure the extent of urban poverty taking into account the real deprivations of the urban poor. The Expert Group classified the vulnerabilities of the urban poor into three categories: residential, occupational and social. The residential vulnerabilities included houslessness or a house with roof and wall made of plastic or polythene, grass thatch, bamboo, mud, unburnt bricks etc. Lack of water connection and non-availability of sewer, drain, electricity, etc. added to these vulnerabilities. The quality of surrounding environment in which the house was located also had a bearing on the score for vulnerabilities. Among occupational vulnerabilities, the Expert Group included begging, rag picking, uncertain employment, irregular wage payments etc. Social vulnerabilities included child headed households, households with all or some members with disability or chronic illness. The Socio-Economic and Caste Census 2011 (SECC) was requested to include a questionnaire in order to generate data helpful to prepare scores on the identified vulnerabilities. The questionnaire was prepared by the Expert Group in association with the Ministry of Housing and Urban Poverty Alleviation (MOHUPA). The SECC was planned to capture the residential status, the physical conditions of dwelling, the number of rooms occupied, provision of civic amenities, occupational status, employment conditions, health and education status, age and social vulnerabilities.

The Group recommended a three-stage identification process: (a) automatic exclusion, (b) automatic inclusion and (c) scoring index. The methodology was worked out keeping in view the data availability. Full data from SECC have not been released. On the basis of some data that was available for testing the methodology, it was reported that the urban poverty in India was 35% in the year 2011–12 (Hashim 2016). Further work is needed in this direction. But what is even more important is to evolve a framework for data collection on poverty related indicators on a periodical basis.

Housing has become one of the acutest problems in the cities, not just for the poor or the immigrants, but also for the middle income people, since the cities have started expanding at a fast rate. The main reason is that the land costs rise when the city grows. Regulatory solutions to the rising costs of land and housing do not work as the economic forces are stronger. Houses that are subsidised for the poor change hands and pass into the possession of better off families. A proximate solution to the housing problem is an efficient system of transportation which makes commutation for work possible from long distances at affordable costs. Middle and high income people are already finding solution to housing in suburbanization, depending on private vehicular transportation. But for the low-income workers, a system of affordable public transportation is the only solution.

Slums in big cities have been the result of unsolved and neglected housing problems over a long period in the past. Conditions of slums have been deteriorating further. Very high density of population, poor conditions of inner roads, nonexistence of drainage, poor sanitary condition, lack of arrangements for waste disposal, poor water supply and lack of other amenities make for unhygienic and unacceptable living conditions. Improvement and replacement of slums is a big issue in urban management. Some experiments have been successfully made in finding solutions with the involvement of the people themselves.

11.5 Urban Crimes

It is a general observation that the crime rates are higher in urban areas than in rural areas and are also higher in larger cities than in smaller cities. On the surface it would appear that larger the cluster of people more crime prone is the environment. Cities are large clusters of people. But the cities have existed for a long time, but were not, in the past, particularly known for crime prone environment. Crime proneness of cities has been increasing over time. Could it be because of the transformation of the society from one with rural dominant ethos in terms of culture, social values and aspirations to one with urban dominant ethos? Questions like this as yet do not have a definite answer, but tentative explanations to this and such other questions offer some testable hypotheses. Even though, cities have existed for a long time in history, the rural population was dominant, and the general ethos of the people had evolved in the rural background which prevailed over in cities. The rural ethos was characterized by joint family system and strong community bonds. The family and the community had traditionally evolved a value system which anchored the behavior of each member of the community. The family and the community exercised a control on the behavior of the individual even when the person was away from the family. With growth in urbanization the traditional value system starts becoming weak. With the fast urban growth and the spread of urban influence the rural ethos gives way to a newly evolving urban ethos moulded by compulsions of urban living. The urban ethos is characterized by individualism, anonymity, consumerism and aspirations of high living. With fast communication and with increasing interaction between the city and its hinterland the urban ethos begins to dominate even the rural areas. Urban lifestyle, consumption pattern and aspirations start influencing the rural society as well. This could be termed as 'creeping urbanism'.

In the period between 1920 and World War II, sociologists associated with the University of Chicago began to construct explanation concerning why cities might have higher crime rates than the hinterland (Urban Crime 2020). At that time, it was generally believed that crime in the city was caused by the influx of immigrants and especially those from 'crime prone ethnic groups.' Chicago school scholars concluded that it was not criminogenic characteristics of ethnic groups that contributed to higher rates of crimes, but the nature of the 'urban ecology' in which they lived. It was the poor neighbourhood with great levels of poverty, racial heterogeneity, transience and family disruptions which informed the crime ecology in which mostly the poor immigrants lived (Urban Crime 2020).

The roots of modern theories of urban crimes can be traced to theories of the Chicago school. The most prominent among these is the sociological disorganization theory which also appears to be more relevant to India. Sociological disorganization theory implicates mainly the racial and ethnic heterogeneity, residential mobility, low socio-economic status, family disruption and relative poverty. It points to a lack of solidarity and cohesion and absence of a shared sense of community's capacity for informal social control (Urban Crime 2020).

The American sociologist, Robert Nisbet reorganized these ideas in 1966 and identified four processes that were particularly significant (Sijuwade 2014): (1) conflict of institutions, (2) social mobility, (3) individuation and (4) anomie. In the pre-industrial urban communities' kinship and religion were the dominant institutions. Since the advent of industrialization, economic, political and social service-oriented institutions have become competing institutions. Rise of new institutions as the corporate bureaucracy, public education and welfare, mass communication, political parties, labour unions and the suburban complex, etc., has led to pluralism characterized by a wide range of competition and social strains, leading to almost endless conflict of goals and behaviour patterns. Rigidly fixed social status has been supplanted by a high degree of geographical mobility and vertical social mobility. Intensive strewing for status becomes an obsession for larger number of people leading to conflicts. Individuation refers to a decline and weakening of the individual's ties to such social groups as the extended family, neighborhood, guild or parish and the maximization of the autonomous individual and the impersonal, atomistic and mechanical relations. Anomie is a state of normlessness, when values become confused, when they are in conflict with one another or when they lose their immediacy to human beings. Anomie with its implicit tensions of moral conflict, alienation and meaninglessness is a notable and persistent aspect of modern urban societies. There are elements in these theories which nearly accurately describe the situation in India, prevalent particularly in the last few decades.

Studies of crimes in India are few. Jean Dreze and Ritika Khera (2000) studied crime, gender and society in India, using data for the year 1981 from 'Crime in India', an annual publication of Government of India. District level data on a range of crimes were available, but the study concentrated on in murder data, which, the authors thought, would be relatively more accurate, since data on other crimes would be subject to high degree of under reporting. The data include both rural and urban crimes. Examining the trend in murder rates (murder per million of persons) from 1950 to 1995, they found that there was a significant upward trend from mid-seventies. In 1995, the murder rate of 41 per million was not particularly high for a country Like India, the authors perceived. Also they found that murder rates were not markedly higher in urban areas. A significant finding of this study was that murder rates in India were highly correlated with male–female ratio in the population——districts with higher female ratio had lower murder rates. The authors give a plausible explanation that low female–male ratios and high murder rates were joint symptoms of a patriarchal environment. The study also suggested that gender-relations, in general had crucial bearing on criminal violence.

Verma and Kumar (2008) have analyzed the major reasons of criminality among convicted offenders in India. Results suggest that growing urbanization, breakdown of extended family system, lure of consumerism and influence of criminal associates are some of the main reasons for these people to fall into a life of criminality. It is also suggested that socio-economic circumstances and failure of the justice system to control social conflicts are important causes of criminal behavior. They also found that there was a preponderance of persons with low education and low income. There were only a little over 2 percent graduates among the criminals. However, it is difficult to functionally relate education or low income with criminality on the basis of the observations in this study considering that fact, that there is a preponderance of people with low education and low incomes in the entire population, urban or rural.

National Crime Records Bureau (NCRB) compiled and collated data on crimes from state records and report in their annual publication Crimes in India. These data are available in a comparable series from 1980 onwards. The data are reported under two major categories. Indian Penal Code (IPC) crimes and Social and Social Law (SLL) crimes. The data are combined for rural and urban areas and are separately presented for metropolitan cities. The compilation of data on crimes follows the 'principal offence rule', which is that among many offences registered in a single FIR, only the most heinous crime (which attracts maximum punishment) will be considered as counting unit. For example, murder with rape will be counted as murder only. Hence the actual count of each crime head (and hence the total number of crimes) may be undercounted. This may be in addition to under reporting of crimes in general, particularly is remote and rural areas. Murder has less chance of being under reported. However, over the years reporting of crimes may have improved on account of certain citizen centric public initiatives like launching of e-FIR facility or women helplines, etc. Subject to these limitations, the data are useful in providing some insights into the level and trends of crimes in India. We will discuss only the IPC crimes in what follows.

In Table 11.3 we present the data on IPC crime rates (all India, rural urban combined) at an interval of 5 years from 1980 onward and annually from 2011 to 2019. The crime rates remained quite stable around 187 (per 100

Table 11.3 Rate of IPC crimes over the years 1980–2019 (all India)

Year	Crime rate	Year	Crime rate
1980	206.2	2012	196.7
1985	184.4	2013	215.5
1990	194.0	2014	229.2
1995	185.1	2015	234.2
2000	176.7	2016	233.6
2005	165.3	2017	237.7
2010	187.6	2018	236.7
2011	192.2	2019	241.2

Note Crime rate = number of crimes per 100 thousand persons. Population is taken as projected population of each year

Source NCRB Reports (Volume I, Statistics), 2019

thousand persons) from 1980 to 2012. There was a sudden spike in the rates from the year 2013 and it continued to rise reaching a rate of 241 in 2019. It is difficult to attribute any definite reason for it. One remarkable thing that happened in India during the second decade of the new millennium was unprecedentedly high penetration of internet. The internet penetration rate went up from just 4% in 2007 to 15% in 2013 and to 50% in 2020. As a result of very high. rate of growth of GDP during the first decade of the new millennium and particularly during the years 2004–05 to 2007–08, a level of prosperity had already been achieved and cell phone communication had already penetrated the rural areas. Subsequently, with very high rate of internet penetration, the urban rural divide in the spread of consumerism was bridged to a large extent. This could be called the penetration of urbanism. Could this also be a factor in leveling the crime rates between the rural and the urban areas and pushing up the overall crime rates. The suggestion is only speculative at present. But it can be said with some degree of confidence that poverty is not the main factor behind rising crime rates. In fact, prosperity could possibly be more convincingly implicated.

Table 11.4 presents data on recent trends in a few selected crimes. The murder rates are seen to have declined slowly over the years. In 2010 the murder rate was 2.8 (per 100 thousand persons) which declined to 2.2 in 2019. Dreze and Khera (2000) had reported a crime rate of around 4 (or 41 per million) for early eighties. Murder rate in India appears to be quite low. Murder rate in USA is reported to be 5 per thousand, which has declined from the peak of 10 during the decade 1975–1995 (Mirabile and Nass 2018). The USA has perhaps the world's highest rate of murder because of the freedom to own guns, perhaps. A traditional crime like robbery does not appear to be increasing. Crimes against women show a tendency of rising over the years. Part of the rise could be due to better reporting, but still the rise in the rate is more than what could be attributed to just better reporting. Theft shows a high and a significantly rising tendency. A good part of theft in recent years is auto theft which has been increasing fast.

Crime rates for 19 mega cities (with 2 million plus population) are presented in Table 11.5. On an average, the crime rates in these cities are more than double the average for the country as whole. There appears to be, however, no relationship between the city size and the crime rates. Mumbai, Kolkata, Chennai and Coimbatore had the lowest crime rates among these cities. Delhi and Jaipur had the highest crime rates. Other cities showing very high crime rates (more than the average) are: Kochi, Patna, Indore and Lucknow. Jaipur, Delhi and Indore show the highest rates of rape, kidnapping and abduction. Surat and Ahmedabad also had high crime rates. While the sample is rather small to reach definite conclusions, there is a tendency for the crime rates to tilt on the higher side in the cities of North and North- West India.

Social and economic factors behind urban crimes are complex. Factors like breakdown of family and community bonds, individual's

Table 11.4 Rates of selected IPC crimes, all India

Crimes heads	2010	2011	2012	2013	2017	2018	2019
Murder	2.8	2.8	2.8	2.7	2.2	2.2	2.2
Rape	3.9	4.1	4.3	5.7	5.2	5.5	4.9
Assault on women	7.1	7.3	7.7	11.9	13.8	13.8	13.6
Kidnapping and abduction	3.2	3.7	3.9	5.3	7.4	8.0	7.9
Cruelty against women	16.4	16.9	18.2	20.1	16.8	16.0	19.3
Theft	27.9	28.2	27.8	30.3	45.7	47.3	50.3
Auto theft	–	–	–	–	17.5	17.7	17.8
Robbery	2.0	2.0	2.3	2.6	2.4	2.3	2.3

Source NCRB Reports, Volume I, Statistics for 2014 and 2019

Table 11.5 IPC crimes rates in mega cities

		IPC cognizable crimes			Selected crimes,2019		
	Mega Cities	2019	2018	2017	Rape	Kidnapping/ abduction	Population (million)
1	Mumbai	221	221	212	4.6	11.4	18.41
2	Delhi	1806	1385	1306	16.2	35.2	16.32
3	Kolkata	136	140	141	0.2	2.7	14.11
4	Chennai	208	232	221	1.0	0.6	8.70
5	Bengaluru	321	362	555	3.5	12.4	8.50
6	Hyderabad	198	185	187	4.9	6.7	7.75
7	Ahmedabad	423	453	288	2.8	3.9	6.35
8	Pune	260	229	300	3.8	13.0	5.05
9	Surat	489	680	264	2.1	7.8	4.59
10	Jaipur	1080	740	693	35.6	27.0	3.07
11	Kanpur	259	256	256	1.8	10.0	2.92
12	Lucknow	565	592	600	5.7	9.3	2.90
13	Nagpur	423	412	453	6.0	21.5	2.50
14	Ghaziabad	388	499	352	5.4	17.3	2.36
15	Indore	656	640	612	14.4	27.0	2.17
16	Coimbatore	119	134	144	0.1	0.3	2.15
17	Kochi	685	963	809	8.6	1.0	2.12
18	Patna	663	574	751	1.4	38.8	2.05
19	Kozhicode	249	195	235	7.2	1.3	2.03
	Mega cities simple average	530	478	462	6.1	13.0	–

Notes
(1) Rate is number of crimes per 100 thousand population (population is taken as in 2011)
(2) Mega cities are cities with 2 million plus population in 2011
(3) Cities in the table are arranged in descending order of population
Source NCRB Report, Volume I, 2019

detachment from traditional value systems, huge urban sprawls creating anonymity, social heterogeneity, etc. would gains further dominance as urbanization grows. Economic factors like unemployment and inequality appear to be rising along with economic growth spurred by globalization and new technology. The ICT technology creates new opportunities for cybercrimes. Even the developed countries have found it difficult to resolve all the issues. Very strong social movements and political will is required to create a dent in the crime prone environment. But the crimes can be kept under reasonable restraint and control by good governance and good policing.

A recent study on police adequacy and working conditions in India (CSDS – 2019) brings out some interesting features of the status of policing in India. The study was based on a sample survey of 11,834 police personnel across 105 locations in 21 states of India and was conducted during the year 2018. The picture that emerges is far from being satisfactory in respect of the adequacy of police personnel, their training and their working conditions. Long hours of work, lack of weekly rest, inadequate provision of basic equipment and resources as well as external pressure in work are commonly reported by police personnel across states. The police, at all-India level, worked at 77% of its sanctioned strength. Over the last five years, on an average, only 64% of the police force were provided in-service training, and this included much higher proportion of senior officers. The ground level personnel who come face to face with the public, woefully lacked in education and training. The police stations were poorly equipped. The relatively deprived sections of the society, i.e., SC, St, OBC and minorities as well as women were poorly represented in the force. The police personnel were subject to frequent transfers—the average stay in one posting being less than two years. The police personnel were overworked, working for 14 h a day on an average and without being paid for overtime.

The police is the only organ of society and administration which comes face to face with crime, and hence the perception of the police personnel in this regard may offer some useful insights. The general perception of the police personnel (pp) was that the crimes had gone up in recent times. A majority of them also believed that the number of crimes reported was less than the crimes actually committed. It was also believed that the common people were hesitant to approach the police even when there was need. Among those people who believed that crime had gone up, 50% believed that the society related issues were the most important reasons for this. Economy related issues were also mentioned. The society related issues mentioned were population overcrowding, increase in number of migrants, lack of education, new technology, political pressure, corruption and caste-based violence. The economic issues that were mentioned were poverty, inequality and unemployment. A small proportion of respondents believed that inadequate policing was responsible for the increase in crimes. Among those who thought that the crime rates were on the decline, about 50% attributed the decline to better policing. This asymmetric perception of the pp about the role of policing in the rise and fall of crimes is understandable. Quite a significant number believed that pressure from politicians was the biggest hindrance in crime investigation.

It has been noted earlier in this paper that migrants to the cities are at the receiving end in the urban social and economic structure. Police personnel were also more likely to report that migrants from other states are very much naturally prone to committing crimes. This perception was strongest in Maharashtra, Andhra Pradesh and Chhattisgarh. There were other biases as well. Many of the pp also expressed the view that complaints against gender violence were generally bogus and that the people from transgender and minority community were more prone to crime.

Adequate staffing of police force and substantive improvement in their working conditions including the adequate provision of necessary modern equipment to help them perform is important. And even more important is proper training particularly of ground level police personnel in modern ways of effective investigation and in moulding their perception and attitude towards an urban society which necessarily has to welcome more migrants, and which is becoming ethnically, culturally and linguistically more heterogenous. Good governance and policing are of utmost importance for creating a livable and sustainable urban environment.

11.6 Summing Up

Given the nexus between the process of economic growth and urbanization, the pace of urbanization is India will inevitably accelerate and will also create more of urban sprawls and agglomerations. Clustering is of tremendous advantage to growth of certain industries and also to the growth of knowledge and ideas. Attention to planning with a vision, provision of good infrastructure and amenities and facilitating convenient interaction between various satellite units of the agglomeration will go a long way in making these agglomerations work as efficient urban units. Small and medium size towns have been growing at faster rate. They need to be statutorily recognized and their growth needs to be supported with good town planning, financial support and better governance structures so that they are able to attract industries and create employment. They have the potential of absorbing surplus agricultural labour to a great extent.

Migrants contribute to urban growth and are needed for growth of industries and services. They suffer from biases and discrimination in their host locations. A campaign to educate city dwellers about the contributions that migrants make would be useful. Housing problems are the acutest for migrants and the urban poor. A good transport network providing affordable long-distance transport to ease commutation can go a long way in easing the housing problem. For identifying the urban poor and for estimating the extent of urban poverty, a methodology which recognizes the residential occupational and social vulnerabilities of the poor needs to be evolved and a system for collection of relevant data with regular periodicity needs to set up.

Social and economic factors behind urban crimes are complex. Some of the factors which appear to be implicated, like, breakdown of family and community bonds, individual's detachment from traditional value system, huge urban sprawls creating anonymity and social heterogeneity, etc., are likely to gain further dominance with growing urbanization. Economic factors like unemployment and inequality appear to be rising with economic growth spurred by globalization and the new technology. Facilitated by relative prosperity and fast communication, urban ethos is spreading over rural areas also. The overall rate of crimes is rising, and the crimes are taking new forms. Very strong social movements and political will is required to create a dent in the crime prone environment. But the crimes can be kept under reasonable restraint and control by good governance and good policing.

Adequate staffing of police force and substantive improvement in their working conditions is important. Equally important is proper training and education of the police force with particular focus on the low hierarchy police personnel who come face to face with the public. The biases that generally prevail against the poor, the immigrants and certain sections of people can only be removed by proper orientation and training. Training in the use of modern technology and modern ways of effective investigation can go a long way in making the police force more effective. The country has to be prepared for dealing with a society which is culturally, ethnically and linguistically more heterogenous and which is adopting increasingly more to the urban ethos. Good governance and effective policing are of utmost importance in creating a livable and sustainable urban environment.

References

Alagh Task Force (1979) Task Force on Projections of Minimum Needs and Effective Consumption Demand, Planning Commission, GOI

Annez, Patricia Clark, Robert M Buckley (2009) Chapter I in Urbanisation and Growth, Eds., Michael Spence, Patricia Clarke Annez and Robert M Buckley, Commission on Growth and Development, The World Bank

CSDS (2019) Status of Policing in India Report 2019, Police Adequacy and Working Conditions, Common Cause and Centre for the Study of Developing Societies

Dandekar VM, Rath N (1971) Poverty in India, Indian School of Political Economy, Poona

GOI (2019) Handbook of Urban Statistics, Ministry of Housing and Urban Affairs, Govt of India

Hashim Committee (2010) Report of the Expert Group to Recommend the Detaild Methodology for Identification of Families Living below Poverty Line in the Urban Areas, Planning Commission, GOI

Hashim SR (2016) Counting the Poor: Changing Perspective with Reference to Poverty in Urban Areas, in IASSI quarterly -Contribution to India Social Science, Nos. 3 and 4, Volume 35, July-December

Hashim SR (2020) Urbanisation in India, Issues and Challenges in IASSI Quarterly—Contribution to Indian Social Science, Volume 39, No 3, September

Jean Dreze, Ritika Khera (2000) Crime, gender and society in India, Insights from Homicide Data, in Population and Development Review, Volume 26, No. 2, June

Kundu Amitabh (2014) Exclusionary growth, poverty and India's emerging urban structure in Social Change, Volume 44, No 4, December

Lakdawala Committee (1993) Report of the expert group on estimation of proportion and number of poor, Planning Commission, GOI

Mirabile Francesca, Daniel Nass (2018) The TRACE: thetrace.org/2018/04

Mistri Avijit (2015) Estimation of internal migration in India, 2011 Census, based on Life Table Survival Ratio Method, online portal on 26 Jan, 2015, at https://mpra.ub.unimuenchen.de/615771/

NCRB (2019) Crimes in India 2019, Volume I, National Crime Records Bureau, Ministry of Home Affairs, GOI

Rangarajan Expect Group (2014) Report of the Expert Group to Review the Methodology for Measurement of Poverty, Planning Commission, GOI

SECC (2011) Socio Economic and Caste Census, 2011, Census of India, GOI

Sijuwade Philip O (2014) Urban Crime, Unrest and Social Control in Global Journal of Interdisciplinary Social Science

Srivastava Ravi (2020) No relief for the nowhere people, The Hindu, May 4, 2020

Tendulkar Expert Group (2009) Report of the Expert Group to Review the Methodology for Estimates of Poverty, 2009, Planning Commission GOI

The Economist, September 28, 2019

Urban Crime (2020) Urban Crime/ Encyclopedia. Com (updated Dec ember, 12, 2020)

UNFPA (2007) The State of the World Population Report, 2007, Unleashing the Potential of Urban Growth. www.unfpa.org>files>695filenameswop 2007engPDF

Verma Arvind, Manish Kumar (2008) The Etiology of Crime in India, International Journal of Criminal Justice Sciences, Volume 3, Issue I

Wikipedia (2020) Covid—19 Pandemic in India, March, 2020

Working Group (2017) Report of the Working Group on Migration, Ministry of Housing and Urban Poverty Alleviation, GOI

12 Are Smart Cities Inclusive? Issues and Concerns

Shipra Maitra

12.1 Introduction

India's current urban population is around 377 million, likely to be nearly double by 2030. According to McKinsey Report (2014), by 2030, India will need to revive 468 cities with more than one lakh population to cope with fast urbanisation growth. Cities are likely to generate about 70% of national income in near future though supporting around 40% of total workforce. Indian economy has grown by more than 7% on average during last decade. However, employment has not kept pace with the GDP growth. Agriculture still accounts for large proportion of work force despite fast-declining contribution to GDP. Urban poverty is around 27%, mostly concentrating in small and medium towns. The self-employed employment category, with very low earnings has increased in proportion over time while that of regular employment and wage-paid casual labour have decreased. The restructuring and dismantling of large industries in cities for reducing environmental threats, have fostered casualisation of workers. In order to increase competitiveness of the cities, modernisation and capital-intensive techniques has forced casualisation of labour. All these have a negative impact on local resource generation resulting in urban local bodies becoming more and more dependent on external resources.

The 2030 Agenda for Sustainable Development Goals (SDG), approved by the United Nations is a "Plan of Action for People, Planet and Prosperity" adopted by many member countries for implementation. India as a member country of the United Nations and signatory to UN Sustainable Development Summit in September 2015 has adopted the SDG 2030 as the guiding framework for its National Development Agenda till 2030. The prime objective of this universal agenda is to eradicate poverty in all its forms and dimensions including extreme poverty that is the greatest global challenge and an indispensable requirement for sustainable development. The SDG 2030 includes 17 Goals with 169 targets.

The framework is built on the Millennium Development Goals and aims to realise the human rights of all and to achieve gender equality and empowerment of women by stimulating action in the areas of critical importance of humanity: economic, social and environmental. Targets for Inclusive city (SDG 11) are chosen to achieve sustainable urbanisation.

India's effort to achieve sustainable urbanisation is evolving in the spirit of realizing the SDGs, with focus on SDG 11 (Inclusive Cities). It is based on competitiveness, sustainability and inclusiveness with sharp focus on efficient delivery of key public goods and services like

S. Maitra (✉)
Institute for Human Development, New Delhi, India
e-mail: shipra.maitra@gmail.com

A. Alakshendra et al. (eds.), *Development, Inclusion and Sustainability*,
Sustainable Development Goals Series, https://doi.org/10.1007/978-981-97-6863-9_12

health, education, power, water supply and other infrastructure including connectivity. This will help in promoting entrepreneurship and private investment with optimum mix of public–private partnership. The framework identifies some major components of equitable development, i.e., sustainable livelihood, human capital, social development with inclusiveness, governance and environmental sustainability. The flagship programme of the Government of India, the Smart City Mission (SCM) tries to address these challenges to make its cities inclusive.[1]

A smart city is one that uses Information and Communication Technology (ICT) to enhance quality of life. It seeks to improve urban service delivery and engage effectively and actively with its citizens. The SCM tries to achieve urban transformation through wider and deeper application of ICT. The vision is to preserve India's traditional architecture, culture and ethnicity while implementing modern technology to make cities liveable, use resources in a sustainable manner and create an inclusive environment. The promises of the SDM include reduction of carbon footprint, adequate water and electricity supply, proper sanitation including solid waste management, efficient urban mobility and public transport, affordable housing, robust IT connectivity and digitalization, good governance, citizen participation, security of citizens, health and education.

Smart cities will generate several common assets through infrastructural investments. Participatory city governance is imperative for maintenance of such assets in the long run. Cities must own their resources for fair distribution and regulate livelihoods of their residents for inclusive growth. Sustainable spatial and economic development planning is needed to arrest skewed spread of urbanisation and economic opportunities. Uncoordinated and selected development incentives without strong citizen and beneficiary involvement may defeat the objective of SCM to build a competitive yet inclusive city. This paper tries to explore how SCM is achieving its major objective of participatory governance.

This paper attempts to analyse how the smart cities are facing City Challenges with 20 selected Smart Cities. It takes examples of two cities, i.e., NDMC and GIFT City to find out (1) how effective is public participation in guiding city development and (2) scopes of increasing investments through PPP mode. Future course of urban management in India will be significantly influenced by experiences of smart cities, whether they have been able to become inclusive for sustainability.

12.2 Urban Development in India[2]

The process of urbanization is continuing but not accelerating fast in India. It took four decades (1971–2011) to increase the share of urban population in total population by 10% i.e., from 21 to 31.16%. Still now, with a little more than 377 million, it is about a third of India's total population. It is estimated that the pace of urbanization will be a lot faster in the coming decades, from 377 (31%) to 600 (41%) million by 2031 i.e., two decades (UNFPA, 2016). The major indicators of urbanization are demographic and spatial. The demographic components are represented by the proportion of urban population in total population and growth rate of urban population. The proportion of urban population has increased over time as revealed in every Census. But the growth rate of urban population declined from 3.1% in 1991 to 2.76% in 2011.India, having the second largest urban population and experiencing much faster urban growth, compared to earlier decades, faces

[1] Maitra, Shipra. 2021. "Sustainable Cities for Sustainable Development, in NAGARLOK" Vol LIII.(4): 16–37, October–December.

[2] This section borrows heavily from the author's previous work, Maitra, Shipra. 2021. "Sustainable Cities for Sustainable Development, in NAGARLOK" Vol LIII.(4): 16–37, October–December.

extreme challenge of developing economically productive and socially inclusive settlements and strike a balance among spread of development in different size classes of urban areas. The share of population has gone up in metro cities (above 1 million) as well as in the towns with population below 20,000 but other size classes of towns are losing population share. Large UAs are increasing in area while the smallest towns are increasing in number in order to take advantage of some extended infrastructure like road networking market facilities or industrial activities. (NIUA 2020). Overcrowded cities and towns with infrastructure bursting at the seams are familiar scenes in India. This problem will only worsen without proper interventions. Cities as engines of growth will only expand in size and result in spatial concentration. The country's urban population contributes over 65% of India's GDP. This will increase to 75% by 2030. On the other hand, there is little incentive for people to migrate out of cities and contribute to balanced regional development. Earlier attempts at providing better urban infrastructure or at creating new townships have not been able to deal with the issue of liveability satisfactorily. Uncontrolled occupation of space in Indian cities has resulted in proliferation of slums and shortage of infrastructure in urban settlements threatening the quality of life and environmental sustainability.

India's urban experiments since Independence can broadly be divided in three phases. The first fifteen years of planning era were marked by efforts towards housing provisions, slum clearance and rehabilitation. The master planning approach endorsed during this time resulted in low-density urban settlements as land-man ratio was not alarming.

The next two decades initiated a significant departure in policy; from slum clearance to slum improvement and up gradation. Emphasis was given to balanced regional development and development of small and medium towns, while containing the growth of metropolitan cities, making land available for provisioning of services, housing for the urban poor, and control of land prices.

However, these initiatives were not successful in spreading urbanisation evenly. There was no networking among smaller towns and villages or with the next size class in order to make economic activities and social services mutually complementary in space. Bigger cities continued to have strong pull factors both in terms of economic and social opportunities and attracted migration from smaller towns, thereby creating both spatial and economic imbalance.

India's economic liberalisation from 1990's, impacted the urban sector also. Some key developments during this period included the opening-up of the sector to private participation, participatory approach in city planning, strengthening the link between urban growth and economic development and employment generation. The 73rd and 74th Constitutional Amendments (1993) introducing Panchayati Raj and Urban Local Bodies (ULB) as constitutionally accepted third tier of administrations in India pushed the agenda for decentralisation seeking greater accountability for the ULBs and paving the way for state transfers and subsidy based urban infrastructure financing regime to market-based financing regime.

Even before the adaptation of SDGs at the global level, Twelfth Five-year Plan (2013–17) of India highlighted that vision of India's growth must be aligned with the objective of inclusion and sustainability. It observed that good urbanization ensures cities and towns to be free from slums and offering adequate opportunities for productive employment and a decent quality of life to all their inhabitants including the poor. Two major challenges of urbanization in India, as recognized by the Twelfth Plan, are job creation and provision of basic services to urban citizens.

12.3 Concept of Smart Cities

The concept behind developing the smart cities is to create highly advanced urban regions in terms of overall infrastructure, sustainable real estate, communications and market viability. Presently, people are facing new challenges as the urban population becomes richer, enthusiastic to travel and more demanding in terms of happiness such as the lessening or control of pollution due to increasing activity, balancing the ease of individual transport and its consequences on traffic blockage, the need for security while preserving privacy, the need for instant information on any situation all within the reduced taxation system.

At the same time, there are unemployed youths migrating from their home states to technologically developed cities such as Delhi, Mumbai, Surat, Ahmadabad, Chennai and Bangalore to improve their lifestyle and gain prosperity which is causing huge economic problems. Such migration greatly put pressure on the public facilities available in these cities. There are also job requirements to needy people. It is imperative to prepare the cities for speedy inclusive urbanisation.

The interest in smart cities assumed significance globally at the time when the entire world was facing one of the worst economic crises. In 2008, IBM began work on a 'smarter cities' concept as part of its Smarter Planet initiative (Harrison and Donnelly, 2011). By the beginning of 2009, the concept captivated the imagination of various nations across the globe.

Plenty of literature is available world wise on definitions, indices, rankings, project scale and priorities of smart cities thereby creating lots of confusion in what can be termed as smart city. There is no standardised domain for smart cities in the present city development models. For example, the standards related to smart cities on sustainable communities broadly cover indicators on smart urban infrastructure. However, there are others standards related to city services, its built and non-built environment and city's functional cycle, which have not been addressed. There is not much evidence that these are integrated into the standardisation process of smart cities. There is a lack of clarity on the specific scope of each of the standards. There is also, lack of consistency in development of indicators for smart cities. There are several indicators for smart cities based on similar principles; but methodologies for developing the indicators and smart city indexing vary, leading to inconsistency. There is no accepted methodology for ranking cities as smart. Moreover, there are overlapping criteria for ranking cities as smart as well as under various other genres. While all the indicators intend to reflect enhanced quality of life for its citizens, the lack of standardisation across indicators has created a confusion for aspiring cities in choosing the right path to become a 'smart city'. (Cohen, 2014, Hollands, 2008).

The International Standards Organization (ISO) established a new set of city performance indicators for smart cities and a universal approach for measuring them in 2014.These could be categorised into three major groups:

1. Sustainability: primarily environmental sustainability as the major objective to be achieved, quality of life and economy emerging as subsequent attributes.
2. ICT applicability: required outcomes such as city efficiency, management, infrastructure, environment, and quality of life to follow automatically. However, it does not guarantee overall functionality, resilience, city form and urban design of the smart city.
3. City governance: larger understanding of the use of ICT in delivering governance, recognises the critical relevance of human resources, and puts emphasis on quality of life as well as environment.

12.4 The Smart City Mission (SCM)

In the Smart City Mission, the government has tried to develop the concept in the Indian context describing smart city as an urban region that is highly advanced in terms of overall infrastructure, sustainable real estate, communications and market viability. It is a city where information technology is the principal infrastructure and the basis for providing essential services to residents. There are many technological platforms involved, including but not limited to automated sensor networks and data centres.

According to the documents released on the Smart Cities website, the core infrastructure in a smart city would include:

- Adequate water supply
- Assured electricity supply
- Sanitation, including solid waste management
- Efficient urban mobility and public transport
- Affordable housing, especially for the poor
- Robust IT connectivity and digitalisation
- Good governance, especially e-Governance and citizen participation
- Sustainable environment
- Safety and security of citizens, particularly women, children and the elderly
- Health and education
  (Source: http://smartcities.gov.in/)

In short, the Mission attempts the following:

- Widespread use of ICT for governance, infrastructure and resource management.
- Top-down effort to build bottom-up stakeholder participation.
- Tripartite coalition between centre, state and urban local body.
- Competitive, convergent and built on best practices from previous and current missions.

12.5 Spatial Development

The SCM is a redevelopment programme for the large satellite towns. It is focussing on developing one hundred.

Smart Cities as satellite towns of larger cities and modernising the existing mid-sized cities with competitiveness, sustainability and improved quality of life. Competitiveness refers to a city's ability to create employment opportunities, attract investments, professionals and people. Quality of life refers to safety and security, inclusiveness, entertainment, ease of seeking and obtaining public services, cost efficient healthcare, quality education, transparency, accountability and opportunities for participation in governance. The whole programme is based on the four pillars of sustainability which include social, physical, institutional and economic infrastructure. The selected 100 cities are of various sizes and economic strength. The central government is spending 100 crores per city per year with matching investment from the concerned states in PPP mode. Cities have been selected as follows (Table 12.1).

SCM tries to cover both large and smaller cities to break the concentration of investments in larger cities only. It further specifies land use development into both Brownfield and Greenfield ones. Minimum size of land is specified but not maximum limit.

The Mission elaborates that smart city proposals should have a combination of two types of development programmes as specified below:

A. Area—Based Development (ABD)

ABD could be any development of three types of land sites or combination of them.

1. Retrofitting (area > 500 acres) will make areas more efficient and liveable. It will have more extensive infrastructural service levels and

Table 12.1 Types of cities selected in SCM

Type of cities	Criteria	Number
1. Cities with population of 4 million or more	1 satellite city each	9
2. Cities with population between 1 to 4 million	To choose among 44 cities	35
3. State/UT capitals irrespective of population size		19
4. Tourist cities, Religious cities, Special productions	Not included in any 3 groups above	10
Cities with population between 0.5 to 1 million		27
Total		100

Source Ministry of Housing and Urban Affairs, Government of India http://smartcities.gov.in/

a large number of smart applications. It will be identified by the city in consultation with citizens.

2. Redevelopment (area>50 acres) will replace existing build-up environment by co-creating new layout with enhanced infrastructure using mixed land use and increased density. It will be identified by ULB in consultation with citizens.
3. Greenfield (area>250 acres) will introduce a smart solution to the previously vacant area using innovative planning, plan financing and plan implementation tools with housing, especially for the poor. It could be located within limits of the ULB or within limits of local Urban Development Authority (UDA).

B. Pan-City Programme

The Pan-City Programme will provide application of the selected smart solution to the existing city-wide infrastructure. For example, one solution could be intelligent traffic management to reduce commuting time or cost.

Special Purpose Vehicle (SPV).

The Mission throws big challenges to the cities to program for alternative financing other than government transfers. The city can raise resources through property taxes, transfer of development rights in land, use of municipal bonds and the like. It encourages the cities to expedite financial reforms through intermediaries and institutional and governance reform through Special Purpose Vehicle (SPV).[3] The process of implementation of Smart City projects starts with the setting up of the SPV. It is a wholly public limited company registered under the Company's ACT, 2013. The Municipal Commissioner is the Chief Executive Officer of the SPV. The decision-making powers available to the Urban Development Department / Local Self Government department / Municipal Administration department are delegated to the Board of Directors of the SPV in which the State and ULB are represented. The State Level High Powered Steering Committee (HPSC) for Smart Cities is delegated the power of approval from the state. SPV is sufficiently empowered to do all development related works, i.e., plan, appraise, approve, release funds, implement, manage, operate, monitor and evaluate the development projects within the demarcated areas of the city.

Table 12.2 Time line of project completion for cities

Selection period	Number of cities	Completion year
January 2016	20	2020–21
May 2016	13	2021–22
September 2016	27	2021–22
February 2017	30	2021–22
January 2018	10	2022–23

Source Ministry of Housing and Urban Affairs, Government of India, http://smartcities.gov.in/

SCM is to be completed within a period of 5 years. The time line for completion of smart city projects according to the rounds in which they were selected is as follows (Table 12.2).

The projects under the SCM are largely capital-intensive infrastructure projects with long gestation periods and extensive ICT application. Smart cities, though, are not only related to data and technology, but sustaining for the future. It is imperative for the cities to find out how they will cope and compete in terms of being attractive places to live and at the same time, are affordable both for the city authorities and the citizens. The focus is largely on the ABD that carries a weightage of 55% in the selection criteria for smart cities.

Total projects under SCM are 5151 with total estimated investment of 2, 05,018 crore. They are under various stages of completion (Table 12.3).

The implementation process was slow to start in the beginning, but it has gathered momentum since mid-2018, as per the official data. Substantial capital expenditure is pumped in the urban sector to generate financing in the PPP

[3] Maitra, Shipra. 2021. "Sustainable Cities for Sustainable Development, in NAGARLOK" Vol LIII.(4): 16–37, October–December.

Table 12.3 SCM projects under various stages of completion (Cumulative since 2016)

Number of projects	Status	Amount (Rs. Crore)
1636	Completed	27,000
3832	Under implementation	1,25,000
4686	Tender issued	1,65,000

Source MoHUA E-Book, 2020

mode for construction of common assets. It was anticipated that private investments worth at least Rs 4 lakh crore will be generated till 2020 in the 100 Smart Cities, accounting for 80% of the estimated spending on the mission. It seems that it was gross over estimation.

12.6 Experience with the Smart Cities

SCM is in the middle of implementation so far as the time lines are concerned. Smart Cities are still in the formative stage as the projects are not yet completed. It is too early to comment on the level of socio-economic transformations, reduction of income inequality, increase in affordability and sustainability. However, a glimpse at the proposals to meet the city challenges may provide the idea of how the emerging landscape is going to take shape. The Mission emphasises on two major elements to face City Challenges—firstly, encouragement of private sector involvement through PPP mode to raise necessary investment and secondly, citizen participation to identify areas of improvement.

We will analyse the efforts to meet City Challenges of 20 selected Smart Cities and to find out (1) how effective is public participation in guiding city development and (2) scopes of increasing investments through PPP mode. List of the 20 cities is given in the Appendix tables.

12.6.1 Public Participation

Globally, city making has been largely considered as collective endeavour where citizen engagement is an inseparable part of planning and development. There are three aspects of public participation. Firstly, there must be a clear understanding among the participants taking part in a decision-making process whether the participation is voluntary to all citizen or by invitation to selected elites. Secondly, if the participatory process includes discussions and deliberation or only providing information. Thirdly, if the decisions through public discussions are translated through policies. The SCM emphasises public participation as an important component of city management. It mandates citizen engagement in three rounds—visioning the Smart City, identifying ABD and Pan City solutions and during implementation. According to official source, 15 million citizens were consulted during framing of City Challenges process, i.e., to identify major challenges faced by the cities.

The cities involved the citizens both through direct and indirect engagement. These can be broadly categorised into three forms.

(1) Eleven cities tried to collect citizens' opinions directly and extensively through votes or forms/schedules, covering minimum 10% of the total city population.
(2) Five cities undertook representative survey among citizens but the coverage was comparatively low.
(3) Four cities followed the method of consulting the citizens through online survey, ward level meeting or through citizens' representatives. The details of citizens' engagement in shaping vision and goals are given in Annexure Table 12.5.

Some cities took special efforts to solicit opinions of the slum dwellers.

Area-based development (ABD)

Citizens' participation was more limited in choosing ABD. It was done in two ways. First, they could choose a particular area from several options and second, they could choose specific programmes to be undertaken in selected areas.

The citizen could express opinion to select particular areas for ABD through various methods like (i) offline and online voting; (ii) stakeholders and ward level consultations; (iii) only wide-ranging consultations and discussions. In 9 cities, offline voting was main method of choosing particular area. In the rest 11 cities, online opinion was also sought but it was not the major deciding factor in identifying particular area.

Citizens' direct participation was much less in choosing specific programmes in selected areas. Only in 3 cities, citizens' view was explicitly sought through polling by their choice of preference. These cities were Jabalpur, Indore and Udaipur. In other cities it was decided after wide consultations with various stakeholders, expert groups, and urban planners or arrived at from the lists of priorities of major interventions that were chosen by citizens during finalisation of city's vision and goal.

Udaipur followed a structured approach in getting citizens' opinions about city's vision and goal. They prepared structured questionnaire through which they provided 4 options of areas for ABD and sought citizens' opinion about the kind of programmes that should be undertaken in each of these 4 areas. The city authorities were able to identify specific area as well as the list of preferred programmes that is required to be undertaken in the selected area. Chennai followed a different approach for choosing particular area through online polling. They did not provide any options for area but chose T. Nagar through maximum votes among large number of options chosen by citizens themselves.

Smart City Guidelines mentions that the ABD could be Brownfield (retrofitting, redevelopment) or Greenfield (new city) or combination of them.

In the selected ABDs in these 20 smart cities, not a single development was Greenfield type. The reasons are several. First, only in four cities, i.e., Bhubaneswar, Surat, Solapur and Coimbatore, Greenfield options were put forward before citizens. In rest of the 16 cities this option was not part of the shortlisted options. Several arguments were given for excluding Greenfield development from shortlisted projects. New Delhi and Kakinada municipalities argued that there is no large land parcel (250 acre or above) available in or around municipality boundary to consider this option. Other cities like Pune argued that if smart city programme is supposed to be landmark programme that can be replicated in other parts of the city, then Greenfield is not an ideal option. Another argument was that the project cost of Greenfield option is much higher and mobilising additional funds is an issue. Second, in four cities where this option was explicitly considered by citizens it was not chosen as retrofit option or retrofit cum redevelopment option would directly benefit the citizens.

Hence, retrofitting option was chosen by 15 cities, retrofit cum redevelopment option was chosen by 4 cities and only in Bhopal, citizens chose Greenfield development option. Bhopal had the option of large 350 acres underutilised government land available in the heart of the city flanked by two large commercial areas that could be utilised through land monetisation.

Few observations can be made regarding the nature of ABD. First, in 5 city plans, the area chosen were transit/market/commercial areas. In 4 other city plans, the area chosen was Central Business District (CBD) area. In these areas, the major thrust was provision of better civic facilities, facilitation of trade and improving road and traffic management. Business/commercial areas together constituted the ABD in 9 out of 20 city plans. Major thrusts of choosing these areas seem to promote commercial returns for investment. Second, the next category of chosen ABDs was to promote tourism by restoring heritage structures and improving infrastructure and civic facilities in the surrounding areas. These cities are Jaipur, Kochi and Udaipur. The motivation of heritage development programmes is to be to promote growth through tourism. Third, next category of area development projects can be broadly classified into a heterogeneous group of beach/riverfront/lake side development. These cities are Pune, Vishakhapatnam, Coimbatore and Guwahati. Apart from promoting recreational activities of local citizens another objective was to restore ecological balance in these areas. Fourth, Chennai stands out for its ABD as it intends to provide better civic facilities including promotion of non-motorised transport in a single residential cum commercial area of T-Nagar which aims at improving the quality of life of the citizens in single area that can be replicated in other areas in course of time. Last, two other cities Bhubaneswar and Bhopal undertook multiple activities in a single area as they could leverage the value of government land as additional source of fund.

Pan City Programme

In most of the cities, the pan city programme was based on the citizens' priorities that were collected through citizens' engagement exercise in shaping up vision and goal. But citizens' opinion was not explicitly sought at the time of identifying pan city programme.

However, four smart city proposals stood out. These were Bhubaneswar, Jaipur, Udaipur and Ludhiana. They directly sought citizens' opinion through offline polling and the most preferred options were chosen. In three other cities of Pune, Coimbatore and Kakinada they relied on online polling to finalise the pan city programme themes.

Scanning through broad components of pan city projects chosen by the cities, a few observations can be made. First, majority of the cities chose transport as the pan city programme. Smart transport management, intelligent traffic system and non-mechanised transportation provision are the primary thrust areas of 13 among 20 smart cities, selected for analysis. Second, integration of provision of public utilities to citizens either through common smart card or through integrated command and control centre was the next preferred scheme for seven pan city plans. Third, effective water management to reduce wastage or providing equitable supply and efficient solid waste management including conversion to bio energy were given prominence in five smart city plans. Fourth, disaster management including advance flood warning were taken up in the pan city programme of two coastal or flood prone cities. Other components were air quality management, e-educations etc.

Table 12.4 Components of SCM

Programme	Total marks	Marks for citizen engagement
City level	30	10
Area-based development	55	5
Pan city development	15	1
Total	100	16

Source As above

SCM assigns marks for each stage of selection in the following way (Table 12.4).

SCM provides maximum weightage to citizen engagement in framing vision and goals for the city. It decreases as actual project implementation becomes operational. The Mission guidelines states the importance of citizen engagement. But citizen has limited role in carrying the vision forward. That is why many cities have decided the areas in ABD before engaging the citizen. Citizens did not know anything about the Pan City Programme in many cities as it carried the least weightage in preparing the City Challenge proposal. Details of citizen engagement in NDMC clarifies this point further.

12.7 NDMC as a Smart City

New Delhi Municipal Council (NDMC) is the only local body of Delhi, selected as the Smart City. Delhi has eleven districts, of which New Delhi district is under the jurisdiction of NDMC, destined for significant ABD in brownfield development. It is a restricted area with seats of the central government, state government, foreign diplomatic enclave and a strong Central Business District (CBD). It occupies less than 3% of both total area and population of Delhi. It is the most significant address in Delhi, being the seat of the governments and foreign embassies. But it is not free of slums. The squatter (Jhuggi Jhopri) settlers comprise of around 7% of total population (2.94 lakh) of NDMC. Residential and commercial constructions are strictly regulated here.

There is no greenfield land of the size of 250 acres and above in the NDMC area. Also, there is no suitable redevelopment area of 50 acre and more due to heritage characteristics of NDMC area. Hence, retrofitting is the only model available to them. The New Delhi City Centre (NDCC) consisting of Connaught Place and contiguous surrounding areas of approximately 550 acres has been selected for undertaking the retrofitting model of development to transform the area as world Class Urban Area. The area is selected through a continuous process of engagement with the stakeholders like citizen's forum, people's representatives, concerned government officials, urban planners, sector experts and suppliers considering its economic and cultural significance, number of footfalls and ease of implementation. The Pan City programme leverages ICT intervention in basic infrastructure to improve liability and inclusiveness. It addresses the issues of water, power, education, healthcare and governance with special emphasis on E-Education in NDMC schools, senior citizen driven mentoring programme for the underprivileged, skill development and enabling access to the Economically Weaker Sections (EWS) in private hospitals, virtual medical services including online access to blood banks, integrated E-healthcare, air quality monitoring.

NDMC has adopted two types of citizen engagement—(1) High-tech engagement through online consultations, multimedia presentations and mobile polling and (2) Low-tech engagement through meetings with stakeholders, street plays for awareness building. It is asserted that around 95% of the residents have been consulted through RWAs. It also says that 10,800,000 mobile SMS were sent to 600,000persons but only 6700 responses received.

It is generally perceived that RWAs represent the middle class, not the slum dwellers. NDMC tried to cover the gap through direct meetings with them. It has also engaged a system of volunteers for ensuring citizen participation.

SCM provides larger role to the SPV in planning and implementation of projects. NDMC's SPV, namely, "New Delhi Municipal Council Smart City Limited", a public limited company, wholly owned by NDMC, has been incorporated. It is the sole implementing agency with no direct involvement of elected representatives.

12.7.1 Leveraging Private Sector Funding

The public private partnership in smart city project is expected to be one of the major sources of fund. It is captured in Annexure Table 12.6.The analysis of funding estimates shows that overall contribution of PPP in smart city capital expenditure (CAPEX) was little less than one-fourth (23.8%). Contribution of PPP in different smart city CAPEX was asymmetrical. In 8 out of 20 smart cities, PPP component was not expected or it was negligible. Only in 4 smart city proposals, PPP contributed more than two-fifth of total CAPEX. These cities are Bhubaneswar, Ahmedabad, Indore and Bhopal. In all these four cities, ABD was either retrofit cum redevelopment or redevelopment type. In three out of these four cities where PPP contributed major part of the CAPEX, the city incorporated monetisation of available pubic land in the chosen areas for ABD. These are Bhubaneshwar, Indore and Bhopal. In Bhubaneshwar and Indore, the total CAPEX amount crossed 4,000 crores by virtue of mobilisation of additional fund by capturing the value of public land.

In Ahmedabad, PPP made substantial contribution but land monetisation of public land was not opted. Almost two-third of PPP contribution is likely to come from redevelopment of slum area with contribution from beneficiaries of slum or EWS housing. Similar approach was followed in Surat where PPP contributed 30% of CAPEX and the contribution of affordable housing beneficiaries in PPP was 58%.

In smart city proposals where PPP made noticeable contribution to CAPEX but land monetization option was not chosen, beneficiaries of slum/EWS housing contributed major part of the PPP. Other options like contribution from benefitted commercial entities, greater user charges etc. did not make noticeable contribution to CAPEX. The innovative approach of raising contribution through PPP mode had not been implemented properly till now.

12.8 Experiment with GIFT City

Gujarat International Finance Tec-city (GIFT) city was conceptualised as first Smart City much earlier than the SCM. The glimpse of expenditure on infrastructure required for smart city project is reflected in concept of this city in the Gandhinagar district. It is a greenfield development. Gujarat International Finance Tec-city Ltd (GIFTL) was formed in 2007 to develop the city as a financial CBD. The project has been revived after eight years. Demand for financial services in India has seen a tremendous transformation over two decades. Financial service sector employs about 4 million people and generates approximately 5% of GDP. It is estimated that if India can build an international financial centre with relevant rules and regulations, earnings out of financial service exports will surpass that of IT sector. GIFT city is expected to generate five lakh direct and five lakh indirect jobs within its mammoth infrastructure.[4] This is based on ninety million square feet of real estate development in office space and residential buildings. Along with the state-of-the-art physical infrastructure, it will have social infrastructure including international and domestic schools and hospitals, high quality shopping malls and entertainment hubs. The city has earmarked around 3000 acres for development of integrated township. It has been projected to develop as an International Financial Service Centre (IFSC) to rival those of Dubai, Hong Kong and Singapore.

Over all management for infrastructure development is entrusted with GIFTL while city regulation is entrusted with a dedicated Area Development Authority. There is a single window clearance through GIFT Urban Development Authority and GIFT Notified Committee for city management. The estimated cost for this project is Rs. 80,000 crore to be shared by different companies. The company is a 50:50 joint venture between IL&FS and the state-owned Gujarat Urban Development Co.

GIFT City is being developed as the high-density urban area covering of 886 acres with infrastructure services in Gandhinagar district of Gujarat. It is being planned for development of 62 million square feet of built-up area (BUA). Currently, only 2 million square feet of BUA is ready for occupancy while 2.5 million square feet of BUA is under-construction. At present, approximate workforce is around 7000 persons. The major land-use is commercial; occupied mainly by service sector-oriented institutions.

Total land allotted to GIFTCL till now is approximately 900 acres of which 250 acres are earmarked for multi-service Special Economic Zone (SEZ). It is divided into four parts: (a) Processing area consisting of IFSC with off-shore banking, data centres and others, (b) Non-processing areas that includes service apartments, related commercial buildings, hostels, recreation centres etc. (c) Enclaves—financial, technology, corporate enclaves and world trade centre and (d) Social facilities like schools, clubs, convention centre. The Union Budget 2016 announced several tax sops for SEZ in the city thereby creating a difference between SEZ and domestic Tariff Area (DTA) within the city. Minimum Alternate Tax applicable to SEZ is 9% against 18.5% in DTA. There are tax waivers on foreign currency sale of equity shares, sale of units of equity-oriented funds, sale of units of business trusts etc. that earns foreign exchange. The city is a deemed foreign territory expected to give a big push to foreign currency transactions under special tax regime. Several

[4] "The Glimpse of Expenditure For Smart City Project | Bartleby." Retrieved July 3, 2024 (https://www.bartleby.com/essay/The-Glimpse-Of-Expenditure-For-Smart-City-P3MZ9LQP4P).

banks like the State Bank of India, ICICI Bank, HDFC Bank, Bank of India, Yes Bank and other financial institutions have already started operations. It has sold 12 million square feet in phase I and hopes to create and sell another 22 million square feet in phase II. The SEZ will enjoy all the financial benefits enjoyed by the manufacturing SEZ.[5]

Phase I of the project, costing Rs. 1818 crore has been funded with a debt of Rs. 1157 crore from banks at 11.5% rate of interest, which is lower than usual charge of 12%. The rest of the money has come from equity (Rs. 65 crore) and internal accrual (Rs. 596 crore). Land development rights have been sold at the rate of Rs. 5500 per square feet which is less than existing rate of Rs. 7500–8000 prevailing in Ahmadabad at that time. GIFTCL charges very low maintenance cost of property at Rs. 5 per square feet per year as compared to Rs. 70 for similar property in Ahmadabad. Phase II will cost about four times of Phase I i.e., Rs. 7696 crore to create more than double the earlier land assets.[6]

12.9 Lessons Learnt so Far

A close look at SCM guidelines shows that public participation is sought to define vision and goals of a city but not so much in implementation process. IT infrastructure is not widespread in India. It still creates substantial digital divide. NDMC's experience shows that even with accessibility of digital gadgets, citizen interest is not generated. Public monitoring of services would have generated more interest as the services are closely related to everyday life. Goals and vision seem a little abstract to many. RWAs and other stakeholders are involved in setting city's agenda in a limited way. Management of projects are through public limited companies, who are not accountable to people. NDMC enjoys highest per capita revenue among the ULBs in Delhi with least share of poverty. Only seven% of its total population lives in slums. Its population is less than 3% of total population in Delhi. Hence, it is easier for it to carry out inclusive programmes without feeling any financial pressure. It has included all municipal schools within E-education and is extending quality health services to the economically weaker sections. Construction activities are strictly regulated and unauthorised encroachments are very limited in scope. Development of NDCC will bring very high return through land monetisation. So, even without vigilant citizen participation, it can be inclusive as well as competitive. But this case cannot be replicated as NDMC has a unique status.

Similarly, GIFT city enjoys unique support from the government. It has advantages of an SEZ with huge tax incentives, low user charges for services and low land cost compared to other cities. Quality technology with uninterrupted power and water supply and other facilities increases competitiveness of this green field development. City management rests with the joint venture company and development authority. There is no scope of involvement of elected representatives. GIFT city is inhabited by highly qualified persons who can articulate their demands effectively and get their problems solved with ease. This kind of cities will create highly competitive environment and generate their own resources with significant initial support from the government.

SCM anticipated handsome private investment within four years of launching. Actually, less than half of the anticipated amount has come till now. Social sector programmes are much less in proportion to other infrastructural programmes. Smart city programmes have some pro-poor plans that affect them directly or indirectly. There are programmes like designated vendor zones for informal vendors, promotion of e-rickshaws, provision of e-learning in municipal schools and provision of better

[5] "The Glimpse of Expenditure For Smart City Project | Bartleby." Retrieved July 3, 2024 (https://www.bartleby.com/essay/The-Glimpse-Of-Expenditure-For-Smart-City-P3MZ9LQP4P).

[6] "The Glimpse Of Expenditure For Smart City Project | Bartleby." Retrieved July 3, 2024 (https://www.bartleby.com/essay/The-Glimpse-Of-Expenditure-For-Smart-City-P3MZ9LQP4P).

housing in slums. Very few of them are financed through PPP. SCM relied heavily on this component of financing, but it could not be expected to provide significant contribution in capital expenditure. Cities which could depend on monetization of public land have been able to fund capital expenditure. Private sector investment is less than one-third of total capital expenditure. In housing schemes, beneficiary contribution was significant in some of the cities. In Surat, the share of PPP in capital expenditure is estimated to be 30% while beneficiaries of affordable housing have contributed around 58% (Annexure Table 12.7).

ABDs focus more on the projects that would make the cities competitive, rather than inclusive. More citizen involvement could have made the projects more people oriented. Selection of areas was not focussed on deprivation. Only one city, i.e., Pune selected the areas for development depending on population density. Others depended more on expert groups and other stakeholders to identify areas for development. Economic potentiality was one important criterion for choice keeping resource generation in mind.[7] Bhubaneswar provides a good example of how wards were chosen for ABD. We take this example as it topped the First-Round selection process of Smart Cities. More than one-third (35%) of the city's population lives in slums. Nearly 40% of slum households reported open defecation, 64% did not have drainage facilities and only 37.4% had treated tap water as their main source of drinking water. Any Smart City plan of Bhubaneshwar must include these basic facilities of drinking water provision within premises, clean toilet facilities and minimum facility of disposal of waste water. In a report submitted to UNFPA (2016), we calculated a Deprivation Index based on seven socio-economic and demographic indicators and ranked all 67 wards of Bhubaneswar Municipal Corporation. We categorised them into four groups—Undeveloped wards, mediocre developed wards, reasonably developed wards and better developed wards. None of the seven wards chosen for ABD in Bhubaneswar was from the underdeveloped wards. Inclusiveness of the city was compromised to encourage competitiveness.

PPP as major source of investment funds leads to a major dilemma. PPP will not be possible without remunerative schemes and generally, remunerative schemes are not inclusive. It seems, the SCM is encouraging competitiveness at the cost of inclusiveness and risking sustainable urbanisation in the process.

Appendix

See Tables 12.5, 12.6, 12.7 and 12.8.

[7] Maitra, Shipra. 2021. "Sustainable Cities for Sustainable Development, in NAGARLOK" Vol LIII.(4): 16–37, October–December.

Table 12.5 Nature of citizens' participation in shaping vision and goal

Sl no	Name of city	Citizens participation in shaping city's vision and goal
1	Bhubaneshwar	Citizens connect initiative had 3.1 lakhs participation (32% of population) through offline and 47 thousand online votes. Further community meeting with different sections including slums reached additional 20 thousand persons
2	Pune	Citizens' engagement was done through pan-city manner, where 4 lakhs households, 50% of all households were reached through door-to-door campaign apart from webpage and social media networks
3	Jaipur	It began with 91 ward councillors meeting held twice and then stakeholders meeting, presentations in colleges and disabled school and online opinion and poll questionnaire survey at ward levels
4	Surat	Citizens engagement was done through opinion polls, stakeholders meeting, online opinions etc. and around 50 thousand responses were received
5	Kochi	It was done through ward level meetings and questionnaire survey through ward committee among 5 thousand participants, online, professional forums, industry chambers etc.
6	Ahmedabad	Among various methods, survey forms, both online and offline got more than 10 thousand responses
7	Jabalpur	Total outreach of citizens were 2.56 lakhs or about 20% of city population including online 99 thousand and offline 1.56 lakhs. It included door-to-door survey of slum dwellers that constituted 18% of respondents
8	Vishakhapatnam	Survey forms were distributed to citizens to select priority areas of improvement in order of preference. 5 lakhs offline forms were circulated. 78% citizens wanted improvement in public transport and others chosen priority areas were park, air quality, public toilet in that order
9	Solapur	City vision documents got over 56 thousand suggestions. Further, 12,500 households including 500 slum households were directly contacted. 35% of citizens voted for cleanliness, 32% for road and transport facilities and 23% for water
10	Devangere	For vision and goals, citizens in 41 wards were consulted and these included slum dwellers, hawkers, shopkeepers, student, elected representatives, professionals etc. After analysing the feedback data, key areas were identified
11	Indore	As a whole, 5.91 lakhs citizens were contacted that constitute 25% of city's population. In city's vision and goal, 2.53 lakhs citizens were consulted and 4 areas out of 24 proposed through different forum were shortlisted
12	New Delhi	The whole exercise was based on interaction with 1056 citizen representatives from RWAs, students, slum dwellers, traders, vendors, govt. employees and diplomats
13	Coimbatore	15–20% city's population was directly or indirectly approached through face to face connect, FGDs, a citizen pulse survey (15,000) and citizens forum, social media etc.
14	Kakinada	For shaping of vision and goal two major tools were used. i) Polls (online and survey forms: 50 thousand total submission) and ii) FGDs that reached 1.2 lakhs citizens. 40% of citizens were reached through various means. However, through online polling 38% of citizens wanted solid waste management to keep city clean, 10% chose IT connectivity, another 9% desired affordable housing and 6% gave priority to slum development
15	Belagavi	Through physical survey with prescribed format, public meetings, and online response around 70% of population was reached. Through physical survey citizens gave their options in order of priority and decentralising city centre got 13% of votes, followed by water supply, sanitation
16	Udaipur	In shaping vision and goals, various means of engagement were undertaken and altogether opinion of 1.25 lakhs citizens were collected that consisted of 25% of population. Tourism was topmost option with 44% of votes followed by water & sanitation, public transport and walkability
17	Guwahati	Various modes including online, FGDs and door to door survey was undertaken to identify city's vision and goal
18	Chennai	Citizens' engagement was done to identify smart city initiatives and non-motorised transport was top priority followed by public transport through smart card
19	Ludhiana	Vision and goal were based on 26 questions and priority exercises. Feedback of 75 thousand citizens was collected through filled up questionnaires
20	Bhopal	3 lakhs responses were received and identified priorities were sanitation, safety, energy efficiency, infrastructure and employment

Source Tables generated from Smart City Proposals (Stage: City Challenges) sent by 20 Cities

Table 12.6 Leveraging private sector funding

Sl. No	Name of city	CAPEX[1] total (in Rs. Crore)	Share of ABD[2] in CAPEX	Share of PPP[3] in CAPEX	Share of slum/ EWS[4] beneficiary in PPP	Monetization of public land
1	Bhubaneshwar	**4,537**	90.3	**56.5**	3.5	Yes
2	Pune	2,932	74.9			Yes
3	Jaipur	2,404	59.1	*23.8*		
4	Surat	2,597	69.4	*30.6*	57.9	
5	Kochi	1,323	65.5	18.4		
6	Ahmedabad	2,290	76.5	**41.0**	64.3	
7	Jabalpur	3,808	94.6			
8	Vishakhapatnam	1,602	91.2	10.2		
9	Solapur	2,247	87.6	10.9		
10	Devangere	77	63.8	13.2		
11	Indore	**4,857**	92.0	**72.5**		Yes
12	New Delhi	1,897	35.3	14.8		
13	Coimbatore	1,570	90.9	11.1		
14	Kakinada	1,993	93.6			
15	Belagavi	47	46.8			
16	Udaipur	1,221	72.1			
17	Guwahati	2,181	72.4			
18	Chennai	1,366	64.3			
19	Ludhiana	1,049	85.0			
20	Bhopal	3,093	79.0	**55.3**		Yes
	Share in CAPEX		76.9	23.8	7.7	

Note 1. CAPEX means capital expenditure
2. ABD means area-based development
3. PPP means public private partnership
4. EWS means economically weaker section
Source Tables generated from Smart City Proposals (Stage II) sent by 20 Cities

Table 12.7 Pro-poor approach and provision of better facilities

Sl. no	Name of city	Provision of better housing for slum dwellers	Whether cross subsidized	Does slum dwellers contribute	Slum rehabilitation includes provision of better civic facilities	Include provision of education, health and skill training	Kind of promotions in Pan city Programmes
1	Bhubaneshwar	Yes	Yes, through PMAY and HIG/MIG housing	Yes	Yes	Skill training, Anganwadi centres	None
2	Pune	Yes	Yes, through land monetisation	No	Yes	Schools, hospital, skill training	None
3	Jaipur	No	N.A	N.A	N.A	N.A	Vendor zones e-rickshaws
4	Surat	Yes	PMAY	Not clear	Yes	No	
5	Kochi	Yes	Yes, PMAY and other funds	No	Yes	No	E-rickshaws in Kochi Fort
6	Ahmedabad	Yes	Yes, with EWS housing	Not clear	Yes	Schools, Anganwadi centre	None
7	Jabalpur	Yes	Yes, through PMAY and HIG/MIG housing	Yes	Yes	Health and Education	None
8	Vishakhapatnam	Yes	Yes	No	Yes	No	Electric rickshaw promotion
9	Solapur	Yes	Yes	No	Yes	No	None
10	Devangere	Yes	Yes	Not Clear	Yes	No	None
11	Indore	Yes	Yes, under PMAY and other subsidies	No	Yes	No	Skill development centres
12	New Delhi	No	N.A	N.A	N.A	No	e-education and e-health
13	Coimbatore	Yes	Yes, through other programmes	Yes	Yes	Skill centres	None
14	Kakinada	Yes	Yes, through other programmes	Yes	Yes	e-pathshala and ESI	None
15	Belagavi	Yes	Yes, through PMAY and other programmes	No	Yes	No	None
16	Udaipur	No	N.A	N.A	N.A	Battery operated vehicles, vendors zone	
17	Guwahati	No	N.A	N.A	N.A	No	No
18	Chennai	No	N.A	N.A	N.A	No	No
19	Ludhiana	No	N.A	N.A	N.A	No	e-rickshaw promotion
20	Bhopal	Yes	Yes	Not clear	Yes	No	None

Source Tables generated from Smart City Proposals (Stage: City Challenges) sent by 20 Cities

Table 12.8 Ranking of wards in deprivation index in Bhubaneswar

Socio-Economic and Demographic Index		Basic Amenities Index		Overall index	
ward Number	Index and Status	ward Number	Index and Status	ward Number	Index and Status
Ward No.55	1.74	Ward No.64	0.03	Ward No.64	2.01
Ward No.64	1.98	Ward No.65	0.56	Ward No.52	3.02
Ward No.60	2.00	Ward No.52	0.73	Ward No.55	3.06
Ward No.34	2.03	Ward No.62	1.00	Ward No.49	3.13
Ward No.50	2.03	Ward No.44	1.03	Ward No.65	3.17
Ward No.49	2.09	Ward No.49	1.04	Ward No.50	3.27
Ward No.67	2.10	Ward No.50	1.24	Ward No.42	3.40
Ward No.42	2.11	Ward No.32	1.24	Ward No.26	3.51
Ward No.53	2.14	Ward No.26	1.27	Ward No.19	3.62
Ward No.58	2.23	Ward No.19	1.28	Ward No.32	3.70
Ward No.26	2.24	Ward No.42	1.29	Ward No.62	3.82
Ward No.56	2.27	Ward No.55	1.32	Ward No.44	3.84
Ward No.41	2.27	Ward No.35	1.54	Ward No.58	3.97
Ward No.52	2.29	Ward No.25	1.73	Ward No.25	4.15
Ward No.54	2.32	Ward No.58	1.74	Ward No.34	4.16
Ward No.40	2.32	Ward No.23	1.84	Ward No.22	4.28
Ward No.19	2.35	Ward No.22	1.88	Ward No.40	4.31
Ward No.22	2.41	Ward No.11	1.88	Ward No.18	4.34
Ward No.25	2.42	Ward No.18	1.91	Ward No.53	4.55
Ward No.18	2.44	Ward No.40	1.98	Ward No.41	4.55
Ward No.32	2.46	Ward No.21	2.11	Ward No.21	4.60
Ward No.66	2.47	Ward No.34	2.12	Ward No.54	4.68
Ward No.21	2.48	Ward No.41	2.28	Ward No.35	4.71
Ward No.5	2.55	Ward No.54	2.36	Ward No.11	4.79
Ward No.2	2.60	Ward No.53	2.41	Ward No.23	4.85
Ward No.65	2.61	Ward No.33	2.47	Ward No.33	5.10
Ward No.31	2.63	Ward No.39	2.76	Ward No.39	5.43
Ward No.33	2.63	Ward No.20	2.78	Ward No.56	5.66
Ward No.36	2.65	Ward No.36	3.06	Ward No.36	5.71
Ward No.39	2.67	Ward No.5	3.17	Ward No.5	5.73
Ward No.44	2.81	Ward No.24	3.23	Ward No.2	6.22
Ward No.62	2.82	Ward No.13	3.26	Ward No.20	6.27
Ward No.8	2.83	Ward No.56	3.40	Ward No.13	6.29
Ward No.38	2.87	Ward No.10	3.55	Ward No.31	6.39
Ward No.29	2.90	Ward No.38	3.60	Ward No.38	6.48
Ward No.11	2.91	Ward No.2	3.62	Ward No.66	6.53

Table 12.8 (continued)

Socio-Economic and Demographic Index		Basic Amenities Index		Overall index	
Ward No.30	2.93	Ward No.8	3.72	Ward No.8	6.54
Ward No.27	3.00	Ward No.31	3.76	Ward No.10	6.76
Ward No.23	3.01	Ward No.51	3.99	Ward No.24	6.98
Ward No.3	3.01	Ward No.3	4.05	Ward No.29	6.99
Ward No.13	3.03	Ward No.12	4.05	Ward No.3	7.06
Ward No.57	3.11	Ward No.66	4.06	Ward No.57	7.46
Ward No.28	3.13	Ward No.29	4.09	Ward No.51	7.70
Ward No.35	3.17	Ward No.57	4.35	Ward No.12	7.77
Ward No.59	3.17	Ward No.46	4.73	Ward No.30	7.80
Ward No.10	3.22	Ward No.43	4.73	Ward No.27	8.01
Ward No.9	3.28	Ward No.30	4.87	Ward No.9	8.20
Ward No.37	3.35	Ward No.9	4.93	Ward No.43	8.34
Ward No.45	3.45	Ward No.27	5.00	Ward No.60	8.52
Ward No.20	3.50	Ward No.47	5.01	Ward No.37	8.54
Ward No.7	3.52	Ward No.45	5.15	Ward No.45	8.60
Ward No.43	3.61	Ward No.37	5.20	Ward No.28	8.72
Ward No.4	3.61	Ward No.1	5.28	Ward No.46	8.81
Ward No.51	3.71	Ward No.28	5.59	Ward No.1	9.32
Ward No.12	3.71	Ward No.7	5.99	Ward No.47	9.35
Ward No.14	3.74	Ward No.6	6.19	Ward No.7	9.52
Ward No.24	3.75	Ward No.60	6.52	Ward No.6	9.94
Ward No.6	3.76	Ward No.63	6.62	Ward No.4	11.12
Ward No.61	4.00	Ward No.48	6.78	Ward No.63	11.18
Ward No.1	4.05	Ward No.61	7.47	Ward No.61	11.47
Ward No.46	4.08	Ward No.4	7.51	Ward No.59	11.54
Ward No.47	4.34	Ward No.14	8.07	Ward No.48	11.70
Ward No.63	4.56	Ward No.59	8.37	Ward No.14	11.81
Ward No.17	4.77	Ward No.17	10.45	Ward No.17	15.22
Ward No.16	4.84	Ward No.15	12.25	Ward No.67	15.42
Ward No.48	4.92	Ward No.16	13.22	Ward No.15	17.27
Ward No.15	5.02	Ward No.67	13.32	Ward No.16	18.06

Undeveloped ward
Mediocre developed ward
Reasonably developed wards
Better Developed wards

Note: 1. Colour Schemes of Wards in four Quartiles
2. Chosen Smart City Wards are highlighted
Source Preparing Smart City Development Plan Based on Demographic and Socio-Economic Analysis, UNPFA—IHD, 2016

References

Bartleby. Glimpse of Expenditure for Smart City Project I. Retrieved July 3, 2024 (https://www.bartleby.com/essay/The-Glimpse-Of-Expenditure-For-Smart-City-P3MZ9LQP4P

Cohen B (2014) The smartest cities in the world 2015: methodology, Fast Company

Harrison C, Donnelly IA (2011) A theory of smart cities. Proceedings of the 55th Annual Meeting of the ISSS—2011, Hull, UK, 55(1). Retrieved from https://journals.isss.org/index.php/proceedings55th/article/view/1703

Hollands R (2008) Will the real smart city please stand up? Creative, progressive or just entrepreneurial? City: analysis of urban trends. Culture, Theory, Policy, Action 12:303–320. https://doi.org/10.1080/13604810802479126

Maitra Shipra (2021) Sustainable cities for sustainable development. NAGARLOK LIII(4):16–37, October–December

McKinsey Report (2014) India's ascent: five opportunities for growth and transformation. MGI@mckinsey.com

Ministry of Housing and Urban Affairs, Government of India (2015) Smart cities Mission. http://smartcities.gov.in

Ministry of Housing and Urban Affairs, Government of India (2020) E-Book available at http://smartcities.gov.in

NIUA (2020) Cities insight report, National Institute of Urban Affairs, New Delhi. www.niua.org

Report (2016) Preparing smart city development plan based on demographic and socio-economic analysis, UNPFA—IHD, New Delhi

Twelfth Five-year Plan (2012–2013—2016–2017) Planning commission, New Delhi

UNPFA. 2016. Annual Report available on UNPFA website

Integrating SDGS in Development Planning—Experiences from Countries

13

Sarthi Acharya

13.1 Introduction

13.1.1 Basics

The 2030 Agenda for Sustainable Development, adopted in 2015, has 17 Sustainable Development Goals (SDGs) flowing from it. Both, the (earlier) Millennium Development Goals (MDGs) and SDGs are a part of the larger goal of achieving higher levels of *Human Development.*

The Agenda strives for a more just, equitable and inclusive world, committing different stakeholders to work together for promoting inclusive and sustainable economic growth, and meeting targets in social and human empowerment and environmental protection. This agenda draws upon and supplements several existing (and past) multilateral agendas and treaties, like the Addis Ababa Declaration, the Paris Declaration, the Istanbul Declaration (in regard to least-developed countries), SAMOA Pathway, Rio Declaration/ Agenda 21, MDGs, and the like. The SDGs aim to mainstream them in a larger totality.[1]

The importance of SDGs stems from the fact that many countries across the world have problems with human capital formation, and with increasingly less demand for unskilled workers, poverty, unemployment, inequality and inequity are real threats. The last two decades have also seen global warming to be threat. The need to intervene collectively for ensuring a sustainable human development has thus become paramount.

This paper is an attempt towards illustrating how national development goals and the SDGs have been integrated across different countries. Hence, it illustrates on how to promote *2030 Agenda* through integrating SDGs in the national planning processes. It suggests entry points to integrate the *2030 Agenda* in aligning with the **existing** plans, and in integrating it into the **up-coming** long term (vision), medium term (typically, five-year) plans, and (short-term/annual) budgeting cycles, using different approaches laid out in the Mainstreaming, Acceleration and Policy Support approach (MAPS) of the United Nations.[2] *This is not an original paper but a compilation of existing knowledge in one place with commentaries and explanations.*

[1] For SAMOA Pathway, see https://www.un.org/ohrlls/content/samoa-pathway; and for Rio Declaration, https://www.un.org/en/development/desa/population/migration/generalassembly/docs/globalcompact/A_CONF.151_26_Vol.I_Declaration.pdf.

[2] See, https://sdgintegration.undp.org/maps-mainstreaming-acceleration-and-policy-support. Also, see papers presented at the conference on *Sustainable Development Transition Forum 25–27 October 2016, Bangkok.*

S. Acharya (✉)
Institute for Human Development, New Delhi, India
e-mail: sarthiacharya@gmail.com

A. Alakshendra et al. (eds.), *Development, Inclusion and Sustainability*,
Sustainable Development Goals Series, https://doi.org/10.1007/978-981-97-6863-9_13

Table 13.1 Sustainable development goals

(1)	(2)
Goal 1	End poverty in all its forms everywhere and revitalise the Global Partnership for Sustainable Development
Goal 2	End hunger, achieve food security and improved nutrition and promote sustainable agriculture
Goal 3	Ensure healthy lives and promote well-being for all at all ages
Goal 4	Ensure inclusive and equitable quality education and promote lifelong learning opportunities for all
Goal 5	Achieve gender equality and empower all women and girls
Goal 6	Ensure availability and sustainable management of water and sanitation for all
Goal 7	Ensure access to affordable, reliable, sustainable and modern energy for all
Goal 8	Promote sustained, inclusive and sustainable economic growth, full and productive employment and decent work for all
Goal 9	Build resilient infrastructure, promote inclusive and sustainable industrialisation and foster innovation
Goal 10	Reduce inequality within and among countries
Goal 11	Make cities and human settlements inclusive, safe, resilient and sustainable
Goal 12	Ensure sustainable consumption and production patterns
Goal 13	Take urgent action to combat climate change and its impacts
Goal 14	Conserve and sustainably use the oceans, seas and marine resources for sustainable development
Goal 15	Protect, restore and promote sustainable use of terrestrial ecosystems, sustainably manage forests, combat desertification, and halt and reverse land degradation and halt biodiversity loss
Goal 16	Promote peaceful and inclusive societies for sustainable development, provide access to justice for all and build effective, accountable and inclusive institutions at all levels
Goal 17	Strengthen the means of implementation

Source UN-DESA

13.1.2 SDGs—An Exposition

SDGs (earlier MDGs) aim to end poverty and promote equitable and sustainable growth.[3] The 17 SDGs, seen in Table 13.1, have 169 targets, but each country is expected to define and/or prioritise their goals, targets, indicators, etc.

The first seven goals are extended MDGs (Table 13.1). Next, Goals 8 to 12 are the drivers of first seven goals, i.e. the means to achieve the latter. Finally, the goals listed in 13–17 are more in the genre of common resources: they are indivisible and require global cooperation and collective action.

There are as well, links between goals within each grouping of goals. Thus, education, health and gender equity (Goals 3, 4 and 5) help alleviate poverty through the human empowerment route and also strengthen each other. There are many others, to the extent that when fully drawn, the interconnectivity is enormously complicated.[4]

[3] See, Nilsson M, D Griggs, M Visbeck and C Ringler 2016, *A Draft Framework for Understanding SDG Interactions, International Council for Science*, http://www.icsu.org/publications/reports-and-reviews/working-paper-framework-for-understanding-sdg-interactions-2016/SDG-interactions-working-paper.pdf (accessed Dec. 27, 2017).

[4] See for interlinks between these goals, Nilsson M (2016), Integrated Decision Making: Concepts and Approaches, Regional keeping in view that all planning Knowledge Exchange, UNDP at, http://www.asia-pacific.undp.org/meetTheSDGs.html, and Howells M 2016, Climate, Land, Energy and Water Systems, at https://un-desa-mhumanodelling.github.io/.

13.2 SDGs into Planning: Initials[5]

Two tools useful to find entry points are presented here with illustrations from different countries in the last several years. These are:

1. Gap Analysis; and
2. Awareness Building.

13.2.1 Gap Analysis

Gap analysis identifies the distance between the current state of development in a particular sector/target area, and the ideal aimed for, in this case the SDGs. Gap Analysis is also essential for setting *realistic intermittent targets* and also *raising resources*. Tools like SERVQUAL are useful for undertaking gap analysis.[6]

Malaysia—Gap Analysis

Malaysia has carried several gap analyses to determine gaps between the actual achievement and set/desirable target. In addition to gaps regarding absolute targets, the country has also conducted *comparative analyses* to judge gaps in social targets between Malaysia and the OECD countries. Table 13.2 presents the results succinctly.

Background: Malaysia is one of the few countries in developing Asia that appears to have overcome its early basic development impediments and it now aims to match-up with the OECD countries. Also, international agendas fit into its quest to improve human capital, which it has pursued since the 1970s.

Cambodia: Gap Analysis[7]

The Cambodian government—in partnership with the UNDP—has in the recent past conducted annual reviews of the progress it makes on its medium-term plan (5-year) targets and the gaps therein. Until recently, the reviews explicitly consisted of measuring progress on the MDG targets (later SDGs). Since the plan targets and MDG/SDG targets in that country are similar, this annual review had served as a good gap analysis tool. Cambodia uses both, survey data and administrative information for evidence and analysis.

Other than for discussions within the government, the results of the gaps analysis are discussed in Working Groups composed of donor agencies and the government to assess the situation and finding out how the targets can be furthered.

Background: Cambodia being a post-war transitional economy conducts gap-analysis with several parallel purposes, the principal ones being good governance (for a post-civil war country this is critical), property rights (this is central to people getting settled and having stakes), poverty alleviation and employment generation, and forestation (it lost it forests very fast in the 1990s). It is developing databases such that it gets quick information on the above issues and people's wellbeing at short intervals. International agendas like the SDGs appeal to it as they provide strategic directions to their development plans.

[5] This section draws extensively upon, Learning Manual Module 6 (UNITAR (2015), *Post-2015 National Briefings Package*, United Nations Institute for Training and Research (UNITAR), available at: https://www.learnatunitar.org/course/view.php?id=517&section=1; and CARE (2010), *Toolkit for Integrating Climate Change Adaptation into Development Projects*(Version 1.0), International Institute for Sustainable Development, Winnipeg.

[6] SERVQUAL, in simple language Service Quality, is a multi-dimensional research instrument designed to capture consumer expectations and perceptions of a service along the five dimensions that are believed to represent service quality. See, Parasuraman, A, Ziethaml, V. and Berry, L.L., 'SERVQUAL: A Multiple-Item Scale for Measuring Consumer Perceptions of Service Quality, *Journal of Retailing*, Vo. 62, no. 1, 1988, pp 12–40 online:https://www.researchgate.net/publication/225083802_SERVQUAL_A_multiple-_Item_Scale_for_measuring_consumer_perceptions_of_service_quality.

[7] Source: Ministry of Planning, Royal Government of Cambodia, Phnom Penh.

Table 13.2 Summary of Malaysia's performance in comparison to OECD countries on select SDG indicators—a comparative gap analysis

SDG #	Goal	Indicator	Malaysia's performance	OECD average
1.	Poverty	1.1. Incidence of relative poverty	19.3%	11.5%
		1.2. Poverty gap	NA	NA
2.	Agriculture and nutrition	2.1. Gross agricultural nutrient balances	NA	NA
		2.2. Obesity rate	13.5%	18.4%
3.	Health	3.1. Healthy life expectancy	64	77
		3.2. Life satisfaction	5.8	6.6
4.	Education	4.1. Upper secondary attainments	90.0%	76.3%
		4.2. PISA results (average score)	413	497.2
5.	Gender equity	5.1. Share of women in national parliaments	10.4%	27.7%
		5.2. Gender pay gap	8.4%	15.5%
6.	Water	6.1. Freshwater withdrawal as percent of total internal resources	2.3%	25.6%
		6.2. Population connected to wastewater treatment	99.3%	88.8%
7.	Energy	7.1. Energy intensity	6.4	6.0
		7.2. Share of renewable energy	0.4%	17.9%
8.	Economy and labour	8.1. GNI per capita	10,660	37,515
		8.2. Employment to population ratio	65.0%	66.9%
9.	Infrastructure and innovation	9.1. Gross fixed capital formation	27%	20.9%
		9.2. R&D expenditure	1.1%	2.0%
10.	Inequality	10.2. Palma ratio	2.2	1.3
		10.2. PISA Social Justice Index	NA	NA
11.	Cities	11.1. Particulate matter	NA	NA
		11.2. Rooms per person	NA	NA
12.	Consumption and production	12.1. Municipal waste generated	410.2	483
		12.2. Domestic material consumption per capita	16.8	18.9
13.	Climate	13.1. Production-based CO_2 emission	9.5	8.4
		13.2. GHG per GDP	400	352.1
14.	Oceans	14.1. Ocean health index	NA	NA
		14.2. Overexploited fish stocks	10.9%	12.3%
15.	Biodiversity	15.1. Terrestrial protection areas	NA	NA
		15.2. Red list index of birds	NA	NA

(continued)

Table 13.2 (continued)

SDG #	Goal	Indicator	Malaysia's performance	OECD average
16.	Institutions	16.1. Homicides/10,000 habitants	0.2	1.9
		16.2. Transparency corruption perceptions index	52.0	69.2

Source Institute of Strategic and International Studies 2015, Gap Analysis for an Implementation Framework in Preparation of Proposed Goals and Targets of Sustainable Development for Post-2015 Development Agenda, EPU/UNDP, Kuala Lumpur

Bhutan: Gap Analysis

Presently the penetration of cash in the economy is not too deep and there is also large out-migration of educated and skilled people. The Bhutanese government has identified financial inclusion, migration, gender and social protection as four areas where it needs to fill gaps. It has proposed the following:

1. Recognise the value of domestic/unpaid work.
2. Promote social protection policies.
3. Improve the efficiency of water use.
4. Strengthen the capacities of domestic financial institutions.
5. Facilitate orderly, safe, regular and responsible migration and mobility of people, including through the implementation of well-managed migration policies.
6. Provide universal access to safe, inclusive and accessible, green and public spaces, in particular for women and children, older persons and persons with disabilities.

Background: Bhutan is transiting to a democracy and a market economy. Its authorities are aware that an unplanned transition would result in its natural resources being ravaged and people getting marginalised. Hence, its principal concern is to ensure the transition process as smoothly as possible, and the *2030 Agenda* is showing the pathway to it.

13.2.2 Awareness Building

Starting Point

Entries into development plans cannot be made without the national stakeholders agreeing to espouse the goals. Example: The construction of millions of public toilets by the Indian government in the 1980s mostly remaining unused, since there was little public awareness about who they belonged to, who would be in-charge of their upkeep, where was the water to be sourced from, how the waste would be disposed, etc. The *Swachh Bharat* (Clean India) Campaign in India, in contrast, is being promoted with getting maximum stakeholders on board. The following are proposed for raising awareness[8]:

1. An introductory series of workshops to be conducted to sensitise government officials and stakeholders at all levels to the *2030 Agenda* and SDGs.
2. A public awareness campaigns to be carried out to communicate the *2030 Agenda*/SDGs to the general public, including indigenous peoples, women, marginalised sections, others.
3. Weave-in SDGs in all meetings: business, civil society, media, others.

Bangladesh, Albania and Others[9]

Bangladesh, Albania, Honduras, and Kenya (amongst others) developed detailed advocacy strategies that considered carefully who to

[8]Source: UNITAR (2015), *Post-2015 National Briefings Package*, United Nations Institute for Training and Research (UNITAR), available at: https://www.learnatu-nitar.org/course/view.php?id=517&section=1.

[9]Source: UNDG 2015, *Mainstreaming the 2030 Agenda for Sustainable Development*, New York.

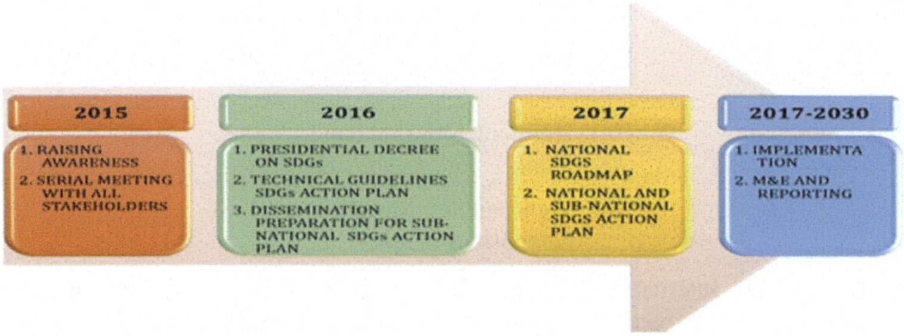

Fig. 13.1 Indonesia's SDG milestones. *Source* Presentation made at the Regional Knowledge Exchange Workshop, Oct. 24–26, 2016, UNDP Bangkok

reach, why they were important to communicate with, and various means to do it. The advocacy was particularly effective in mobilising communities around the MDGs and informing them about the 2030 *Agenda*. Albania has provided a unique example whereby local landmarks, renowned personalities, songs, imagery, and other aspects of the region were used in advocacy materials to make the goals relevant to the local reality and promoting *localisation of the goals*.

Indonesia: UN-Government Partnership in Raising Awareness About SDGs[10]

UNDP Indonesia (with support from the Ford Foundation, Tanoto Foundation, and the SDG Philanthropy Platform Partners) has committed to provide support to the Ministry of National Development Planning of Indonesia in rolling out the SDGs. This partnership will support the preparation of the Presidential Instruction for the implementation of SDGs, drafting of the National SDGs Roadmap, and establishment of an SDG Secretariat. At the sub-national levels, UNDP Indonesia will support the localisation of SDGs in Riau Province through mainstreaming, acceleration and policy support.

UNDP Indonesia, its funding partners, and the local provincial governments ensures that the project is implemented in a participatory manner by involving broad-based multi-stakeholders. UNDP especially ensures that local civil society organisations representing different marginalised and poor communities participate in the planning, development, implementation and monitoring-evaluation of the project.

To reinforce the localisation commitment to the SDGs, UNDP supports the Provincial SDGs Secretariat to develop a Provincial SDGs Roadmap. This long-term planning document sets milestones, objectives and strategies to attain the SDGs, and identify responsibilities of different actors. UNDP and the Provincial SDG Secretariat have also worked together in translating the global and national sustainable development agenda into local development agenda.

The milestones that the country has charted for itself for implementing SDGs could be seen in Fig. 13.1.

By 2017 the country rolled out the roadmap at both national and sub-national levels and the period thereafter, until 2030 is the implementation (and simultaneous evaluation) phase. For raising awareness and capacity building the Indonesian authorities have held a series of meetings with different stakeholders including the media at the sub-national levels for capacity building and has also hoisted a website. A sketch of interrelationships between the SDGs and national planning in Indonesia could be seen in Fig. 13.2.

Background: Indonesia has been committed to achieving social goals for especially for reaching out to the far-flung provinces and

[10] See, https://unstats.un.org/unsd/unsystem/Documents-Sept2018/BG-Item6-MAPS.pdf.

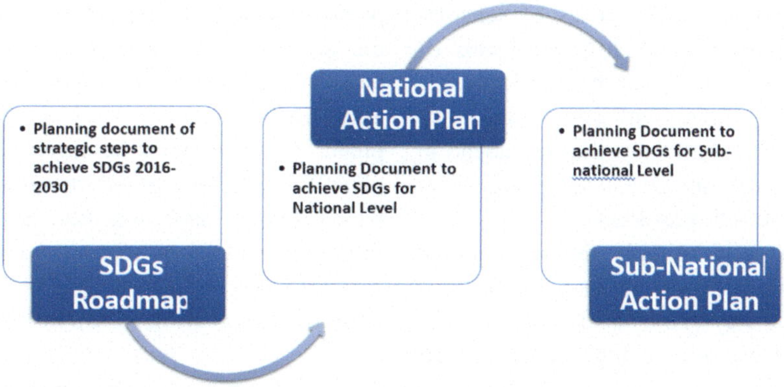

Fig. 13.2 Mainstreaming SDGs into national development Agenda—Indonesia[11]

peoples in its 6,000+islands. Indonesia, competes with Singapore, Malaysia and Thailand in economic wellbeing. However, being highly diverse (though resource/forest-rich), the distribution of gains is not always equal. It hence embraces policies, strategies and approaches like the *Agenda 2030*, that would help promote conservation as well as equality along with development.

Uganda: UNITAR's National Post-2015 Development Agenda Briefing Package[12]

The Ugandan government together with the UN Country Team and two training experts has piloted a briefing package on the post-2015 Agenda. The event was led by Ministry of Foreign Affairs, the Ministry of Finance, Planning and Economic Development, and the National Planning Authority. Experts from UNDP and the UN Department of Economic and Social Affairs (DESA) also contributed to the facilitation by presenting a global perspective on the evolving issues of global partnership, financing for development, review, follow-up, and SDG synergies.

The exercise highlighted how the government has already made significant progress in integrating the proposed SDGs into national planning. Specifically, the proposed National Development Plan II (NDP-II) includes many of the SDGs and a significant portion of the proposed targets have been adjusted to the national context.

Background: Uganda has faced political uncertainty and turmoil in the past few decades, but is now embarked upon a (sustainable) development path. Its endorsement of the *2030 Agenda* is a logical step in this direction.

Rajasthan (Sub-National): MDG Awareness Building Workshops, 2005[13]

The Government of Rajasthan (in India), in partnership with the UNDP, had organised eight workshops for its officials in the capital city of Jaipur and at the division/district levels to make the officials aware of the MDGs: goals, progress made, and challenges ahead, in 2005. UNDP's implementing partner, the Institute of Development Studies at Jaipur, organised these

[11] Sources: https://sdgs.un.org/partnerships/support-roll-out-sdgs-indonesia, and Presentation made at the *Regional Knowledge Exchange Workshop, Oct. 24–26, 2016, UNDP Bangkok.*

[12] Source: UNITAR (2015), *Post-2015 National Briefings Package*, United Nations Institute for Training and Research (UNITAR), available at: https://www.learnatunitar.org/course/view.php?id=517&section=1.

[13] Source: Institute of Development Studies, Jaipur.

workshops with the twin purposes of sensitising government officials at different levels of disaggregation on the importance/performance of MDGs in Rajasthan (with inter-state comparisons made) and challenges that lay ahead in implementing/achieving MDGs and their integration in the state's 5-Year Plans.

The officials were initially defensive about the status of development in their work area/sector, but soon they recognised that the gap-finding exercise was not to find faults in their performance but to raise efforts for bridging the gaps. They eagerly opened up and offered suggestions on how to better implement poverty alleviation schemes, bring about greater gender equity, and improve the drinking water and sanitation situation. The final results were seen in the next 5-Year Plan of the state where references to poverty alleviation, gender issues (child education with clear mention to the girl child), water supply and sanitation were made, with explicit mention to the MDGs.

Background: Rajasthan being a desert-state, its interest in promoting a social development agenda stems from a realisation that its economic development cannot happen without improvement in conservation (water, forests), and human development (education, health, gender).[14]

13.3 Adapting and Integrating the SDGs at National Level[15]

Each government requires deciding upon integrating of SDGs into national plans and strategies as they are applicable.[16] The existing development plans/strategies require to be scrutinised at both, the national and sub-national levels—including those at the sectoral and sub-national levels—for assessing the extent to which they are SDG compliant. It would then be possible to formulate plans for matching ambitions and commitments with resources and capacities. Experiences from different countries are reproduced below[17]:

Bhutan: Integrating SDGs in Planning

The Bhutanese government has gone ahead with indigenising the SDGs, following an additional dimension in the trio of economic-social-environmental, namely 'happiness' for which it has defined a Gross National Happiness Index (GNH). The country identified the following three goals as vitally important:

Goal 1. End poverty in all its forms everywhere (priority: high urgency).

Goal 13. Take urgent action to combat climate change and its impacts (to remain carbon–neutral).

Goal 15. Protect, restore and promote sustainable use of terrestrial ecosystems, sustainably manage forests, combat desertification, and halt and reverse land-degradation and halt biodiversity-loss (showcase leadership in biodiversity).

A comparison of Bhutan's plans with the SDGs suggests the following:

- National Key Result Areas (NKRAs): 16 outcomes, 61 key performance indicators (KPIs) and four dimensions (including happiness).
- Sustainable Development Goals: 17 goals, 169 targets, three dimensions.

Fifteen of the SDG goals are already integrated into the 16 NKRAs (Table 13.3). Next, of the 169 SDG targets, some 134 are found relevant for Bhutan and were integrated in its 11th Five-Year Plan (2013–2018).

[14] *Human Development Report of Rajasthan—An Update 2008*, Government of Rajasthan, Jaipur.

[15] This section draws upon *Learning Manual Module-6 (UNITAR (2015), Post-2015 National Briefings Package*, United Nations Institute for Training and Research (UNITAR), available at: https://www.learnatunitar.org/course/view.php?id=517&section=1; and CARE (2010), *Toolkit for Integrating Climate Change Adaptation into Development Projects*(Version 1.0), International Institute for Sustainable Development, Winnipeg.

[16] Centre for Global Development, https://www.youtube.com/watch?v=s6qJBhQ_3Ag.

[17] The overall analysis could also be seen at, *Mainstreaming the 2030 Agenda for Sustainable Development*, at https://unsdg.un.org/resources/mainstreaming-2030-agenda-sustainable-development-reference-guide-un-country-teams.

Table 13.3 SDGs and National Goals—A Snapshot in Bhutan

SDGs	NKRAs (Bhutan's national goals)
Goal 1: No Poverty	NKRA 3: Poverty and Inequality Reduced
Goal 2: Zero Hunger	NKRA 8: Food and Nutrition Security Ensured
Goal 3: Good Health and Well-being	NKRA 14: Healthy and Caring Society Enhanced
Goal 4: Quality Education	NKRA 7: Quality of Education and Skills Improved
Goal 5: Gender Equality	NKRA 10: Gender Equality Promoted, Women and Girls Empowered
Goal 6: Clean Water and Sanitation	NKRA 15: Liveability, Safety and Sustainability of Human Settlement Improved
Goal 7: Affordable and Clean Energy	NKRA 6: Carbon Neutral, Climate & Disaster Resilient Development Enhanced
Goal 8: Decent work and Economic Growth	NKRA 11: Productive and Gainful Employment Created
Goal 9: Industry, Innovation and Infrastructure	NKRA 2: Economic Diversification and Productivity enhanced
Goal 10: Reduce Inequalities	NKRA 3: Poverty and Inequality Reduced
Goal 11: Sustainable Cities and Communities	NKRA 15: Liveability, Safety and Sustainability of Human Settlements Improved NKRA 4: Culture and Tradition Preserved and Promoted
Goal 12: Responsible consumption and production	NKRA 5: Healthy Ecosystem Services Maintained
SDG 13 Climate Action	NKRA 5: Healthy Ecosystem Service Maintained NKRA6; Carbon Neutral, Climate and Disaster Resilient Dev Enhanced
SDG 15 Life on Earth	NKRA 5: Healthy Ecosystem Maintained NKRA 6 Carbon Neutral, Climate Resilient Development Enhanced
SDG16 Peace, Justice and Strong Institution	NKRA 9: Efficiency and Effectiveness of Public Services Improved NKRA 12. Corruption Reduced NKRA 13 Democracy and Decentralization Strengthened NKRA 16: Justice Services and Institutions Strengthened
SDG17 Partnership for the Goals	NKRA 2: Economic Diversity and Productivity Enhanced

Source Drawn from https://www.ncbi.nlm.nih.gov/pmc/articles/PMC10105210/; Actual presented at the *Regional Knowledge Exchange Workshop, Oct. 24–26, 2016, UNDP Bangkok*

Belize: Merging SDGs in National Planning Process

Belize, a small low-income country, has its economy primarily composed of agriculture, tourism, and services. There is high income inequality and poverty. Belize is committed to implementing SDGs in totality, for which it is collaborating with the UNDESA and UNDP. The country has merged its *Growth and Poverty Reduction Strategy* and the *National Sustainable Development Strategy* into one unified and coherent strategy and planning process, which is called the *Growth and Sustainable Development Strategy*(GSDS). The consultation process for the GSDS could be seen in Fig. 13.3.

Background: Belize is committed to the SDGs through strong advocacy and its quest to become a vibrant democracy after a long period of colonial rule until 1980.[18]

[18] See, http://www.nationalwomenscommissionbz.org/what-are-the-sustainable-development-goals/.

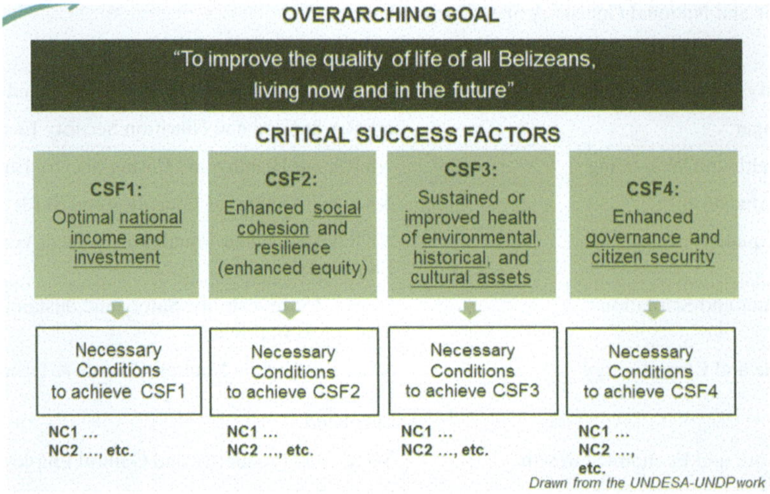

Fig. 13.3 Structure of GSDS—Belize

Cambodia: An MDG/SDG-Centred Plan[19]

Cambodia is a post-war, transitional, open market economy and is among Least Developed Countries (LDCs). Most major private investments save a few in hotels and restaurants are foreign investment-dependent, and public projects are to a great extent Overseas Development Assistance-funded. Planning to that extent is largely indicative, since the major determinants of investments are external.

Cambodia's Five-Year Plan 2014–2018 was so formulated for it to be *MDG-compliant*. Since the SDGs are similar to Cambodia's own goals in many respects (at least the first seven goals), several of its components de facto became SDG-compliant. An abridged version of the plan matched with SDGs could be seen in Table 13.4. The table shows that effort has been made in the plan to promote universal education, health, green technologies, infrastructure, industrialisation, and gender issues.

Cambodia has fully integrated the SDGs in its present plan cycle, 2019–2023.[20] It is localising the framework of the *2030 Agenda* (Cambodia Sustainable Development Goals—CSDGs), which the government has endorsed.[21] To quote Prime Minister Hun Sen, '*…the Royal Government of Cambodia considers that all the 17 SDGs, which were built on the MDGs, are very relevant to Cambodian contexts especially in strengthening its achievements made over the past 15 years'*.[22]

Uganda: Integrating SDGs in Planning

The Ugandan government, in partnership with the UN-DESA, UNDP and UNITAR carried out an exercise to mainstream SDGs. In its second National Development Planning (NDP) process (2015/16–2019/20) in support of the country's Vision 2040, the government recognised that the SDGs offered an initial framework through which

[19] This author was one of the principal advisors to Cambodia's Plan (2014–2018).

[20] https://docs.wfp.org/api/documents/WFP-0000112436/download.

[21] See, Bonnak P 2016, Sustainable Development Goals and their Adaptation to Cambodia Context, presentation made at *Sustainable Development Transition Forum 25–27 October 2016*, Orakai Songdo Park Hotel, Incheon; and The Sustainable Development Goals and their Adaptation at Country Level—Cambodia(Ministry of Planning. Phnom Penh).

[22] Speech delivered at the *United Nations Summit for Adoption of the Post-2015 Development Agenda*, September 25–27, 2015.

Table 13.4 Mapping NSDP with SDGs in Cambodia

Select NSDP 2014–2018 Goals	SDGs
Promote **crop diversification** and **livestock**; adapt to climate change; invest in irrigation; ensure farmers getting stable prices; extend credit facilities, improve farmers' skills	SDGs 1, 2 and 13
Demarcate and rehabilitate **forests**; conserve **fisheries**; minimise **water pollution**; conserve soil quality, wetlands and biodiversity; implement mechanisms for **carbon trading**, **climate change** adaptation and environmental financing	SDGs 2, 13 and 15
Improve access to **safe-drinking water** and **sanitation**; encourage innovations in sanitation in localised environments	SDG 6
Construct **rural roads**; improve connectivity by multi-modal, cross-border transport and logistics systems; expand railways; and seaport and inland infrastructure	SDG 9
Expand irrigation systems with flexibility to adapt to **climate change** and for flood control; harmonise development of irrigation, hydropower and transportation	SDGs 1, 2 and 15
Expand electricity production from '**clean' sources**; realise "by 2020, all villages have access to electricity from grid or other sources"; participate in energy cooperation in the region	SDG 7 and 13
Establish industrial clusters; and promote entrepreneurship, productivity, technology transfer, access to finance, business counselling centres	SDG 8
Promote **Education/TVET Plan** on R&D; promote science and technology, esp. in agriculture, livestock, aquaculture, industry technologies, energy, ICT, healthcare and environment	SDGs 4 and 8
Construct **referral hospitals, health centres and health posts**; create health insurance; improve MNCH and awareness about WASH; reduce morbidity/mortality	SDG 3
Promote women's economic role thru technical and vocational training and entrepreneurial skills; fight gender violence; trafficking and sexual exploitation of women	SDGs 4 and 9
Fight corruption, reform administration and decentralise governance	SDG 16
Prepare a strategy for development cooperation and partnership	SDG 17

Source Presentations made at Workshop on "Sustainable Development Transition Forum 25–27 October 2016, ESCAP Bangkok"

different sectors could trace their linkages to the national priority areas. SDGs were included in NDP-II. An important share of SDG targets was adjusted to national circumstances and included in the NDP-II Results Framework. The government is planning to further incorporate SDG targets in a more detailed results framework and refine it using the SDGs structure. A pictorial presentation of the consultative process in formulation of the NDP could be seen in Fig. 13.4.

Indonesia: Country Experience in Integrating SDGs in Planning

The Indonesian government has entered into partnership with the UNDP for furthering the SDGs (with support from the Ford Foundation, Tanoto Foundation, and the SDG Philanthropy Platform Partners). The government has matched its development goals with the SDGs, a brief of could be seen in Table 13.5.

Malaysia: Integrating SDGs in Planning

As is evident from the statements of its prime minister on September 27, 2015 at the United Nations in New York, Malaysia is committed to implementing the *2030 Agenda*.

> Malaysia is absolutely committed to the Post 2015 agenda—the Sustainable Development Goals and our aspirations to transform the world by 2030; and.
>
> Malaysia is now about to embark on its next five-year development plan—the Eleventh.
>
> Malaysia Plan, from 2016 to 2020. In many aspects, the plan mirrors the multi-dimensional nature of the SDG, and has been formulated with the people at the centre of all development efforts.

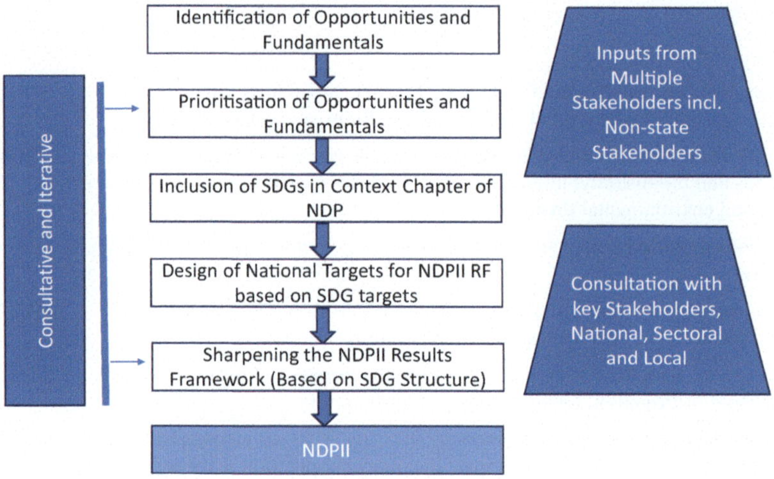

Fig. 13.4 Mainstreaming SDGs in NDP II, Uganda[23]

Table 13.5 SDGs and National Targets Alignment—Indonesia

Pillar/goal	Global targets	National targets	National priorities
(1)	(2)	(3)	(4)
SOCIAL (1.2.3.4.5)	47	27	Poverty reduction, improved welfare; enhanced food security; implementing Smart Health Indonesia programme; protection of children; women and marginalised groups
ECONOMIC (7,8,9,10,17)	54	39	Energy security, acceleration of manufacturing industry, improved labour competitiveness, building national connectivity, well-balanced development, implementing free and active foreign policy
ENVIRONMENTAL (6,11,12,13,14.15)	56	31	Water security, housing and residential development, climate change adaptation and mitigation, development of marine-based economy, protection of natural resources, environment and disaster management, conservation and sustainable use of bio-diversity
LAW AND GOVERNANCE (1)	12	8	Improved quality protection of citizens, enhanced law enforcement, foster transparent and accountable government

Source Yusuf G 2016, *Integrating SDGs to Development Plan*, Presentation made at the Regional Knowledge Exchange Workshop, Oct. 24–26, 2016, UNDP Bangkok

The 11th 5-Year Plan (2016–2020—11-MP) has coincided with the launching of the SDGs. The initiating point of Malaysia Plan (11-MP) was *Anchoring Growth on People*, which at one level matches the one of *2030 Agenda*, which says

[23] See also, https://link.springer.com/referenceworkentry/10.1007/978-3-030-91260-4_31-1#:~:text=Localization%20of%20SDGs%20was%20integrated%20in%20the,(NDP%20II)%202015/2016%E2%80%932019/2020%20of%20Uganda%20(GoU%202015d).

Table 13.6 Integrating SDGs into planning framework—Malaysia

(1)	(2)
Strategic thrusts	Matched SDGs
ST 1: Enhancing inclusion towards an equitable society	1, 2, 5, 6, 10, 11
ST 2: Improving wellbeing of all	3, 11, 16
ST 3: Pursuing green growth and resilience	4, 8
ST 4: Accelerating human capital development for becoming an advanced nation	6, 7, 12, 13, 14, 15
ST 5: Strengthen infrastructure to support economic expansion	7, 9
ST 6: Re-engineering economic growth for greater prosperity	2, 8, 9, 17
Transforming public service for productivity	16

Source Misran H 2016, Integrating the 2030 Agenda: SDG Roadmap Malaysia, *Presentation made at the Regional Knowledge Exchange Workshop*, Oct. 24–26, 2016, UNDP Bangkok

Leave No One Behind. Hence, it is perfectly placed to incorporate these into the national medium-term plan (11-MP).

The 11-MP has defined in the form of Strategic Thrust (ST) areas. A matching of these with the SDGs could be seen in Table 13.6.

In brief, the roadmap that Malaysia charts for itself looks as below:

- Phase I (2016–2020)—prioritising SDG according to 11MP
- Phase II (2020–2025)— focus on post 2020 goals and targets
- Phase III (2025–2030)—remaining goals and targets in line with Malaysia's capacity and global role.

In Malaysia, the laws require public participation in the process of preparing development plans (structure plans and local plans) in Peninsular Malaysia (except Kuala Lumpur). Development plans guide the location of development with criteria-based policies. The public has the right to know and participate in making decisions. Effective and successful public-participation programme allows members of the community to have an active voice in the process and to have free access to important information.

Mongolia: SDG-Planning Harmony[24]

Mongolia has prepared a Sustainable Development Vision 2030 (SDV), and has medium-term development plans, the Government Action Plans (GAP). The SDV or the GAP are not expected to be an operational plan (as elsewhere) and the actual authority rests in the annual plans, but the latter are guided by the SDV and the GAP.

The SDV-2030 is fed into by the GAP/SDGs. A pictorial depiction could be seen in Fig. 13.5. This figure suggests that the SGV is supreme and all other plans and goals are expected to feed into it/fed by it. The SDGs have 169 (intermediate) targets, and the GAPs are the means to achieve the SGV goals.

The broad goals of the GAP for 2014–2020, listed below, have all the three dimensions of the SDGs, though the actual allocations are made in the annual plans.

Actions for economic stabilisation: Promote production, attract investment, ensure fiscal

[24]This section draws heavily on *Mapping of the Sustainable-Development Vision 2030 of Mongolia and the Green Development Policy/Action Plan against the Global SDG Targets*, at https://www.un-page.org/knowledge-hub/mapping-of-the-sustainable-development-vision-2030-of-mongolia-and-the-green-development-policy/ and on the contents of Mainstreaming the SDGs: Policy planning and implementation in Mongolia, https://www.slideshare.net/slideshow/mainstreaming-the-sdgs-policy-planning-and-implementation-in-mongolia/79319570.

Fig. 13.5 Harmony between SDG and SDV—Mongolia. *Source* Ministry of Finance Mongolia 2016, Mainstreaming the SDGs: Policy Planning and Implementation in Mongolia, *Presentation made at the Regional Knowledge Exchange Workshop*, Oct. 24–26, 2016, UNDP Bangkok

policy stability, create a favourable tax environment, and intensify mega-projects.

Ensure sustainable economic growth: Promote real economic sectors, improve city planning, ensure safe living environment for residents, provide reliable sources of energy for power, and expand logistics and transportation.

Social development: Improve human resources in the health sector, prevent communicable and non-communicable diseases, encourage innovation, develop knowledge-based production, promote employment, improve social welfare, scale-up youth education and employment.

Environment and green policy: Preserve natural balance of the ecosystem, protect natural resources, support environmentally friendly technologies, develop renewable energy production, and improve tourist attractions.

Governance: Ensure transparency, accountability, decentralisation, and delivery of high-quality services.

Background: Central-planning was discontinued in 2014 and the Ministry of Finance fully controls development financing. Central line ministries implement the plans. However, under the law, Aimags (provinces) and Soums (districts) have their own jurisdiction and financial allocation for planning and implementation. Also, within the Finance Ministry, there is a planning department whose personnel are the same who worked for the erstwhile Ministry of Economic Development.

Nepal: Integrating SDGs

Nepal aims to pursue the *2030 Agenda* on the back of what it completed and what remained from the MDGs. The SDGs have been taken up to fill-in the infrastructure gaps with a major focus on social sectors, along with the environmental agenda,. The SDG implementation coincides with new three-year planning cycle. SDGs had been mainstreamed in the approach paper prepared in mid-July 2016, and the budget for financial year 2016–2017.[25]

The starting point was taking up activities on awareness, advocacy and orientation, with the goals and targets translated in the local language, and discussions held with parliamentarians and local development bodies during the plan preparation process. A vision document entitled "Envisioning Nepal: 2030" with SDGs, at centre point is being prepared. NGOs and civil society organisations have been brought on board.

Key SDG-priority areas have been identified by the central government and local bodies;

[25] Later information is not available.

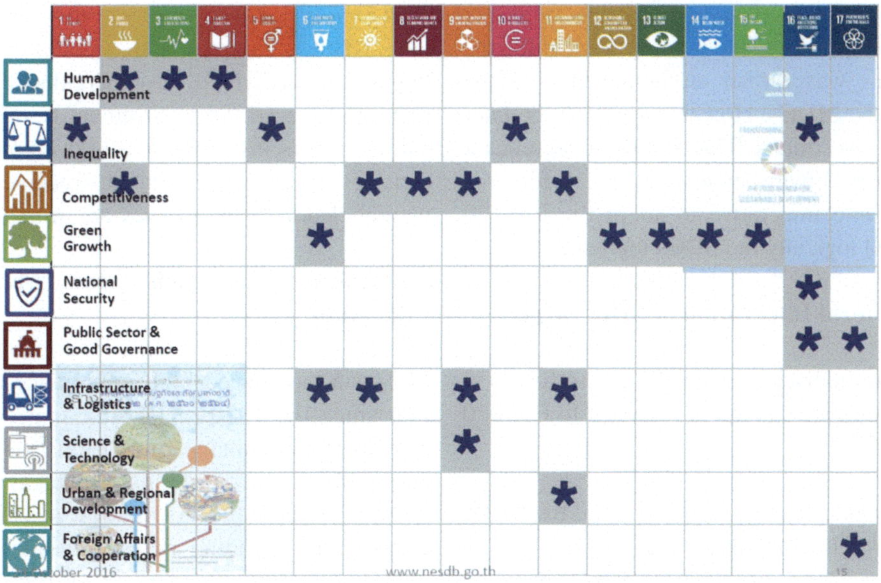

Fig. 13.6 Integration of SDGs into NESDP. *Source* Kumpa Ladawan (2016), Agenda 2030 sustainable development goals: Thailand's perspectives, presentation made at the regional knowledge exchange workshop, Oct. 24–26, 2016, UNDP Bangkok

along with, programming and budgeting guidelines have been undertaken. A Medium-Term Expenditure Framework has been exercised for project prioritisation with a given budget-ceiling. When seen from an SDG perspective, in 2016–2017, 29.5% of the budget was allocated for addressing SDGs 1 to 6 (improvised forms of MDGs), 65% for SDG 7, 9 and 11 (mainly infrastructure) and 5.5% for the other eight SDGs. In terms of the numbers of development projects, 48.5% were on SDGs 1–6, 35% on SDG 7, 9, and 11, and 16.5% on other SDGs.

Nepal has a strong central-planning system with significant decentralisation in a market framework. Yet, authorities in Nepal find that SDGs alignment with the national plan and annual programmes (horizontal) is easier compared to the more challenging work of alignment at the sub-national levels (vertical). Mainstreaming SDGs at the local levels is arduous since local governments do not have periodic plans, implementation rigour, motivation, incentives and capacities. This is a task it plans to undertake in the times to come.

Background: Nepal is categorised as a Least Developed Country. Since it had already embraced the MDGs, it was logical that it embraced the SDGs. The task became smoother since environmental issues are central concerns in the country.[26]

Thailand: Integrating SDGs into National Planning

Thailand has developed a mechanism for integrating SDGs via mapping these goals, both in its long-term vision spanning up to 2036, as well as medium-term planning (National Economic and Social Development Plan—NESDP). A general roadmap of integrating SDGs in NESDP could be seen in Fig. 13.6. The figure is self-explanatory, encompassing government policies and medium- and long-term plans, and the challengers that the SDGs pose.

[26] See, Jha SG 2007, '*A Brief Appraisal of Existing Main Environmental Issues in Nepal and Potential Intervention to Solve the Perceived Problems*', Banko Janakari, Vol 17, No. 1.

Background: Thailand already has a strong social development agenda dating back since the 1980s. This was further strengthened with the advent of the MDGs. However, through the later part of the last century, Thailand faced significant damage to the environment—air, water, forest—and it became imperative to adopt the Rio+20 agenda, and logically the SDGs thereafter.[27]

India: Sectoral Integration of SDG Targets

India's Nationally Determined Targets for 2030 regarding climate change in response to Rio+20 (SDG 13):

1a. Renewable energy target: 40% by 2030; achieved in 2020: 36%.
1b. Achieve 40% of electric power installed capacity from non-fossil fuel by 2030.
2. Create additional carbon-sink of 2.5–3 billion tonnes of CO_2 equivalent through additional forest and tree cover.
3. Reduce by 20–25%, emission intensity of GDP by 2020, compared to the 2005 levels.
4. Enhance investments in development programmes in various vulnerable sectors, including agriculture, water resources, forestry, health and disaster-management.
5. Build capacities while adopting and deploying new energy efficient technology and other technologies to reduce carbon emissions.
6. Mobilise resources to execute plans for combating climate change across sectors.
7. Adopt and promote low carbon-intensive lifestyles on a mass scale through sustainable living based on traditions and values of conservation and moderation.

[27] See Chap. 1 of, 'Overview of Environmental Issues and Environmental Conservation Practices in Thailand', *Overseas Environmental Measures of Japanese Companies (Thailand)*, Ministry of the Environment, Government of Japan, Mar 1999. Retrieved 2 Mar 2015; Choeypun K n.d. 'Thailand Environmental Fund', at https://www.cbd.int/financial/trustfunds/Thailand-ef.pdf; and 'Institutional Aspects of Sustained Development in Thailand', at http://www.un.org/esa/agenda21/natlinfo/countr/thai/inst.htm.

Background: There are strong environmental NGO groups and Community-based Organisations, which have heralded the movement and forced the government to take action. National laws now require all development projects, especially in mining, to receive environmental clearance before they are initiated.

Malaysia's Approach to Mainstream Education

Malaysia has placed high importance to education since the 1970s. The state expenditure on education and human capital formation has been among the highest in terms of percentage of the GDP, at about 5–6%. As in 2011, at 16% of total government spending in education, it was not only higher than the OECD average of 8.7% of total public-spending but also at par with or more than the top-performing countries. On *2030 Agenda*, its approach is summarised below:

1. Understanding the current performance and challenges of the Malaysian education system, for improving access to education, raising standards (quality), closing achievement gaps (equity), promoting unity amongst students, and maximising system-efficiency.
2. Establishing a clear vision and aspirations for individual students and the education system as a whole up to 2030.
3. Outlining a comprehensive transformation programme for the system, including key changes to the Ministry of Education, which will allow it to meet new demands and rising expectations, and to ignite and support overall civil service transformation.

It is attempting to bring closer, the different schools operating in the country—like those operated by ethnic groups. The government participates in PISA examinations (international competitive educations) and the government uses the results for judging and thereby improving children's performance in language, maths, etc.

Uttarakhand: Approach to Setting Targets for SDGs (Sub-National)[28]

Uttarakhand, a northern province in India, began to prepare a development plan for the province's development based on SDGs, starting 2017. They are developing a Mid-term Plan up to 2023 (seven years) and a Long-term Plan up to 2030.

The process consists of a four-step approach to identify what is to be done. First is to document the current situation; second, setting targets—eradicate poverty, provide safe water, safe sanitation, legal shelter to all, provide education up to secondary school to all, preserve the environment, and so on; third is to conduct a gap analysis to find out the extent of work that remains; and the fourth is to develop a strategy based on the first three steps. The plan aims to achieve interim targets by 2023, and final targets by 2030.

The government, in partnership with the Institute of Human Development (IHD) in New Delhi, has developed a roadmap to complete the task of completing the process of identifying the indicators for 2030 and for the interim periods (2020, 2023 and 2026). It identified data sources and responsibilities of different departments by 2016.

13.4 Institutional Issues

Institutional mechanisms for delivering results are central to achieving the targets of SDG (or any other).The nuances of these are discussed here.

1. Creating horizontal and vertical coherence: This is essential for ensuring that goals that can only be achieved through efforts of more than one ministry/department are so done; and for this, the implementation is synchronised across ministries/departments and also between different levels in the government.
2. Stakeholder Analysis: All stakeholders have to be on board for optimal results, and for this, not only is their identification essential, their agreeing to work in congruence is equally important.

13.4.1 Creating Horizontal and Vertical Policy Coherence

For creating horizontal and vertical linkages some key elements are, formalised institutional mechanisms in the form of inter-agency coordinating groups for creating horizontal and vertical policy coherence, integration and partnerships.[29]

Bhutan: Gross National Happiness Commission[30]

Bhutan's *Gross National Happiness (GNH) Commission* is an example of an inter-agency coordinating body designed to foster horizontal coherence, integration and partnerships across government sectors. The GNH Commission is the Government of Bhutan's Planning Commission and is charged with ensuring that GNH is mainstreamed into government planning, policy making and implementation. The GNH Commission coordinates the country's Five-Year Plan (FYP) process and is composed of all ministry secretaries with planning officers that provide links between individual ministries and the GNH Commission.

Finland: Commission for Sustainable Development[31]

In Finland, which is a high-income country, the inter-ministerial secretariat of the *National Commission for Sustainable Development*

[28] Source: Government-IHD Workshop on SDGs, Dec. 14–15, 2015, Government of Uttrakhand, Dehradun.

[29] See, Rasheed N 2016, Integrated Approaches: Health and Environment, presentation made at the Regional Knowledge Exchange Workshop, Oct. 24–26, 2016, UNDP Bangkok.

[30] See, GNH Centre, at www.gnhcentrebhutan.org.

[31] See also, https://kestavakehitys.fi/en/commission.

(FNCSD) is an example of an inter-agency coordinating body that facilitates horizontal policy coherence, integration and partnerships. Steered by the Ministry of the Environment, the secretariat comprises of members from different ministries, each taking the lead in preparing themes within their area of expertise. The secretariat has facilitated horizontal coordination over the years, including establishing a sub-committee for integrating multiple strategies from across government and other stakeholder groups.

Switzerland: Example of Integrated Policy Analysis[32]

Switzerland, which is a high-income country, has had a long history of applying integrated policy analysis methods in the form of 'sustainability assessment' (SA). The Federal Office for Spatial Development provides guidelines and tools for SA, which are intended as instructions on how to evaluate the Federal Government's initiatives (laws, programmes, strategies, concepts and projects) to determine their compliance with the principles of sustainable development (i.e. ARE 2015).

The Swiss ARE has collaborated with representatives from 30 Swiss cantons and local municipalities to prepare guidelines for assessing project sustainability at cantonal and municipal levels. The guidelines are available online and describe the benefits of assessment, how the sustainability assessment process can be initiated and provides assistance for choosing the right assessment tool.[33]

[32] Switzerland, the Policy In, at https://link.springer.com/referenceworkentry/10.1007/978-3-030-90434-0_54-1.

[33] See. https://www.esdn.eu/country-profiles/detail?tx_countryprofile_countryprofile%5Baction%5D=show&tx_countryprofile_countryprofile%5Bcontroller%5D=Country&tx_countryprofile_countryprofile%5Bcountry%5D=28&cHash=cf3aea74ff18a9d6e446fee5e193d434.

Malaysia: Vertical Integration

Malaysia has 13 states (provinces) and two federal territories, and over 60 districts. The central government decides policies and implementation approaches, while different ministries ensure their translation into practice, through the state- and district councils.

National Action Council: This decides overall policies and ensure implementation of development strategies; evaluates the implementation of strategies; and coordinates development issues related to the environment. [→ (HORIZONTAL LINK)].

Action Committee of the Ministry: This ensures implementation and coordinate programmes and projects of the NAC that fall under the ministry; reports the progress of programme and project implementation to the implementation unit in the prime minister's department; and monitors and evaluates the impact of the project implemented under the ministry.

↓ (VERTICAL LINK).

State Action Council: This ensures that all decisions of the NAC are carried out; streamlines and expedites the implementation of NAP and policies at the state level; and monitors and evaluates the effectiveness of the government in implementing NAC's decisions.

↓ (Vertical Link).

District Working Committee: monitors and evaluates the impact of the implementation of development programmes and projects at the district/divisional levels; identifies problems of implementation that cannot be solved at the district level and these be brought to the attention of the SAC; and ensures that the target group receives maximum benefits from programme/project. [→ (HORIZONTAL LINK) + ↓ (VERTICAL LINK)].

India: Centre-Province Vertical Integration in Policy

In India—where the provinces are fairly autonomous and plan for SDGs or other activities independent of central direction—there is a rule-based division of responsibilities between the national and provincial authorities for each sector. For implementing projects, (flexible) formulae are drawn-up to split the revenues between the national and sub-national/regional authorities since most tax revenues are collected by the centre. Both, the jurisdiction and the financial sharing are reviewed periodically.

1. Sectors/activities under the national jurisdiction are identified and agreed upon; like trans-province infrastructure, currency and foreign affairs.
2. Sectors/activities under provincial jurisdiction are identified and agreed upon; like education, provincial roads, health, agriculture, and land issues.
3. Sectors/activities under a concurrent list (joint, central jurisdiction plus provincial) are identified and agreed upon. This is discussed in the parliament and with the provincial assemblies and reviewed every five years; like taxes, education, science and technology, specialised health institutions, and culture.

The jurisdictions are agreed upon in the parliament and state assemblies. There are constitutional guarantees for such arrangements.

13.4.2 Stakeholder Analysis

Consultations take place at various levels depending upon the scope of policy, time frame, and resource availability.

1. Consultation at the local community level in planning and implementation: This would apply for local projects like local connecting roads, common land use, streetlighting, etc.
2. Inter-ministry participation: This is mainly for cross-cutting issues.
3. Inter-stakeholder participation (research institutes, international development partners, private sectors, civil society, media) in larger and more complicated issues.

Germany: Engaging Stakeholders[34]

In Germany, a high-income country, in 2014 government asked its Council for Sustainable Development (RNE) to assess how a national implementation of the SDGs would impact the structures and institutions of the country's sustainability policy. In 2015 the RNE submitted its statement to the government. The content of the recommendation report included the following areas:

- *General:* Two recommendations dealing with the publishing of the recommendations and broadening the public debate on sustainable development.
- *Sustainability as a principle of action*: Three recommendations identifying the existing national SD strategy as the appropriate instrument for implementing the SDGs, and addressing monitoring and review and global financial assistance.
- *Global partnerships*: Four recommendations for advancing the use of national SD strategies around the world, including in the EU.
- *Germany's responsibility*[35]: Twelve recommendations relating to new perspectives for a 'German SD Strategy', including a proposal for all goals to relate to 2030 as a rule, and for broad participation and involvement in implementing the strategy.
- *Redefinition of Germany's sustainability architecture*: Eight recommendations relating to sustainability as a law-shaping principle and institutional interfaces between global and national.[36]

---

[34] See, http://www.nachhaltigkeitsrat.de/en/home/.

[35] Report on the implementation of the 2030 Agenda for sustainable development, at https://sustainabledevelopment.un.org/content/documents/279522021_VNR_Report_Germany.pdf.

[36] See, https://www.bundesregierung.de/breg-en/issues/sustainability/germany-s-sustainable-development-strategy-354566.

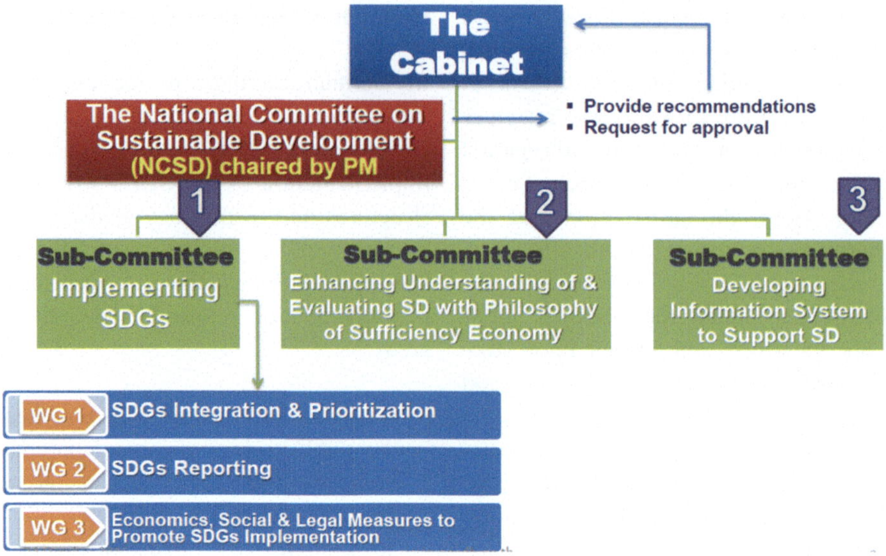

Fig. 13.7 Thailand's institutional SDG mechanism. WG stands for Working Group. *Source* Kumpa Ladawan (2016), Agenda 2030 Sustainable Development Goals: Thailand's Perspectives, *Presentation made at the Regional Knowledge Exchange Workshop, Oct. 24–26, 2016*, UNDP Bangkok

- *Structure of the 2016 German Sustainability Strategy*: Eight recommendations relating to monitoring and peer review.[37]
- *Germany's 2016 goals for sustainable development*: Four recommendations relating to the integration of SDGs in the national SD strategy.[38]

Thailand: Institutional Framework for SDG Implementation

The Thai government has developed a multi-stage framework for implementing SDGs, which has a number of working-groups and sub-groups to advise on the whole process. A sketch of the broad institutional structure is reproduced in Fig. 13.7. This figure shows the aggregate picture, but there are a large number of stakeholders associated with the working groups, which include non-profit organisations, the private sector, and legal bodies among others.

Somalia: Dialogues

In 2013, with the adoption of a new Constitution Somalia was presented with a new window of opportunity for a new phase of stabilisation and peacebuilding. In order to help manage the transition process, the Government of Somalia (FGS), with its civil society, parliament and other Somali stakeholders and the international community, agreed to develop a Compact, guided by the principles of the New Deal.

As part of the Compact, which was produced on the basis of a "Fragility Assessment", the government and the international community defined mutual roles and responsibilities

[37] See also, German Sustainable Development Strategy, Update 2021, at https://www.bundesregierung.de/resource/blob/974430/1940716/1c63c8739d10011eb116fda1aecb61ca/geman-sustainable-development-strategy-en-data.pdf.

[38] Ref: https://www.bundesregierung.de/breg-en/issues/sustainability/germany-s-sustainable-development-strategy-354566#:~:text=Intergenerational%20justice%2C%20quality%20of%20life,%2D%20and%20long%2Dterm%20targets.

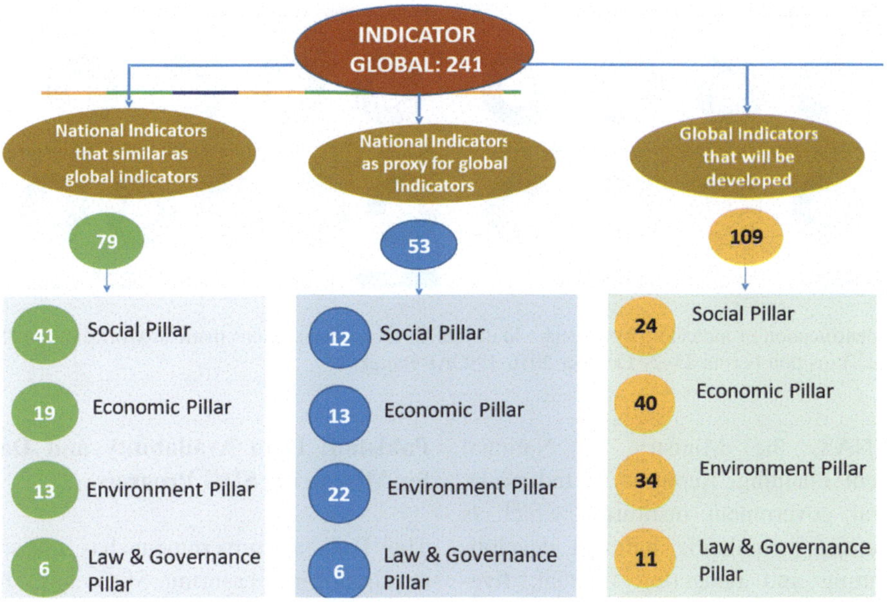

Fig. 13.8 Mapping of SDG indicators that are similar, have proxy, as well as will be for development in Indonesia. *Source* Presentations made at Workshop on "Sustainable Development Transition Forum 25–27 October 2016, ESCAP Bangkok"

including a financing architecture and setting up of overall framework for advancing peacebuilding and state-building in Somalia. The compact was the result of an inclusive process and strong partnership between the government, the United Nations, the World Bank, EU and donors and other key partners. The government and partners made sure the there was strong alignment between international assistance and the Somali Compact priorities and partnership principles. With support from the UN, the government established an Aid Coordination Unit for effective coordination and implementation of the compact.

Background: Somalia is a poor, post-civil war country, which has now embarked upon development.

13.5 Monitoring and Evaluation

13.5.1 Country Experiences[39]

Indonesia: Developing Indicators for SDGs

Figure 13.8, which maps the national development indicators with the SDG indicators, suggests that the ultimate purpose of the authorities is to bring the SDG goals and national goals as close to each other as possible. The 169 targets of SDGs are matched with 96 targets given that not all global targets are likely to find place in the local context. The Indonesian government has also split the SDGs into four pillars, the fourth being law and governance.

[39]The data in this section have been taken from the presentations made at the presentations made at the Regional Knowledge Exchange Workshop, Oct. 24–26, 2016, UNDP Bangkok, for other than Cambodia. For Cambodia, the 5-Year Plan for that country has been consulted; see, http://mop.gov.kh.

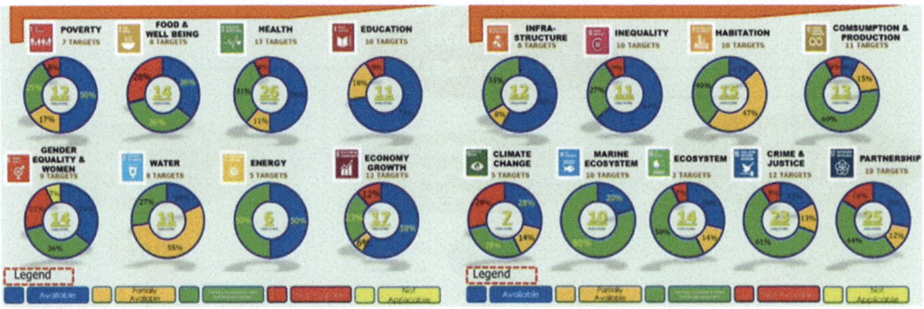

Fig. 13.9 Identification of indicators by targets—Malaysia. *Source* Presentations made at Workshop on "Sustainable Development Transition Forum 25–27 October 2016, ESCAP Bangkok"

BAPPENAS, the Ministry of National Development Planning, Republic of Indonesia, is a central government institution which is responsible for formulating national development planning and budgeting (annual, five-years, and long term). BAPPENAS has also a responsibility to coordinate international development (bilateral, unilateral and multilateral) cooperation. As for the SDGs, they are implemented through the existing mechanisms through directions provided from the central authorities.

Malaysia's SDG Indicators

Malaysia has conducted an exercise to identify indicators by SDG goals and targets and has classified them by those fully available, fully available, partially available (and need development), and not available (Fig. 13.9). A classification of this kind helps identify the unfinished task in regard to SDGs.

Thailand: SDG-Wise Data Availability and Challenges

Thailand identifies 241 indicators relevant for measuring SDG progress in its conditions (Fig. 13.10) when matched with the UN's listing. It finds that 119 indicators could be drawn from the existing databases, another 103 require checking, 14 need revision, and five are not relevant in the Thai conditions.

Pakistan: Data Availability and Data Gaps for Measuring SDG Progress

The Pakistan government has classified data required for measuring SDGs into four: data available in Pakistan on regular basis and standards are available; data available but not on regular basis, though standards are available; data not regularly produced and standards need to be developed; and indicators are not applicable/relevant), hence data not required. A chart classifying data availability in the context of the SDGs is seen in Fig. 13.11.

Issues and gaps regarding availability of quality data increase as one transits from the national aggregates to provinces, districts and more disaggregates levels. At the district level, data for 18 indicators can be made available by adding few questions to existing tools. Additional 37 indicators will require adding new modules existing tools. New surveys to be introduced such as general social surveys and climate/natural resource to reduce gaps further.

Background: Pakistan is a highly diverse low-middle income country in South Asia, which has experienced economic upheavals and high population growth in the recent decades. It is committed to achieve SDGs.

The Philippines: Data Challenges

The Philippine government has classified data availability in a somewhat similar fashion as that in Malaysia and Pakistan. It reads as:

SDG	UN Setting	Existing	Need Revision	Need Checking	Not Relevant
1	12	8	1	3	-
2	14	11	1	2	-
3	26	17	-	9	-
4	11	5	-	6	-
5	14	8	1	5	-
6	11	7	2	2	-
7	6	4	1	1	-
8	17	8	1	8	-
9	12	10	1	1	-

SDG	UN Setting	Existing	Need Revision	Need Checking	Not Relevant
10	11	6	-	5	-
11	15	6	2	6	1
12	13	3	-	10	-
13	7	4	-	1	2
14	10	3	3	3	1
15	14	3	-	11	-
16	23	10	-	13	-
17	25	6	1	17	1
GOALS	241	119	14	103	5

Fig. 13.10 Current status of SDG indicators. *Source* Presentations made at Workshop on "Sustainable Development Transition Forum 25–27 October 2016, ESCAP Bangkok"

Criteria	Availability of data				Total Indicators (N)
	Global	National	Provincial	District	
1- Data available on regular basis and standards are available		45	14	30	89
2- Data available but not on regular basis, however standards are available		6	14	14	34
3- Data not regularly produced and standards need to be developed		8	21	76	105
Total Indicators (N)	13	59	49	120	241

Fig. 13.11 Initial Findings of SDG Data Gap in Pakistan. *Source* Taken from Power Point presentation from *Regional Knowledge Exchange Workshop, Oct. 24–26, 2016, UNDP Bangkok*

Tier 1—With established methodology, regularly collected (41%).

Tier 2—With established methodology, data not regularly collected (21%).

Tier 3—With no established methodology, methodologies are being developed (30%).

Tier 4– Not Applicable/being verified (8%).

(Total: 241 indicators).

Figure 13.12 shows a distribution of those indicators that fall in Tier 1 category by goals and their adequacy. It is evident from this figure that more than half the job is yet to be completed.

Cambodia: Hierarchy of Indicators for MDG-Centred Plan 2014–2018

The Cambodian 5-Year Plans earlier pivoted on the MDGs and now SDGs. Its M&E Framework has a hierarchy of indicators, each level meant for a set of policymakers. It believes that the top policy/political leaders, being generalists and

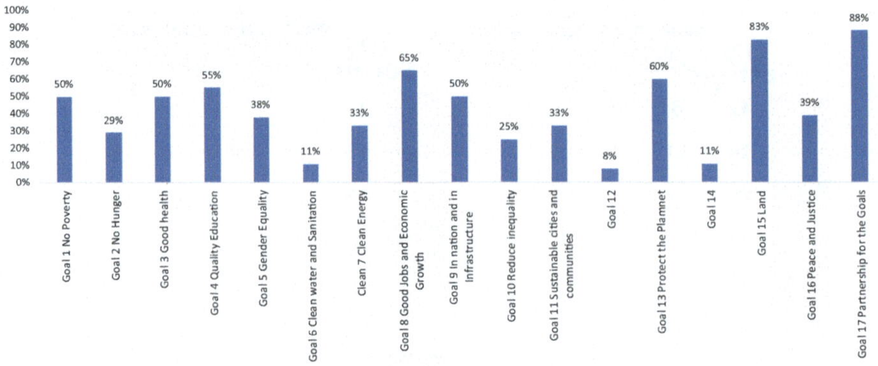

Fig. 13.12 SDG indicators classified as Tier 1, by goal. *Source* Presentations made at Workshop on "Sustainable Development Transition Forum 25–27 October 2016, ESCAP Bangkok"

also too busy to wade through several data files, would be interested only in key goals. These were identified to be 19 in number. At the outcome level, they have identified 48 indicators, and for output and input levels the responsibility has been left to the concerned line/implementing ministries and agencies (Table 13.7).

13.6 Budgeting for the Future

Budgeting for SDGs entails several aspects: effective budgeting of SDGs for the whole period, 5-year periods, or annual; next, there are sector, programme and location/region-specific budget; then there are target-specific budgeting like gender-budgeting, etc. The next step is to enter these into the Medium-Term Expenditure Framework to determine the expenditures and revenues and the short-falls there-in, and make efforts to raise more revenues. Examples of different types of budgeting from different countries are presented here.

Lao PDR: Costing MDGs[40]

Lao PDR had carried out an MDG-costing exercise in later 2009 with assistance from the UN Country Office. There were some 200 interventions identified, spread across eight sectors/eight ministries. These were, agriculture, education, energy, environment, gender, health, transport (including urban planning and zoning), and water and sanitation. Some typical interventions were: expanding irrigation, improving crop varieties, facilitating agricultural trade, developing non-farm/on-farm income generation activities, building capacities among officials in a range of fields, making opportunities for women, strengthening climate change control systems (incl. increasing forest cover), rural electrification and use of renewable fuels, extending potable water and sanitation, scaling-up early childhood and building/renovating primary schools and introducing school nutrition programmes, strengthening and staffing health centres, having more attended deliveries and pre- and post- natal care, controlling malaria, AIDS, etc., developing national, provincial and commune roads, protecting river banks, and district planning.

The total annual price tag for these interventions worked out at about US$1,293 million for the period 2010–2015, which worked out to about US$192 per capita annually. Of this, $800 million was expected to be the government share in fixed investments; $194 million, the government share in recurring expenses; $183 million, the non-government agencies' share in fixed/one-time investments; and $88 million, the non-government share in recurring expenses. If

[40] Source: MDG Costing—A Brief for Lao PDR, Ministry of Planning and Investment Lao PDR and UN Country Office. This author was the Team Leader in this exercise.

Table 13.7 Cambodia: hierarchy of M&E indicators for the plan 2014–2018

GOAL indicators (19)	**(1) Sustainable Growth**: Real GDP Growth Rate, Per-capita GDP, Inclusive Growth (Index), Structure of GDP, %Forest Cover, Total Cumulative Areas Cleared of Mines/ERW **(2) Poverty Reduction:** Poverty Rate, Gini Coefficient of Consumption Inequality **(3) Stability**: Investment Rate **(4) Human Development** *Education and Skills*: Completion Rate Grade 6 (by sex), Completion Rate Grade 9 (by sex), Skill Level (Index) *Gender*: % Women Holding Decision-making Positions in Public Sectors *Health:* Attended Births, IMR, % Stunted Children, MMR *Water and Sanitation*: % Population Having Access to Safe Water, % Population Having Access to Improved Sanitation
Outcome indicators (48)	**(1) Macroeconomic Management**: Budget Revenues and Expenditures, Inflation Rate, FDI, Current Account Balance, Two-way Trade, International Assistance **(2) Sectoral Growth and Diversification**: Sectoral Growth Rates, Paddy Yield Rate, Tourist Arrivals, Distribution of Employment by Sectors, Irrigated Area (All Crops), Land Under All Crops, Crop Diversification Index, Marketable Surplus Index, Annual Victims Caused by Mines/ERW **(3) Management of Natural Resources**: Surface of 23 Protected Areas, Community Protected Areas **(4) Infrastructure**: Length of Paved roads, Railway Track Rehabilitated and Constructed, TV/Radio, Print Medium, Per-capita Electricity Use **(5) Governance**: Rate of Issuance of Land Certificates to Farmers, Conflict on Land/Property, Criminal Offences **(6) Human Development Details** *Poverty:* Poverty (rural), Child Poverty, Food Poverty *Education*: Net Primary Level Enrolment Rates (boys and girls), Gross Lower Sec. Enrolment (boys & girls), Proportion of Children in schools Aged (6–17), Proportion of Persons Graduating from University (age 18–35 years) *Gender:* Proportion of Women in Public Sectors, Female share in Wage Employment (sector-wide), Net Enrolment of girls in Hr. Sec. and Higher Education, % Girls to Total Completing TVE, % Women Visiting Antenatal Care At least Twice, No. of Families Having Problems of Violence at Home *Health*: U5MR, Malaria Fatality Rates, HIV Prevalence, Prevalence of All Forms of TB, Birth Registration, Children Underweight, Women 15–49 years with Anaemia, Children (6–59 months) with Anaemia *Water and Sanitation:* Access to Safe Water, Improved Sanitation (by rural/urban)
Output indicators	A list is developed by the implementing ministry is expected to define these as per their sectoral programs and projects
Input indicators	Project and Programme-level Indicators to be maintained by implementing ministries and agencies as per the RF developed for each programme

Source Presentations made at Workshop on "Sustainable Development Transition Forum 25–27 October 2016, ESCAP Bangkok"

the economy grew at 8% annually and revenue collection is 18–19% of the GDP by the end of 2015, additional external assistance of about $500 to the treasury was expected to be sufficient to meet the targets.

Performance-Based Budgeting and Budget Mainstreaming—Some Country Experiences

Performance-based budgeting (PBB) system prioritises the allocations to ensure that relevant

Table 13.8 Evidence of effects of social transfers in food consumption, health status and education

	Effect	Country
(1)	(2)	(3)
Food Consumption		
1	Improving food security: 7.8 million people	Ethiopia
2	Percentage of transfer spend on food: ~75%	Malawi
3	Increase in food consumption: 165%	Bolivia
4	Change in food share: [1.9 to 4.3] percentage points higher	Ecuador
Health Status		
1	Increase in health visits: 18%	Mexico
2	Increase in children immunisation: 30%	Peru
3	Increase in immunisation coverage: 45% to 98%	Bangladesh
4	Expected average gain in height: 3.5 cm	South Africa
5	Reduction in malnutrition: 1.7 times greater than national trend	Nicaragua
6	Reduction in starvation: from 19 to 48%	Lesotho
7	Gain in children weight: 0.17 kg for age	Bangladesh
8	Reduction on the incidence of illness: 12.5% (children)	Malawi
9	Reduction on the incidence of diarrhoea: 10.5 percentage points (children under 24 months)	Colombia
10	Reduction on maternal mortality: 11%	Mexico
Education		
1	Increase in school enrolment: 10%	Ecuador
2	Increase in school enrolment: 11 percentage points (girls between 11–14 years)	Pakistan
3	Increase in school enrolment: 5 percentage points (children between 6–17 years)	Malawi
4	Increase in school enrolment: 30 percentage points (secondary school)	Cambodia
5	Increase in pass rate: 24 percentage points	Bangladesh
6	Reduction in one day absence incidence: 20%	Brazil
7	Reduction in drop out incidence: 63%	Brazil

Sources Unicef (2012), Estimation of Rate of Return on Social Protection Instruments in Cambodia, Phnom Penh Country Office; and, Schady, N. and M.C. Araujo. 2006. Cash transfers, Conditions, Scholl Enrolment and Child Work: Evidence from a Randomised Experiment in Ecuador, *World Bank Policy Research Working Paper, 3930*

programmes and services are meeting their outcomes. Next, budget mainstreaming promotes integration of specific issue-areas into fiscal budgets. Mainstreaming of gender or the child are important examples of this approach. Table 13.8 provides some evidence on these have been measured and could provide meaningful insights.

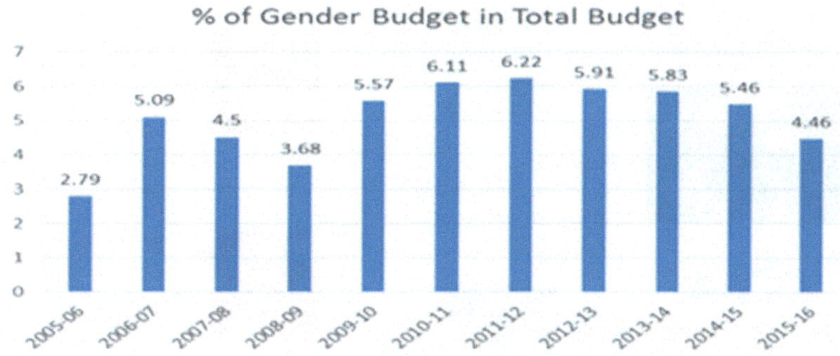

Fig. 13.13 Financing gender development in India. *Source* Chakravarty L (1016), Integrating SDGs in National Budgets, Presentation made at the Regional Knowledge Exchange Workshop, Oct. 24–26, 2016, UNDP Bangkok

India: An Example of Gender-Budgeting

An example of gender-disaggregated budgeting in the aggregate national budget, computed for a decade, could be seen Fig. 13.13. In another study on gender budgeting in the province of Maharashtra, it was found that the budget was generally 'gender blind'.[41] These and such studies have created a public opinion among policy makers to become sensitive to this and such issues.

Thailand: Linking Climate Change and Budgeting[42]

Objective: To build a sound business case for increased investment in conservation, sustainable use and equitable sharing of benefits of ecosystems and biodiversity, with a focus on identifying and filling-in finance needs.

Benefits: Policy planning and budgeting process towards sustainable economic and social development; financial mechanisms for investment and resource management on biodiversity; enhancing National biodiversity strategies and Action Plans as reference point outlining biodiversity issues and plans.

Structures: National Steering Committee; governments, led by National Economic and Social Development Board and co-led by environment policy; secretariat team; working group; UNDP BIOFIN, Thailand; other experts.

Outcomes: Policies, institutions, expenditure reviews; resource mobilisation, financing; 20 Aichi Targets, Convention on Biodiversity (CBD); and Sustainable Development Goals (SDGs) (Fig. 13.14).

Annual Budgeting: Select Cases

- Until 2014, Mongolia's government drew up plans akin to medium-term plans. Each year there is approval for the public-investment projects, based on a Medium-term Budget Expenditure Framework, which is ratified once in three years, though amended every year.
- In some cases, there are legal compulsions placed in regard to budgetary commitments on specific items: Maharashtra (India) is legally bound to provide (unskilled) work at the prescribed minimum wages in rural areas, on demand. Thus, the budget is required to make provisions for the employment programme.

[41] Parikh, A and S Acharya (2004): 'Impact of Maharashtra's Agriculture Policies on Women Farmers: A Gender Budgeting Analysis', a project report prepared for UNDP, Delhi.

[42] This sub-section is been taken from, Kumpa Ladawan (2016), Agenda 2030 Sustainable Development Goals: Thailand's Perspectives, Presentation made at the Regional Knowledge Exchange Workshop, Oct. 24–26, 2016, UNDP Bangkok.

Fig. 13.14 Integrated financing framework—SDG achievement (Thailand). *Source* Presentations made at Workshop on "Sustainable Development Transition Forum 25–27 October 2016, ESCAP Bangkok"

- In Malaysia, there is a strong commitment to education as stated in the medium-term plan as well as the long-term vision. The planning process, coming under the prime minister's office, has the mandate to get approval from the finance ministry.

Finding New Sources of Revenue—A Brief from Different Countries

The Addis Ababa Action Agenda of 2015 emphasises the need to mobilise and make effective use of domestic resources. Some approaches:

a. Improving revenue administration through Internet/mobile technology, progressive and efficient collection tax systems, and a progressive tax policy. India has taken steps to record every transaction through banks for taxation purposes.
b. Allocating more resources towards SDG-oriented sectors—Reallocation has been done in the state of Assam in India.
c. Ushering greater efficiency is the use of existing resources, like sharing resources across more than one purpose—Use of electronic means has ushered efficiency in India.
d. Raising ear-marked taxes: these are taxes levied for specific purposes—The Maharashtra Government (India) does this for their employment programmes.
e. Raising more taxes: This is possible in developing countries where there is an informal sector, which is sparsely taxed and many in this sector have the capacity to pay—many countries, including the developed countries have done this from time to time.

13.7 Conclusion

This paper presents experiences of select countries on how they have attempted to integrate SDG/MDG goals into their development plans. The process not only involves inclusion of goals but also getting different stakeholders on board for better success, weaving the SDGs in the institutions that deliver results, an M&E processes and finally considerations in regard to funding the goals. This compendium, simple as it seems, is a helpful exercise in informing the reader of what it entails while adopting different global development goals locally, the challenges, pitfalls and impediments in the implementing processes. Equally importantly, the paper points the need to bring and keep different stakeholders on board to achieve the results.

The manufacturer's authorised representative in the EU is Springer Nature Customer Service Centre GmbH, Europaplatz 3, 69115 Heidelberg, Germany. If you have any concerns regarding our products, please contact [illegible]ctSafety@springernature.com

[illegible] and bound by CPI Group (UK) Ltd, Croydon, CR0 4YY

01/10/2026

02225934-0020